DESIGN WORKBOOK

Using SOLIDWORKS 2026

Ronald E. Barr, Ph.D.
Professor

Thomas J. Krueger, Ph.D.
Senior Lecturer

Davor Juricic, D.Sc.
Professor Emeritus

Alejandro Reyes CSWE, CSWI

SDC PUBLICATIONS

SDC Publications
P.O. Box 1334
Mission, KS 66222
913-262-2664
www.SDCpublications.com

Examination Copies

Books received as examination copies are for review purposes only and may not be made available for student use. Resale of examination copies is prohibited.

Electronic Files

Any electronic files associated with this book are licensed to the original user only. These files may not be transferred to any other party.

Trademarks

SOLIDWORKS is a registered trademark of Dassault Systems. Microsoft Excel / Word are registered trademarks of Microsoft Corporation. All other brand names or trademarks belong to their respective companies.

Disclaimer

The author makes a sincere effort to ensure the accuracy of the material described herein; however, the author makes no warranty, expressed or implied, with respect to the quality, correctness, reliability, currency, accuracy, or freedom from error of this document or the products it describes.

The author disclaims all liability for any direct, indirect, incidental, or consequential, special, or exemplary damages resulting from the use of the information in this document or from the use of any products described in this document. Data used in examples and sample data files are intended to be fictional.

Download all needed model files from the SDC Publication website (www.SDCpublications.com/downloads/978-1-63057-781-0).

ISBN-13: 978-1-63057-781-0

ISBN-10: 1-63057-781-2

Printed and bound in the United States of America.

Table of Contents

1. Design Workbook Lab 1: Basic 2D Sketching

Introduction to SOLIDWORKS; Screen Layout; Menus; FeatureManager Tree; View Orientations; Sketching Toolbars; Sketch Planes; Starting a New Part; Part Units; Basic Dimensioning; Extruded and Revolved Parts.

2. Design Workbook Lab 2: Advanced 2D Sketching

Review of 2D Sketch Entities; Advanced Sketching Tools; Sketch Editing Tools; Linear and Circular Repeats; Extruded and Revolved Parts.

3. Design Workbook Lab 3: 3D Modeling Part I

Adding Sketch Relations; 3D Features Toolbar; Advanced Extrusion and Revolution Operations; Create Reference Geometry; 3D Mirror Feature; Create Linear and Circular 3D Patterns.

4. Design Workbook Lab 4: 3D Modeling Part II

Creating Advanced 3D Features: Draft, Shell, Dome, Loft, Sweep; Advanced Extrusion and Revolution Operations.

5. Design Workbook Lab 5: Assembly Modeling

Practice 3D Part Modeling; Creating a New Assembly; Assembly Toolbar; Adding Parts to an Assembly; Move and Rotate a Component; Mate Parts Together.

6. Design Workbook Lab 6: Part Evaluation and Configurations

Measure Tool; Component Mass Properties; Mass Properties Units; Editing and Modifying a Solid Model; Design Table Basics; Entering Design Table Parameters; Configuration Manager.

7. Design Workbook Lab 7: Static Stress and Thermal Analysis

Introduction to Finite Element Analysis Using SOLIDWORKS Simulation; Definition of FEA Terms; Basic FEA Stress Analysis; Applying Loads and Constraints; FEA Mesh Creation; Analyzing the Model for Stress Distribution; Printing the von Mises Stress Distribution; Design Changes Based on Analysis Results.

8. Design Workbook Lab 8: Animation, Detailing and Rapid Prototyping

Introduction to the SOLIDWORKS Animation Wizard; Assembly Exploded View; Creating the Animation; Animation Controller; Editing the Animation; Saving an .AVI File; Animation Motion Elements; Introduction to Rapid Prototyping.

9. Design Workbook Lab 9: Section Views in 2D and 3D

Viewing a 3D Section View of a Solid Model; Printing 3D Section View; Changing Drawing and Hatch Pattern Options; Projecting Orthographic Views; Making a 2D Section View.

10. Design Workbook Lab 10: Manufacturing Detail Drawings

Drawing Sheet Options; Projecting Orthographic Views in a Drawing; Adding Centerlines; Importing Annotations from the 3D Model; Dimensioning the Drawing; Adding Manual Annotations.

APPENDIX A – Drawing Sheet Template

NOTES:

Design Workbook Lab 1
Basic 2D Sketching

INTRODUCTION TO SOLIDWORKS

SOLIDWORKS© is a 3D parametric solid modeling software used for mechanical design. Most designs start with a 2D sketch, where dimensions and geometric relations (or constraints) are applied between sketch elements, to fully define its geometry. The sketch is then turned into a 3D model using the Extrude, Revolve, Sweep or Loft command. A part is modeled by adding additional 2D sketch and 3D features, including bosses, cuts, fillets, chamfers, etc. Each new 3D feature either adds or removes material from the part, until the component's design is complete. This process is called Feature Based modeling.

If a part needs to be changed, its design can be modified at any time by editing any 2D sketch or 3D feature, where you can change dimensions, sketch geometry, etc.

In this first chapter, you will learn different 2D sketching techniques to create simple extruded or revolved parts, using four different exercises.

SOLIDWORKS TOOLBARS SETUP

When you first launch SOLIDWORKS the screen will look like the following image. To get started, touch the arrow next to SOLIDWORKS in the upper left corner to expand the menu bar, and click in the Push pin on the right to keep the menu visible. De-selecting it will cause the menu bar to automatically hide.

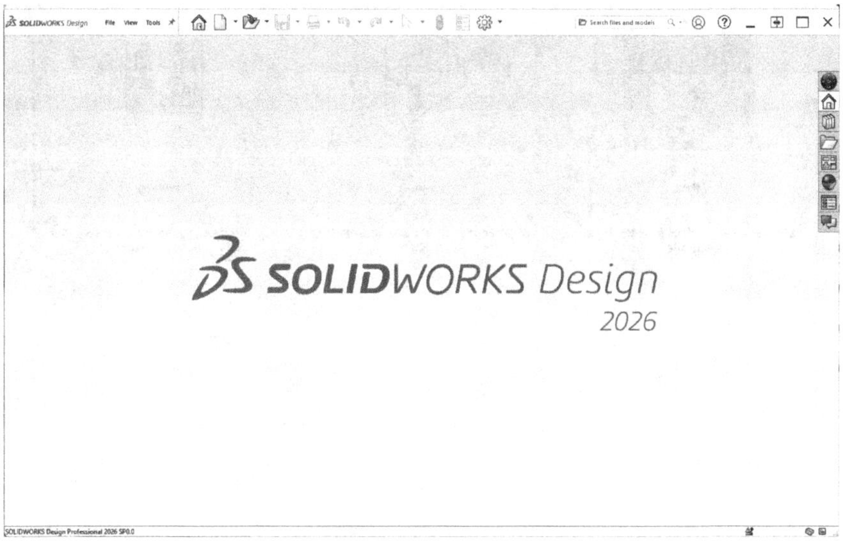

SOLIDWORKS has three different work environments: Part modeling, Assembly and Detail Drawing. The appropriate command toolbars and menu options are automatically activated when you switch to a different document type. If needed, you can change toolbars' visibility to fit your individual needs.

To change a toolbar's visibility on or off, select the menu **View, Toolbars**, and select/de-select the desired toolbar. Optionally, you can right click on any toolbar to display the same visibility options, as shown next.

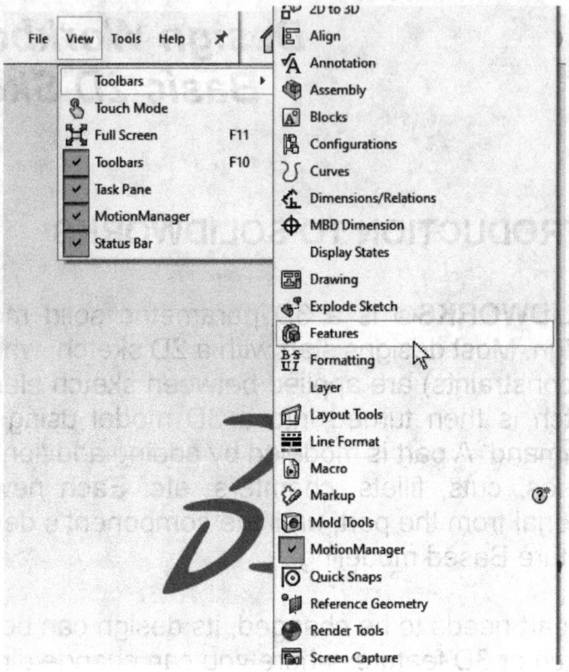

STARTING A NEW PART

To start a new part in SOLIDWORKS, select the menu **File, New…** or click in the main toolbar in the **New** document command. The New Document dialog will be shown where you can start a new Part, Assembly, or Drawing. See **Figure 1-3**. If you cannot see the three options shown, click on the Novice button on the lower left corner.

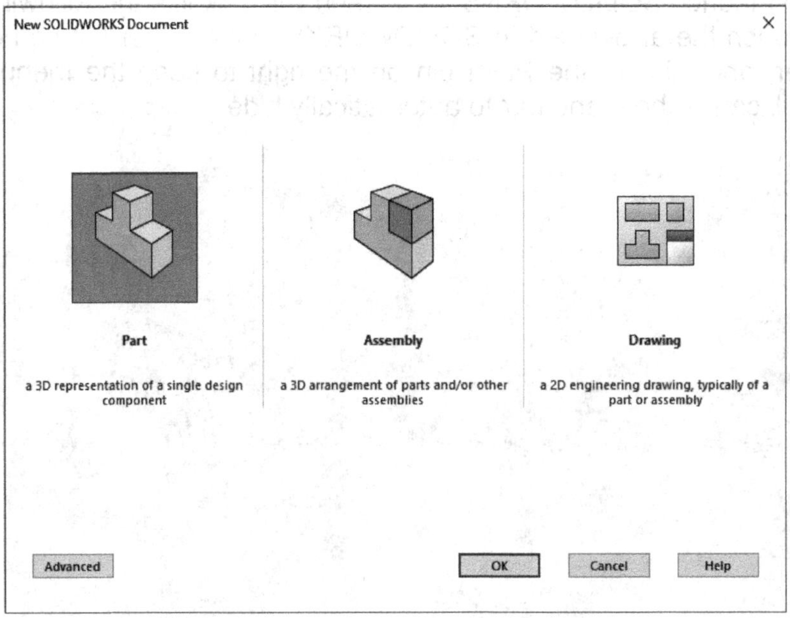

THE SOLIDWORKS SCREEN LAYOUT

After selecting **Part,** click **OK**. A new part file is created, and the part modeling interface is shown. The **CommandManager** bar is enabled by default and located at the top of the graphics area; it combines multiple toolbars in a single location using tabs for a more efficient use of screen space as shown.

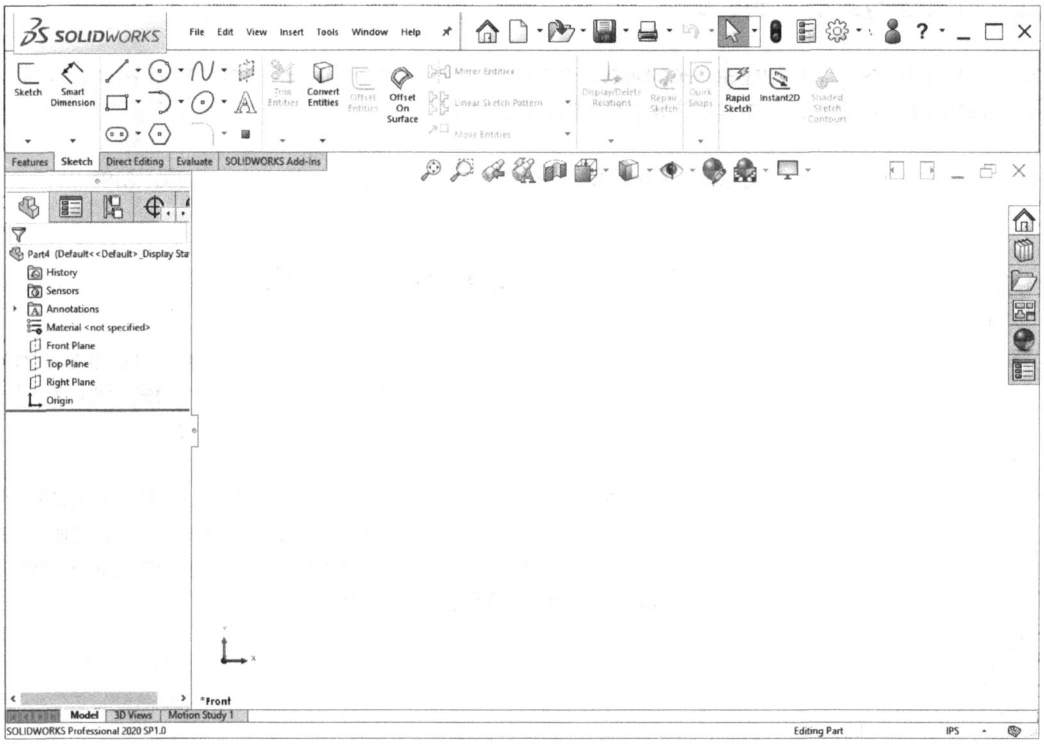

If needed or desired, toolbars can also be shown or hidden separate from the CommandManager. A toolbar's visibility is maintained in each of the Part, Assembly and Drawing environments.

The graphics area at the center of the screen is where you will be doing most of your design work. Each of the menus and toolbars will be described next.

MAIN MENU – to prevent this menu from automatically hiding, select the Push pin on the right of the menu bar.

File is where you can create New files and Open, Save, Print Preview, and Print existing files.

Edit is where you can find commands to modify existing features, Undo the last command, Cut, Copy, and Paste Entities. It also has the Rebuild command to update the model when you make changes.

View is where you can change the model's Display parameters, change the View Orientation, and select view manipulation commands like Zoom, Rotate, and Pan.

Insert is where you find most feature creation commands, including Extrusions, Cuts, Sweep, Loft, Fillets, etc.

Tools include multiple commands to help you create, modify, analyze, and change document and system options.

Window is where you can select between multiple open documents and arrange them to tile them vertically or horizontally on the screen.

Help is where you can access SOLIDWORKS' user help and online tutorials.

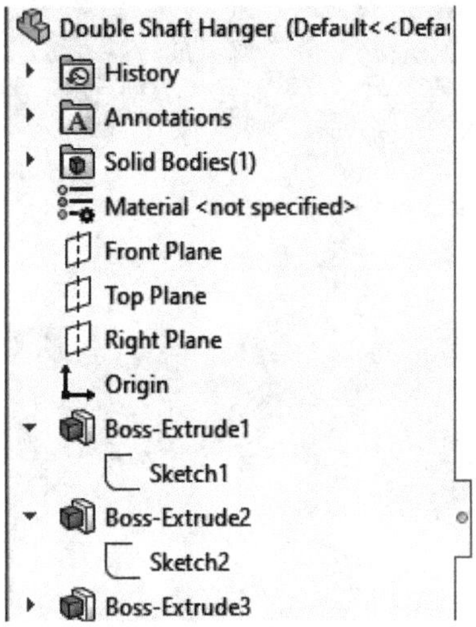

FEATUREMANAGER TREE

The SOLIDWORKS **FeatureManager** Tree is the pane on the left side of the screen used to keep track of the solid modeling process and operations.

It lists every model's features, including their sketch, in chronological order. This is where you can access and edit every feature of your model if needed and is part of every SOLIDWORKS document.

THE REBUILD BUTTON

The **Rebuild Button** is located on the Standard Toolbar and looks like a red/green traffic light.

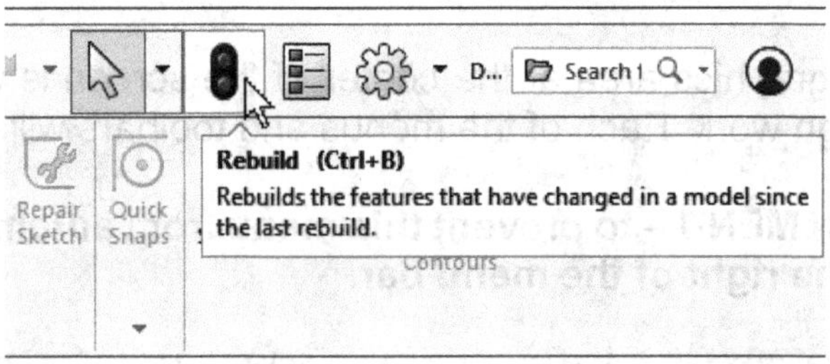

When you select it, the model is rebuilt and updates the geometry with any changes that have been made. When a model is rebuilt, the current command is automatically finished.

TOOLBARS

Toolbars include the commands available to create and modify models and are grouped by functionality. To see all the available Toolbars, right click on any toolbar, or select the menu **View, Toolbars**.

The Heads-Up **View Toolbar** includes the commands needed to manipulate the model's view including Zoom, Pan (move) and Rotate. Using the drop-down command options, you can change the model's display style to Hidden Lines Removed, Hidden Lines Visible, Shaded, etc.; options to change the model's orientation to Front, Back, Left, Right, Top, Bottom, Isometric and perpendicular to a selected plane or face; and visibility controls to hide or show auxiliary geometry.

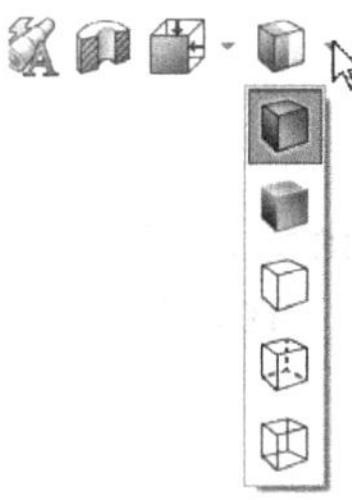

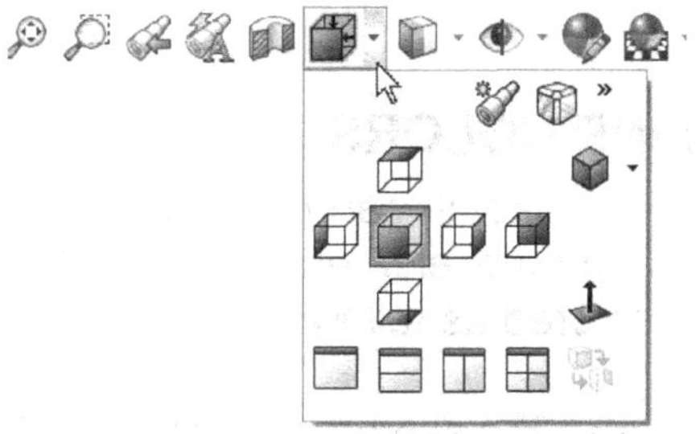

SKETCH TOOLBAR: The sketch toolbar includes the tools needed to create sketch geometry.

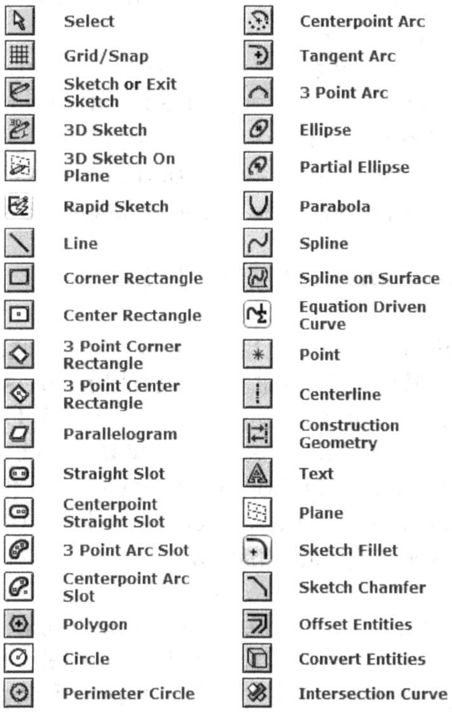

The Dimensions and Relations toolbar has the tools to add relations and dimension between sketch entities.

The sketch toolbar includes multiple commands to create 2D geometric elements including **Line, Arc, Circle, Rectangle**, etc., as well as different tools to modify your sketch, like **Trim, Mirror, Fillet**, and **Offset**.

MODEL PLANES

Before you start a 2D sketch in SOLIDWORKS, you must select a sketch plane.
Every part includes three default orthogonal planes, which match the three standard orthographic views: **Front**, **Top**, and **Right**.

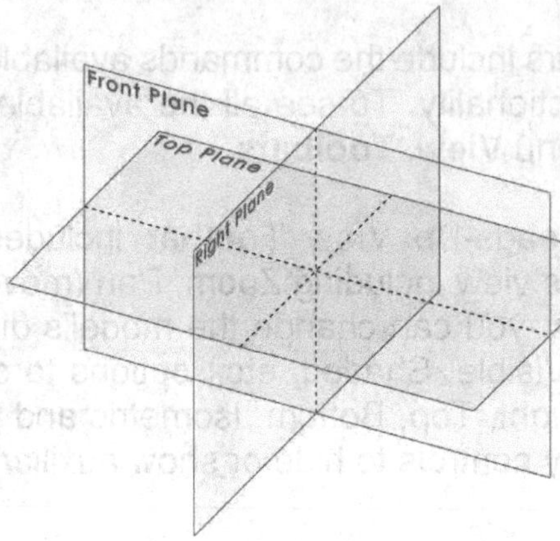

After a 3D model is created, the 3D model's flat faces can also be used to add a sketch. The model's face does not have to be parallel to one of the three principal planes to be used for a sketch.

SKETCH LINE COLORS

A SOLIDWORKS sketch can be in any of many different states, and you can identify them by a geometric element's color. When sketching 2D profiles, the geometric elements are colored as follows by default:

- **Cyan** indicates the sketch element is **Selected**.
- **Blue** indicates the sketch elements are **Under Defined**, which means they don't have the necessary dimensions and/or geometric relations needed to define them (undesirable).
- **Black** indicates the sketch elements are **Fully Defined** and have the necessary dimensions and/or geometric relations needed to define them (preferred).
- **Yellow or Red** indicates the sketch elements have conflicting dimensions and/or geometric relations and are **Over Defined** (undesirable).
- **Gray** indicates these entities are not in the active sketch.

STANDARD TEMPLATES

Before you start any exercises in SOLIDWORKS, it is a good idea to establish some standard 3D Part and 2D Drawing templates. One set of templates will use the American National Standards Institute (ANSI) standard in Inches and the other for Metric (IS) dimensions.

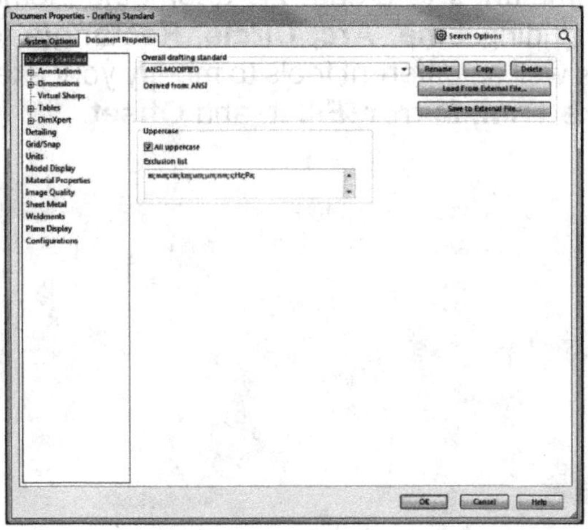

To start, go to the menu **File, New**, select **Part** and click **OK**. Go to the menu **Tools, Options,** select the **Document Properties** tab, make sure the **Dimensioning Standard** is set to **ANSI**.

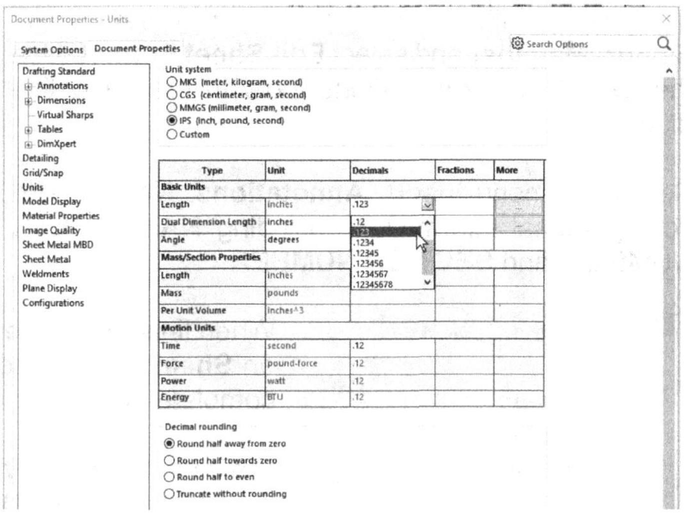

Next, select **Units** in the left-hand column of the dialog and activate **IPS (Inch, Pound, Second)**.

You should also set the Decimal Places to three **(3)** as indicated. Click **OK** to close the "Document Properties" window.

Now go to the menu **File, Save As,** change the **Save as Type** selection to **Part Templates (*.prtdot)**, select the folder where you wish to save this template, and name the template **ANSI-INCHES**. **Make sure that you are saving this into your personal folder**.

Repeat the above process, only this time change the **Units** to **MMGS (Millimeter, Gram, Second)**, and two decimal places. This should be sufficient for metric measurements. Save this template to your folder. Remember to change the **Save as Type** option to **Part Templates (*.prtdot)** and set the name as **ANSI-METRIC**. **Make sure the template is saved to your personal folder.**

2D DRAWING SHEET WITH TITLE BLOCK

The next process is to set up a 2D Drawing with a Title Block to document your designs using either of the two dimensioning standards. You may have to draw a title block if one has not been provided.

An example of a drawing with a title block, with suggested dimensions and style, is available in **Appendix A** and included with the book's files.

a) If a 2D Drawing template has not been provided, select the menu **File**, **New**, select the **Drawing** and click OK. Select the size **A-(ANSI) Landscape.** Turn off the **Display Sheet Format** option and click OK. If needed, click on the red **X** to cancel the **Model View** command.
b) If a 2D Drawing template with Title Block has been provided, go to the menu **File** and **Open** that drawing file.

Go to the menu **Tools**, **Options**. In the **Document Properties** tab make sure the **Dimensioning Standard** is set for **ANSI**. Next, select **Units,** activate **IPS (Inch, Pound, Second)**, and set to three **(3)** decimal places; this will have enough accuracy unless you do a high tolerance drawing.

To draw the Title Block, right click on **Sheet1** in the **FeatureManager** and select **Edit Sheet Format**. Use the Sketch and Annotations tools to draw the title block.

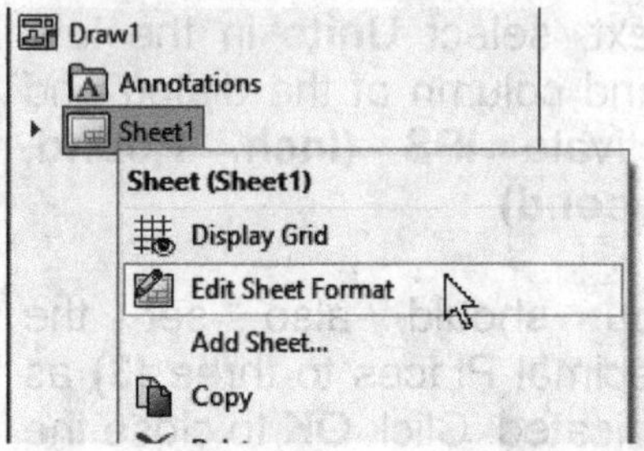

Select the menu **Insert, Annotations, Note** to add your personal information, including **NAME, DESK NUMBER**, and **SECTION NUMBER**.

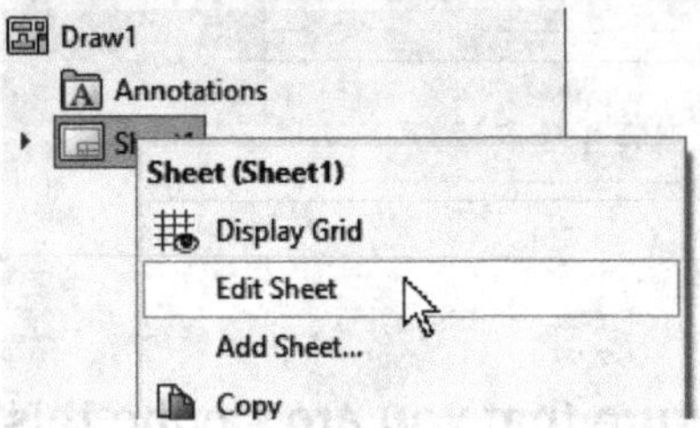

When finished, in the **FeatureManager Tree** right click on **Sheet1** and select **Edit Sheet**. With the title block complete, select the menu **File, Save As**.

From the **Save as Type** drop-down list select **Drawing Template (*.drwdot),** select the folder where you wish to save this template, and save your template as **TITLEBLOCK-INCHES**. Make sure that you are saving it to your personal folder.

Repeat the above process, only this time, in the **Units** section select **MMGS (Millimeter, Gram, Second)**. Two **(2)** decimals should be sufficient for metric measurements. Add your personal information such as **NAME, DESK NUMBER**, and **SECTION NUMBER**.

In the **FeatureManager Tree**, right click on **Sheet 1** and select **Edit Sheet**. Select the menu **File, Save As**. From the **Save as Type** drop-down list select **Drawing Template (*.drwdot),** select the folder where you wish to save this template, and save your template as **TITLEBLOCK-METRIC**. Make sure that you are saving it to your personal folder.

ADDING A 2D DRAWING VIEW OF A 3D MODEL PART

For each of the assignments you will be asked to submit a printed copy of the part that you built in SOLIDWORKS. The following will be the procedure to accomplish this requirement.

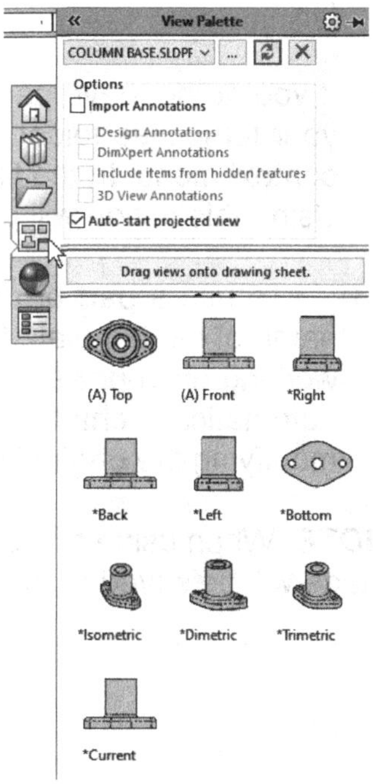

A. Build the solid model according to instructions and save that part in your folder. A part's file extension is **.sldprt**.
B. Open the drawing template to be used (see **Figure 1-12** for example). Use the inches or metric drawing template as needed. If you choose the incorrect template, remember you can change the units of measurement later in the menu **Tools, Options, Document Properties, Units**.
C. From the **Task Pane** toolbar, click on the **View Palette**, select the part you want to add to the drawing from the open files list, or browse to open a new part.
D. Click and drag the Isometric drawing view from the list onto the sheet. In unit nine and ten you will learn how to make a complete detail drawing using multiple orthogonal views. If you do not see the model view after dragging it to the sheet, it means you are editing the **Sheet Format**. In this case, right click on **Sheet1** in the **FeatureManager Tree** and select **Edit Sheet**.
E. To change the scale of a drawing view, select the view; in the **Scale** section of the **PropertyManager** pane select **Use Custom Scale** to change the scale of the view as needed. In the **Display State** section select **Shaded with Edges** for a shaded view.
F. In the Title Block area of the drawing sheet, add the necessary notes using the menu **Insert, Annotations, Note**. Provide the exercise number in the upper right box. Your Title should be placed in an open area of the Title Block using a larger font size, i.e., **20 to 24 PT or .25"**. Your name, desk, sec and exercise number should use a font size of **12pt or .125"**.

HOW TO USE A TEMPLATE

To use the previously created templates, there are two options, depending on the permissions allowed by the system administrator: You can add the files to a SOLIDWORKS templates folder (preferred option) or open the template files individually.

1. *To use the SOLIDWORKS templates functionality, you need access to modify SOLIDWORKS system settings.* Create a folder named "**My SW Templates**" in your personal folder and copy your templates to this folder. In SOLIDWORKS go to the menu **Tools, Options**, and in the **System Options** tab select **File Locations**.

In the drop-down list select **Document Templates**, click **Add** and browse to the "**My SW Templates**" folder.

To use the templates, click on the **New** command or the menu **File, New**. If you cannot see the "**My SW Templates**" tab, click on the **Advanced** button in the lower left. Select "**My SW Templates**" tab, click on the template you want to use and click OK.

2. If you do not have permission to modify system settings, the best option is to open your templates folder, right click on each template and select **Properties** from the context menu. In the file's properties turn on the "**Read Only**" attribute and click OK. Using this option will prevent you from accidentally modifying your templates.

   To use a part or drawing template, open it in SOLIDWORKS; you will see the name of the template followed by **[Read Only]** in the title. At this time, you can start working on your new part or drawing. When you click on **Save**, SOLIDWORKS will automatically change the file type to part (*.**sldprt**) or drawing (*.**slddrw**), without modifying the original template.

NOTE: When using this option, if the file's **Read Only** attribute is not set, any changes made will be saved to the template.

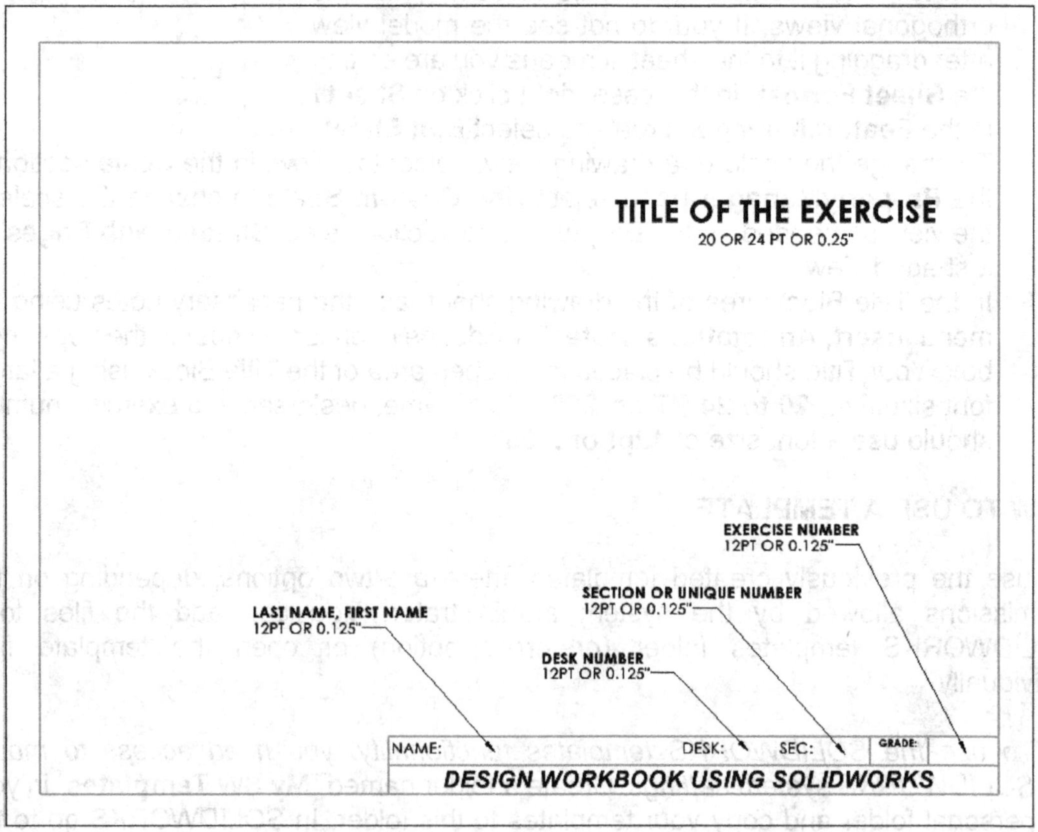

Exercise 1.1: CARBON FIBER GASKET

In this first exercise, you will build a Carbon Fiber Gasket part. To complete this part, you will create a 2D sketch and extrude to create a 3D part. In this exercise you will use many of the 2D sketching tools available.

If you can use the SOLIDWORKS templates folder option, click on the **New** command, select **ANSI-INCH** from the "**My SW Templates**" tab and click OK.

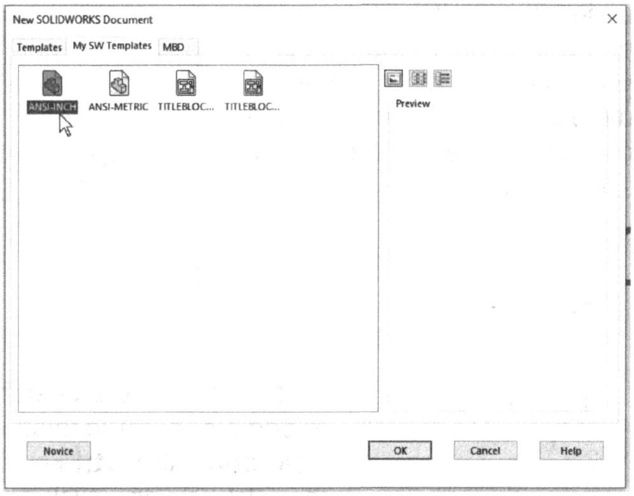

Otherwise, click on the **Open** command, browse to your templates folder, and open the **ANSI-INCHES.prtdot** file. Using either approach, click on **Save** and name your part **CARBON FIBER GASKET.**

The new part's units are already set to inches with 3 decimal places, as our template. If you need to change them, select the menu **Tools, Options,** select the **Document Properties** tab, and click on **Units**. Change the units of measure if needed and click **OK** to finish. The 2D sketch for the gasket will be created in the **Front Plane.** In the FeatureManager tree select the **Front Plane** (it will be highlighted in the screen).

Select the **Sketch Tab** in the **CommandManager** and click on the **Sketch** icon.

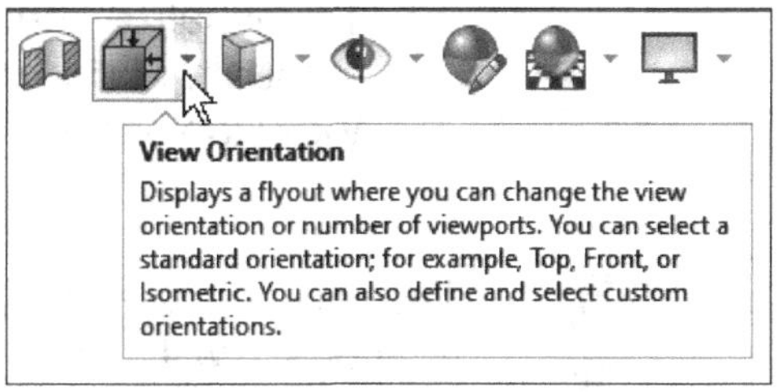

After the sketch is created, your part will be automatically reoriented to a Front view, showing the **Front Plane** parallel to the screen. If needed, you can use the view orientation using the Heads-Up View toolbar.

Select the **Line** tool from the sketch toolbar and draw the profile indicated next starting at the origin. Draw the profile approximately as shown. When drawing either vertical or horizontal lines, SOLIDWORKS automatically shows guides to let you know a horizontal or vertical geometric relation will be added to the line, constraining its orientation. Use either click-click, or click-and-drag method.

When finished press the **ESC** key to turn off the **Line** tool. At this time, you can use the left mouse button to drag the blue geometric elements, because their size has not been defined yet. To fully define the sketch geometry, you will now add geometric relations and dimensions.

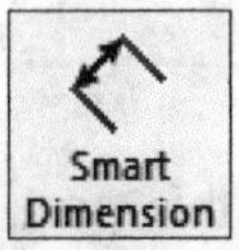

Now, select the **Smart Dimension** tool to add the dimensions shown. Select the endpoints to dimension, locate the dimension on the screen, and enter the value indicated. If you need to change a dimension's value, double click on it, enter the new value, and click OK.

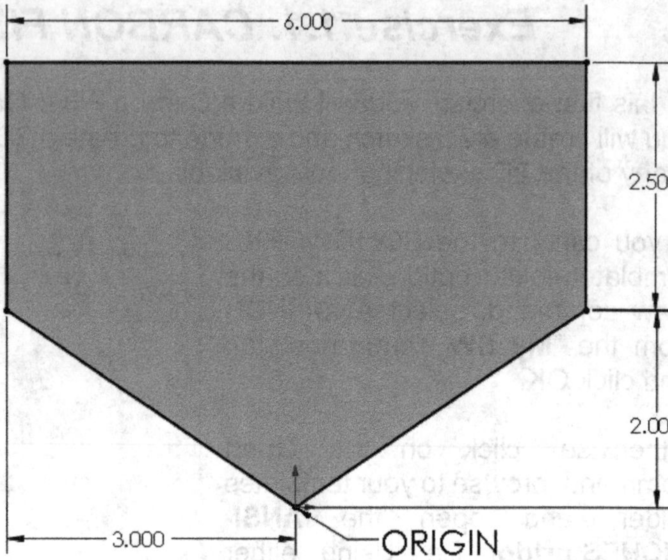

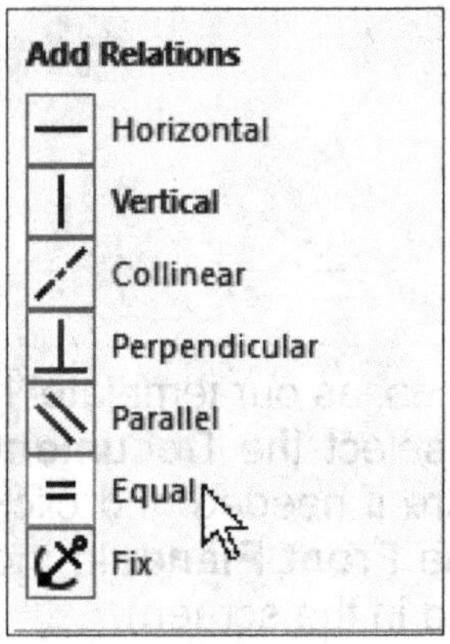

To fully define the sketch geometry up to this point, you need to add a geometric relation to make the two vertical lines equal. Hold down the **Ctrl** key and select both vertical lines, release the **Ctrl** key and click on the **Equal** relation in the **Properties** pane. By making both lines the same length, the sketch is now fully defined, meaning that all necessary dimensions and geometric relations have been entered. Now the sketch has changed from blue to black, letting you know the sketch is fully defined.

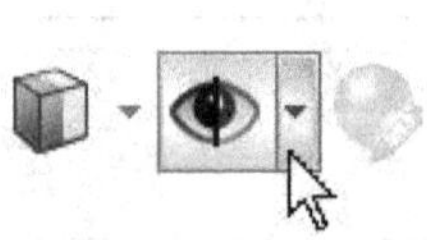

To visualize the geometric relations automatically captured and manually added, click in the **Hide/Show** drop-down command in the **View** toolbar, and turn on the **View Sketch Relations** command.

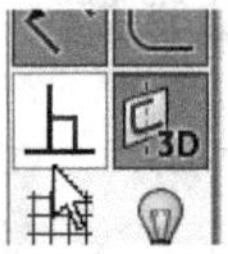

The existing sketch relations will be indicated with the green icons next to each geometric element. This option can be turned on or off as needed.

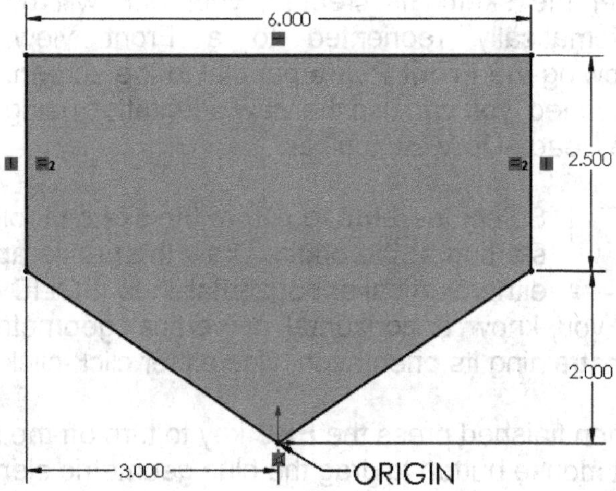

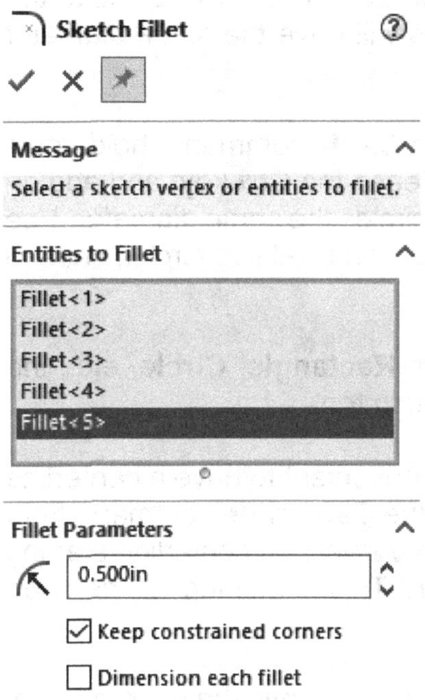

The next step is to round the corners of the gasket. This will be done using the **Sketch Fillet** command. Select the **Sketch Fillet** icon from the Sketch toolbar. The **Sketch Fillet** PropertyManager will be displayed in place of the FeatureManager. In the fillet radius enter **0.500** inches and select each of the vertices (intersections) in the sketch to create a fillet, or the connecting lines.

Since we added an **Equal** geometric relation to two lines, adding the sketch fillet will change the length of the lines, and you will be asked if you want to proceed. Click Yes to continue and select the next intersection. Click OK to finish and close the **Sketch Fillet** command.

Note there is only one fillet dimension added to our sketch. The reason is that all fillets made at the same time will have an **Equal** sketch relation. Therefore, changing this dimension will change the size of all fillets at the same time. The sketch profile should look as follows.

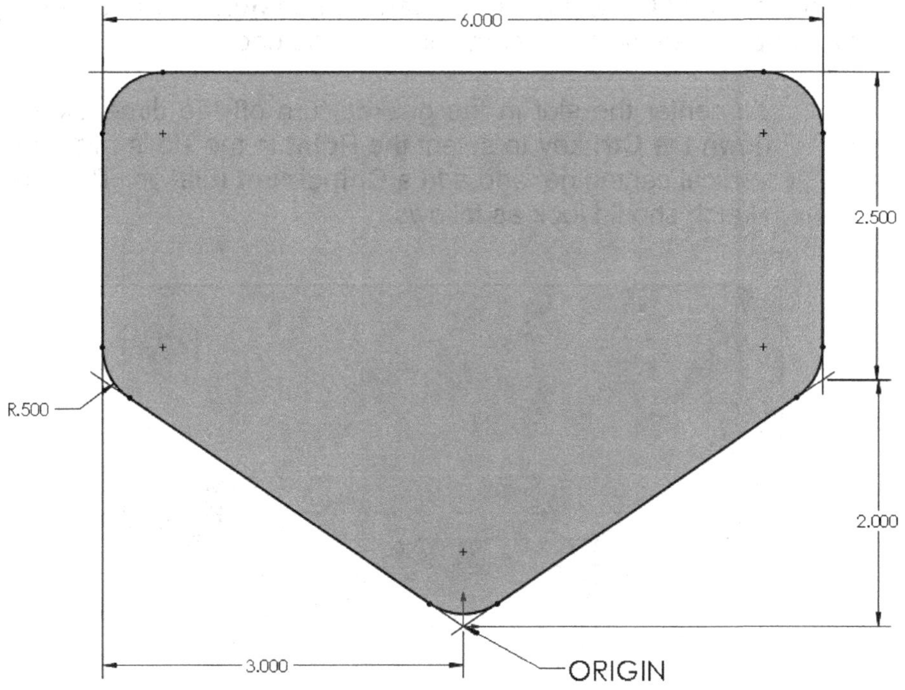

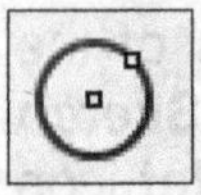

Next, select the **Circle** command, and create the five circles concentric to the fillets (rounded corners) of the sketch. All circles will have the same diameter of **0.375"**.

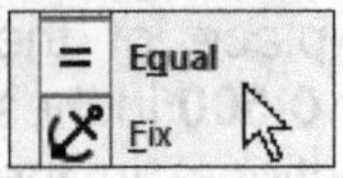

After the five circles are drawn, cancel the **Circle** command, hold down the **Ctrl** key and select the five circles, release the **Ctrl** key, and add an **Equal** geometric relation to make the five circles the same diameter. Use the Smart Dimension tool and dimension one circle **0.375"**. Note all five circles change size at the same time.

Note: While you are using a sketch entity tool like **Line**, **Rectangle**, **Circle**, etc., the cursor will show a small icon that indicates the active sketch tool.

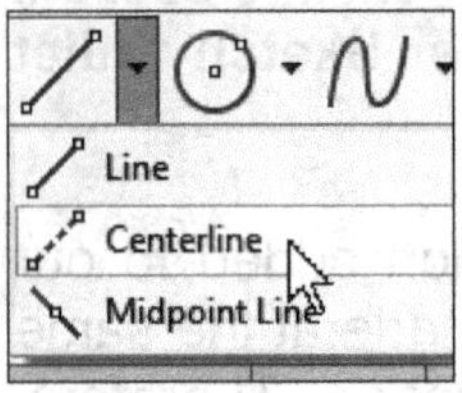

Since the gasket is symmetrical, it is convenient to have a centerline to use it as support geometry. Select the **Centerline** command from the **Line** drop-down command and draw a vertical centerline starting at the origin going up through the part. This centerline will be used in the next two steps.

Select the **Straight Slot** tool and draw a horizontal slot above the bottom circle. If the option Add Dimensions is turned on, the slot will have a width and height dimension added automatically. Double click and change the width to **3.00"** wide, and the height to **0.50"**. Select the **Smart Dimension** tool and add a vertical dimension of **2.00"** between the slot's centerline and the origin.

To center the slot in the gasket, turn off the dimension tool, hold down the **Ctrl** key to select the **Point** in the slot's midpoint and the vertical centerline, and add a **Coincident** relation. The fully defined sketch should look as follows.

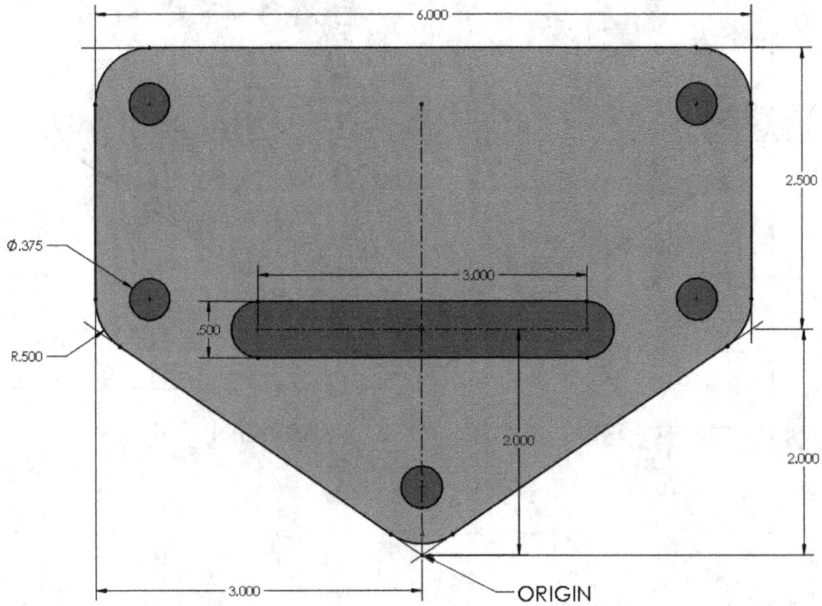

Many models have symmetrical features, and SOLIDWORKS offers tools to make those features easy to duplicate. You will now draw two polygons on the gasket that are symmetrical about a centerline. First, select the vertical centerline previously drawn, and select the menu **Tools**, **Sketch Tools**, **Dynamic Mirror.** The centerline will now have an equal sign at either end to indicate the **Dynamic Mirror** is active. This means that anything drawn on one side of the centerline will be automatically mirrored on the other side.

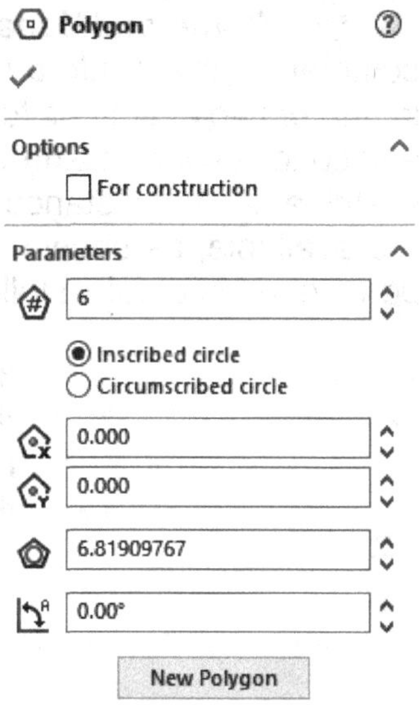

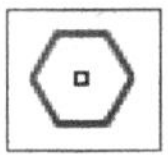

Select the "**Polygon**" command from the sketch toolbar. When polygon's options are displayed, set the number of sides equal to **6** and select the Inscribed circle option.

Draw a polygon on either side of the vertical centerline. First locate the center of the polygon, then move your cursor horizontally to define the size and rotation of the polygon. Select the **Smart Dimension** tool and dimension the horizontal and vertical location of the polygon, as well as the diameter of the inscribed circle.

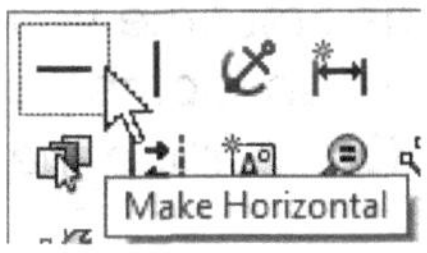

To fully define the sketch, select a hexagon's horizonal line, and from the pop-up menu click on **Make Horizontal** to add a geometric relation to prevent the hexagon from rotating. Your finished sketch will look like the following image.

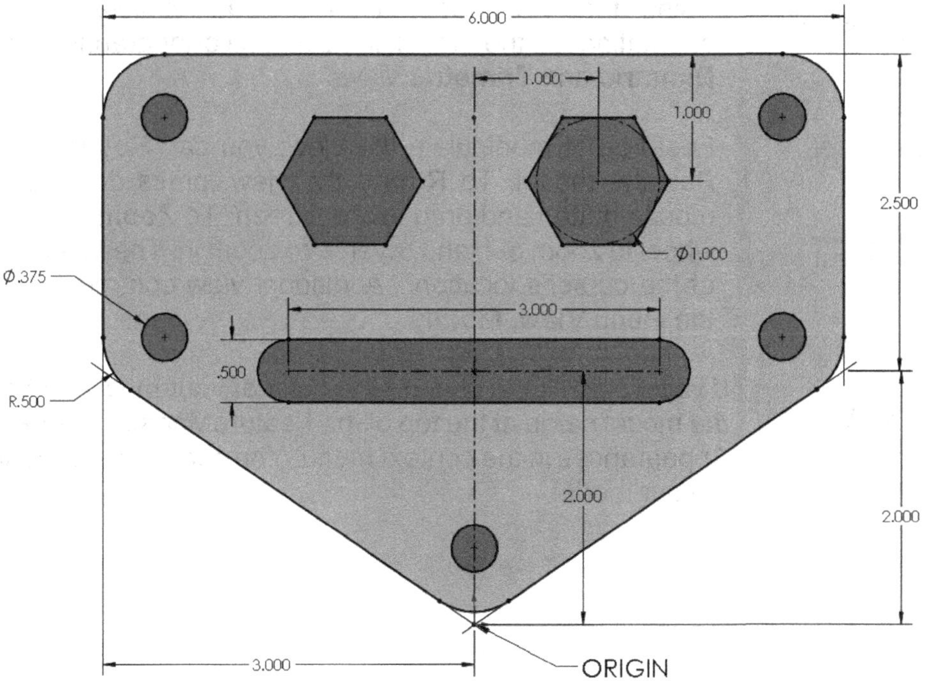

EXTRUDE THE GASKET SKETCH

The **Extruded Boss/Base** command is in the menu **Insert**, **Boss**, **Extrude**, or the **Features** tab of the **CommandManager**. As its name implies, this command will extrude a sketch or selected sketch contours in one or two directions to create a solid feature. The default end condition for an extrusion is **Blind**, which will extrude the sketch a defined distance. Different extrusion end conditions are available, as shown. Many of these end conditions will be covered in detail in following lessons.

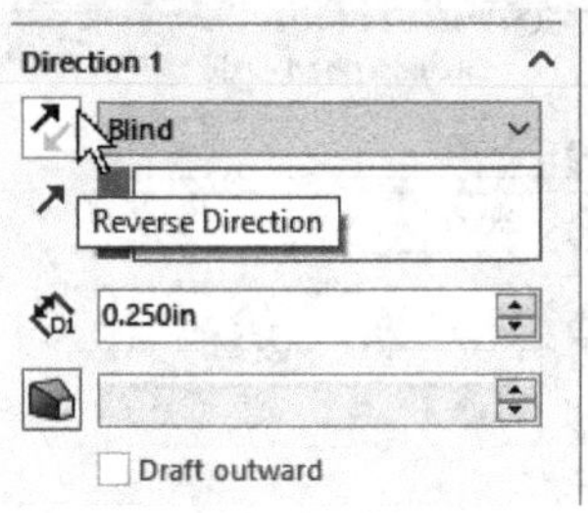

If needed, you can change the direction of the extrusion with the **Reverse Direction** button next to the "**End Condition**" selection box.

Now extrude the gasket to its needed thickness. In the **Boss-Extrude** properties set the end condition to **Blind** and set the extrusion depth to **0.25"** as shown. The first time you create an extrusion in a part, the model's view is automatically rotated to an Isometric orientation, and a preview of the extrusion is shown in the screen. Click OK to finish the extrusion.

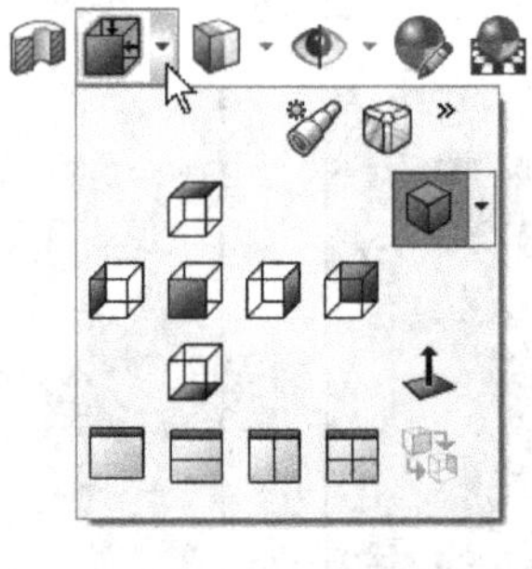

Using the View Orientation toolbar, explore the different orientations of the model, including all orthographic, **Isometric**, **Dimetric** and **Trimetric** views.

Feel free to manipulate the view; you can **Rotate**, **Zoom** and **Pan** the model. To **Rotate** the view, press down the middle mouse button and drag in the screen. To **Zoom** roll the mouse wheel to zoom out, and reverse to zoom in. The model will zoom at the cursor's location. Additional view controls are found in the menu **View**, **Modify**.

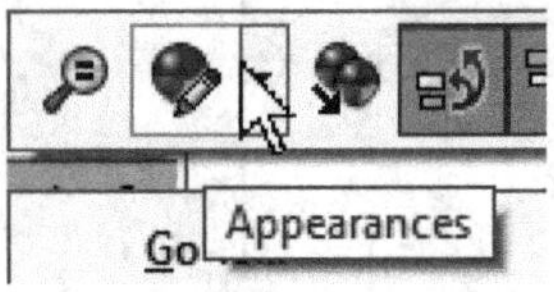

If you would like to change the color of your model, right click on the model name at the top of the FeatureManager tree and select Appearances in the context menu. You can then assign any color to your model.

Save your part as **CARBON FIBER GASKET** in your personal folder. The file type will be automatically set to **Part (*.sldprt)**.

To complete this exercise, you need to print a copy for submission to your instructor. *Check with your instructor for any special printing instructions.*

Using the **TITLEBLOCK-INCHES** drawing template, follow the instructions provided on page 1-9 to add a **Trimetric** view, and change the view's scale to 1:1. Use the menu **Insert**, **Annotations**, **Note** to add the missing notes to complete the drawing as shown. Be sure to use the recommended font and size for the required annotations. Don't forget to add the exercise number in the space to the right of the **GRADE** box.

Save your drawing as **CARBON FIBER GASKET,** using the **Drawing (*.SLDDRW)** file type. Notice that the title of the part and the drawing are the same, but the extension is different. The 3D solid model and the 2D drawing are linked, meaning that any changes made to the solid model will be automatically updated on the drawing.

Before you print your copy, make sure you have set the **File**, **Page Setup** to **Landscape** mode. Use the menu **File, Print Preview** to send it to the designated printer.

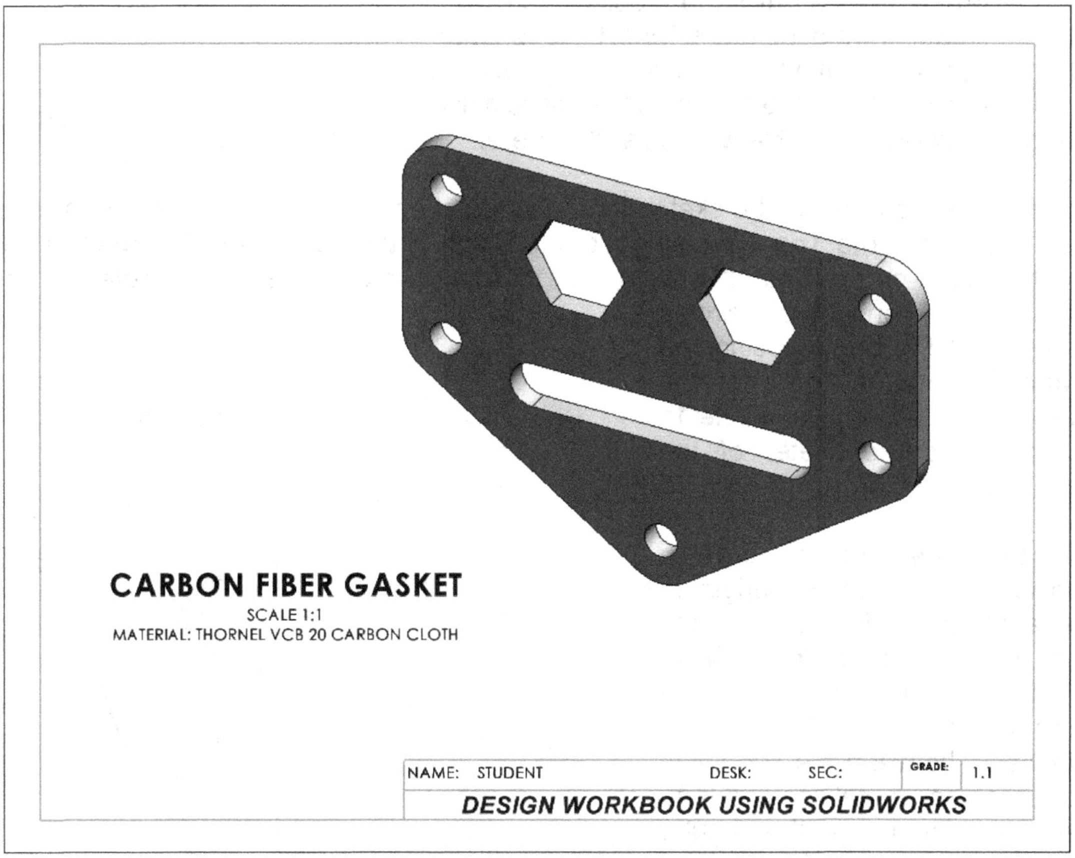

Exercise 1.2: Cover Plate

There are many different approaches to creating a 2D sketch using **SOLIDWORKS**. The approach will depend on the design intent and the designer's own preferences and choices. In the previous Exercise 1.1, the sketch was added to the Front plane. In this exercise you will add a sketch on the Top plane since this is the normal orientation of the cover plate.

Make a new part using the **ANSI-INCHES** template following the instructions on **Page 1-9**. Select the **Top plane** in the FeatureManager, and from the **Sketch** tab in the **CommandManager** select **Sketch**. After adding the first sketch, the part's orientation will change to make the selected sketch plane parallel to the screen, in this case to a **Top** view.

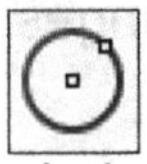

Select the **Circle** command and draw a circle centered at the origin. Use the **Smart Dimension** command and make the circle's diameter **10.00"**. After the circle is dimensioned, the sketch is fully defined, and you are ready to make an extrusion.

From the **Features** tab select **Extruded Boss,** or from the menu **Insert**, **Boss/Base**, **Extrude.** Make the extrusion using the Blind end condition, **0.500"**, and click OK to finish. The cover plate's **Isometric** view is shown here.

The next step is to make the sketch for the indentation of the Cover Plate. Change the view orientation to a **Top** view, select the top face of the part, select the **Sketch Tab** and then click on the **Sketch** icon to start the next sketch. Select the **Circle** tool and draw two concentric circles centered at the origin. Using the **Smart Dimension** tool, set the diameter of the smaller circle to **3.50"** and the diameter of the larger circle to **8.50"**.

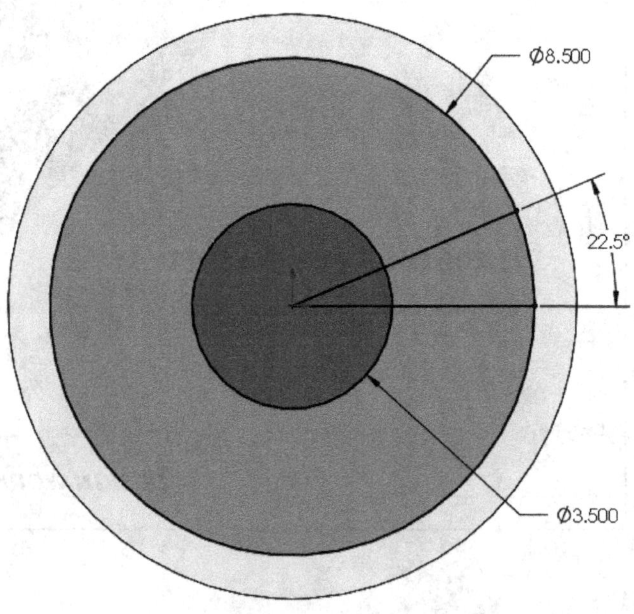

Use the **Line** tool and draw a horizontal line from the origin to the large circle. Next, draw a line from the origin at an angle above the horizontal line, also up to the large circle. Add a **22.5°** angle dimension between the two lines. Select the horizontal line and then the angled line to add the angular dimension as indicated.

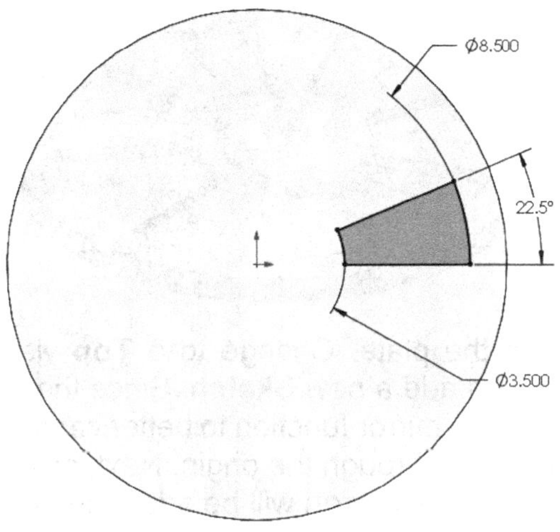

Select the **Trim** tool from the sketch toolbar. In the **Trim command options** in the PropertyManager select the **Trim to Closest** option. Note the cursor has a small scissors icon next to it, letting you know the trim command is active. Use the left mouse button to Trim the lines and circles to match this image.

After the angled line is trimmed, it turns blue, meaning it is no longer defined. To see what this means, click and drag it with the left mouse button; you will see it move up and down. To fully define it again, hold down the **Ctrl** key, select the angled line and the origin, and add a **Coincident** geometric relation.

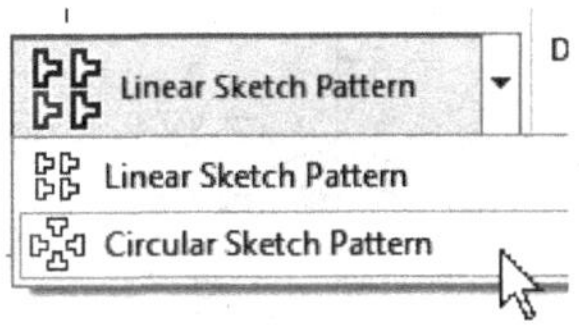

The next step is to add a circular pattern of the sketch profile. From the **Linear Sketch Pattern** drop-down command, select **Circular Sketch Pattern**.

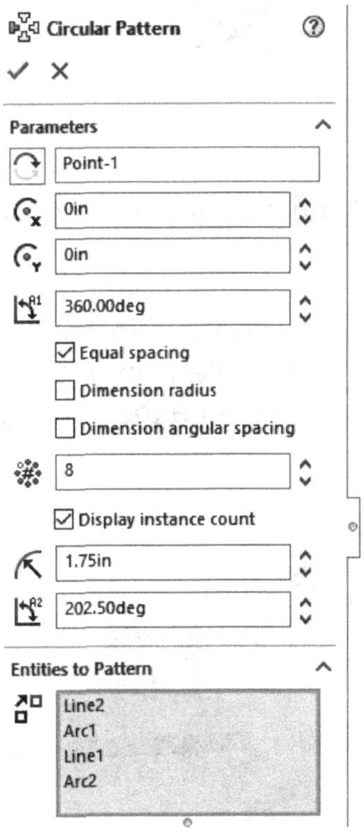

Activate the first selection box (Center of Circular Pattern) and select the Origin, make sure the **Equal spacing** option is checked, and change the number of instances to **8**. Keep in mind this value includes the original. The last step is to activate the **Entities to Pattern** selection box and select the four sketch elements trimmed in the previous step. Review the preview and click OK to finish. The completed sketch.

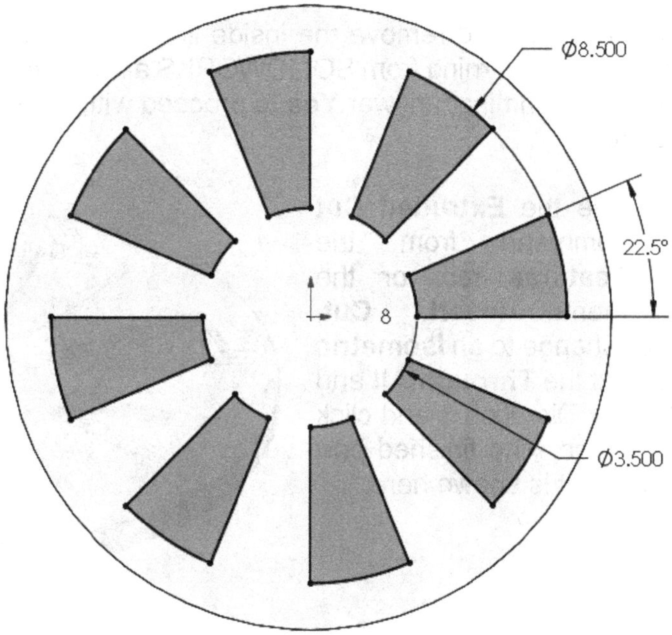

Now you will make a Cut Extrude to make the indentations in the cover plate. At this point, it is helpful to change to an **Isometric** view orientation. From the **Features** tab select the **Extruded Cut** command. In the Cut-Extrude PropertyManager enter set the direction 1 End condition to **Blind** and set the distance to **0.25”**. Click OK to finish.

You will now create the lift hole in the middle of the plate. Change to a **Top** view orientation, select the top face of the cover plate, and add a new **Sketch**. Since the lift hole will be symmetric about the origin, you can use the mirror function to better capture the design intent. Start by drawing a vertical **Centerline** through the origin. Next, select the menu **Tools, Sketch Tools, Dynamic Mirror**. An equal sign will be added to both ends of the centerline.

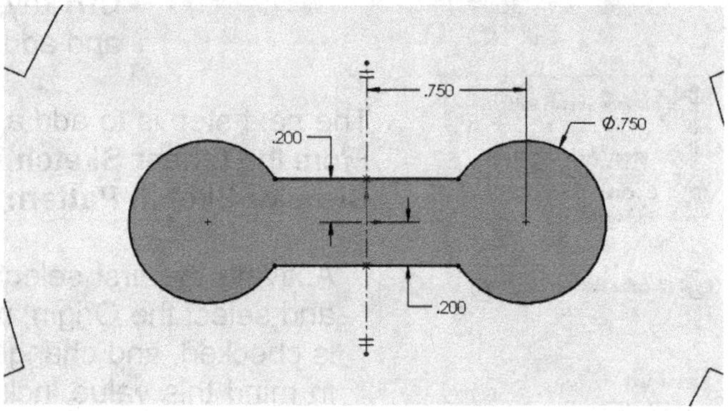

Remember every operation you perform on one side will be duplicated on the other side. Now, draw a small **Circle** to the right side of the centerline and two parallel **Lines** that start at the centerline and go to the circle. Use the Smart Dimension to dimension the sketch as indicated.

Hold down the **Ctrl** key, select the center of one circle and the **Origin**, and add a **Horizontal** geometric relation.

Use the **Trim** tool and remove the inside lines of the circles between the two parallel lines. If you get a warning from SOLIDWORKS about eliminating associated dimensions while you are trimming, answer **Yes** to proceed with the trim, and re-create the missing dimensions.

Use the **Extruded Cut** command from the **Features** tab, or the menu **Insert**, **Cut**, **Extrude**. Change to an **Isometric view**, select the **Through All** end condition for Direction 1 and click the **OK** button. The finished part up to this point is shown here.

To finish the cover plate, you need to add four circular cuts around the perimeter for the guide holes. Change to a **Top** view, select the top face of the cover, add a new **Sketch,** and draw four **Circles**. The circles should be at 90 degrees to each other, and their centers will be located on the perimeter aligned horizontally and vertically with the origin.

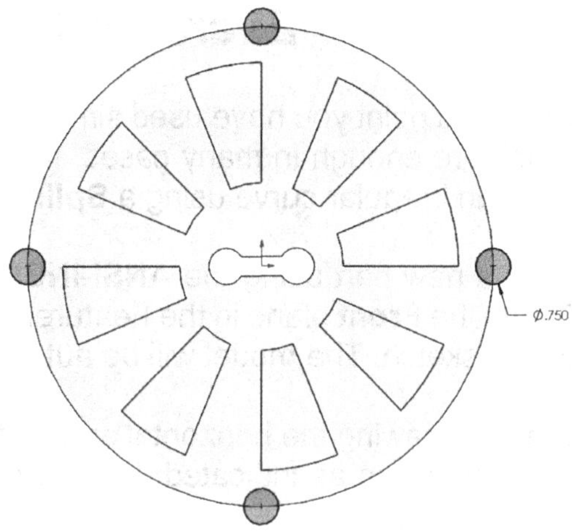

NOTE: For the left and right circles, add a **Horizontal** geometric relation between their center and the origin, and for the top and bottom circles, add a **Vertical** relation to the origin. This way the circles will be fully defined.

After drawing the circles select all four of them and add an **Equal** relation to them. To finish, dimension one of the circles **0.75"**, as shown.

OR

Draw a single circle and use the **Circular Sketch Pattern** tool, using the part's origin as the center of the pattern. Set the number of instances to **4** and click OK to finish.

Now select the **Extruded Cut** command from the Features tab. Set the end condition to **Through All** and click **OK** to finish. To change the part's color, right click in the model name in the FeatureManager, select the **Appearances** command from the context menu, and assign a color to your model.

Save your part as **Cover Plate.sldprt**, make a new drawing using the **TITLEBLOCK-INCHES** template and add an isometric view. Follow the instructions on **page 1-9**.

Remember to add the missing notes to complete the drawing.
Save the new drawing as **COVER PLATE.slddrw**.

Remember the part name and the drawing are the same, except with a different extension. The solid model and the drawing are linked, meaning that any changes made to the solid model will be automatically updated in the drawing.

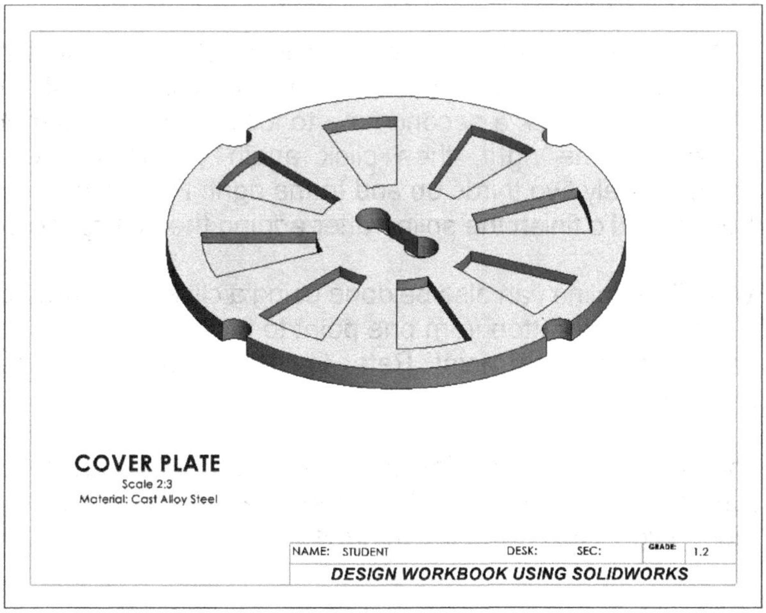

Exercise 1.3: WALL BRACKET

Up to this point you have used simple sketch geometry like lines, circles, and polygons, which are enough in many cases. In this **Wall Bracket** exercise, you will learn how to draw an irregular curve using a **Spline**.

Start a new part using the **ANSI-INCHES** template and save it as **WALL BRACKET.** Select the **Front** plane in the FeatureManager and click on the **Sketch** command to add a new sketch. The model will be automatically rotated to a **Front** view.

Start by drawing the horizontal and vertical lines of the Wall Bracket using the **Line** tool and dimension as indicated. Units are in inches and the part's origin is located in the upper left corner.

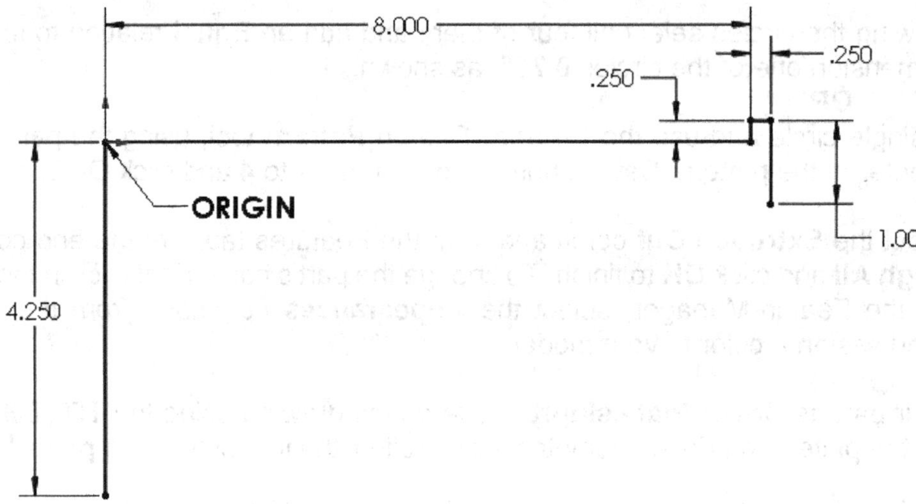

Now you will draw a spline to close the sketch profile. Select the **Spline** command from the Sketch tab, click on the first point at the bottom of the **4.25"** line, click a second time to locate the first intermediate point about one third up and to the right, then click again to locate the second intermediate point approximately two thirds up and to the right. Finally, click on the bottom endpoint of the **1.00"** line. To finish the spline after adding the last spline point, press the **ESC** key.

Note: The spline can also be done using a click-and-drag approach. Click-and-drag with the left mouse button from one point to the next for each spline point; the spline will be finished on the last point. Refer to the following image for the location of the spline points' locations.

The two intermediate points are used to control the shape of the spline. With the **Select** cursor, use the left mouse button to click and drag the intermediate point 1 to create a slight bulge. Next, drag the intermediate point 2 up to create a slight inflection.

Note: The spline can be approximated without exact dimensions for the intermediate points and adjusted as needed to obtain a shape similar to the next image.

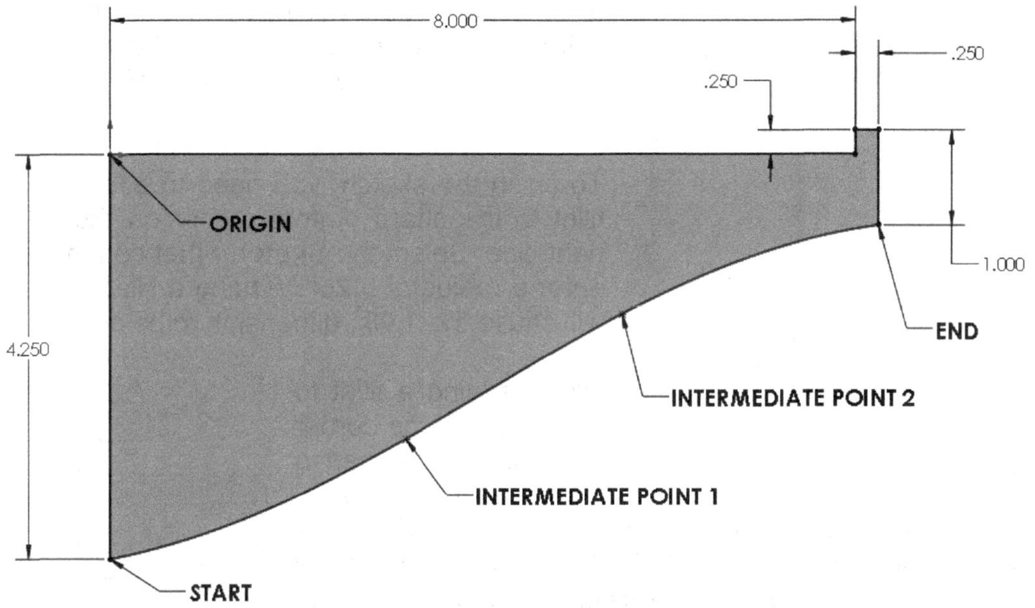

In order to fully define the spline, add the following dimension to the spline points.

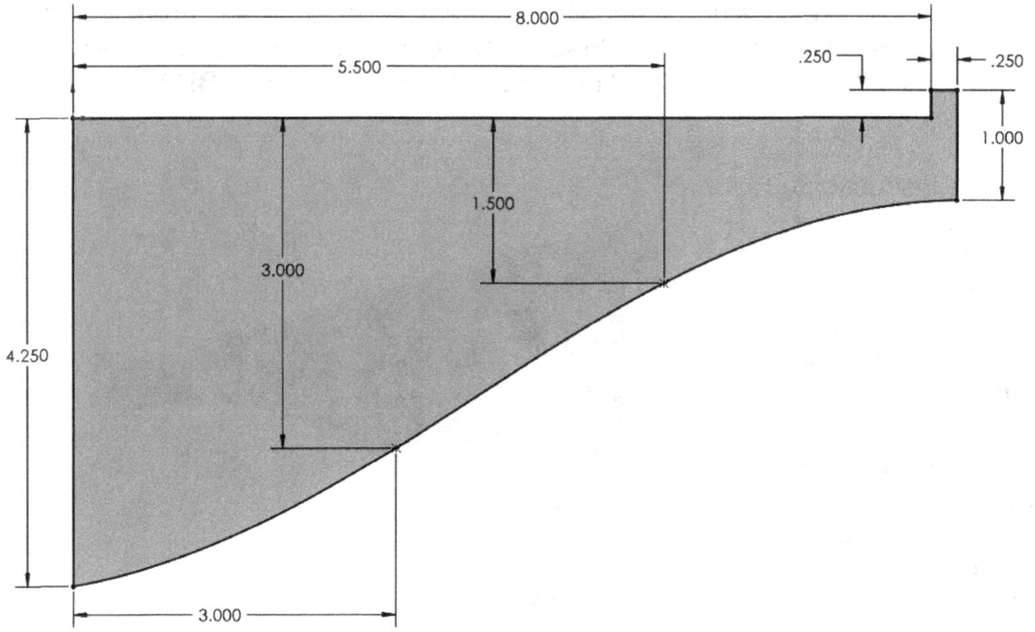

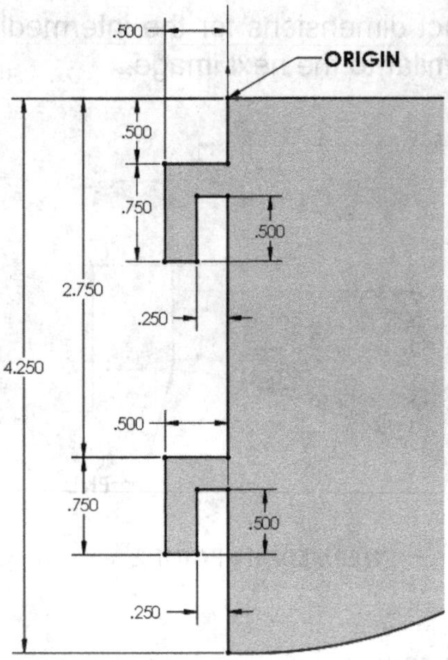

Now you will complete the left side of the Wall Bracket by adding the wall hooks. Add the two hooks using the **Line** tool and **Dimension** as indicated. You will have to use the **Trim** tool in two places to make the wall hooks contiguous with the rest of the profile.

To finish the sketch, you need to add a fillet to the sharp bottom corner on the right side. Select the **Sketch Fillet** command and enter a radius of **0.25"**. Adding a fillet to a spline will cause the **1.00"** dimension to be deleted.

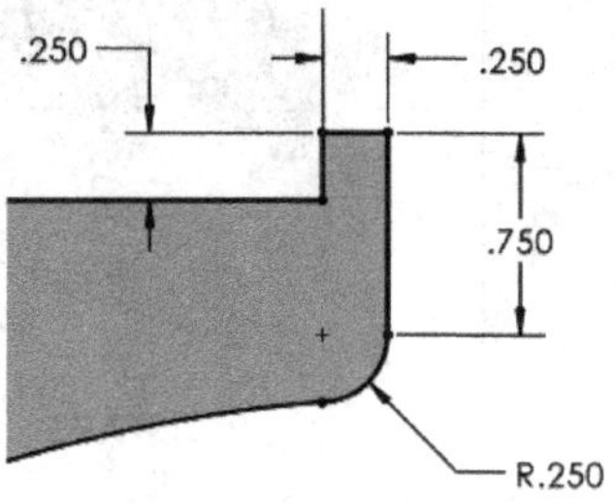

When adding a fillet to linear edges, the corner is usually constrained to maintain the dimension; that is not the case with a spline. When you select the endpoint to add the fillet, **SOLIDWORKS** will give you a warning letting you know the dimension will be deleted. Click Yes to delete it and dimension the vertical line **0.75"**. Add a vertical relation between the fillet's end and its center to fully define the sketch.

The 2D sketch is now complete and ready to be extruded. From the Features tab select **Extruded Boss/Base**, set the end condition to **Blind** with a depth of **0.125"** and click OK to finish. Save the part as **WALL BRACKET**.

Using the instructions from page 1-9, make a new drawing, add a **Trimetric** view with a scale of 1:1 and fill in the missing annotations.

Save your drawing as **WALL BRACKET.slddrw** and **Print** a hard copy for your instructor.

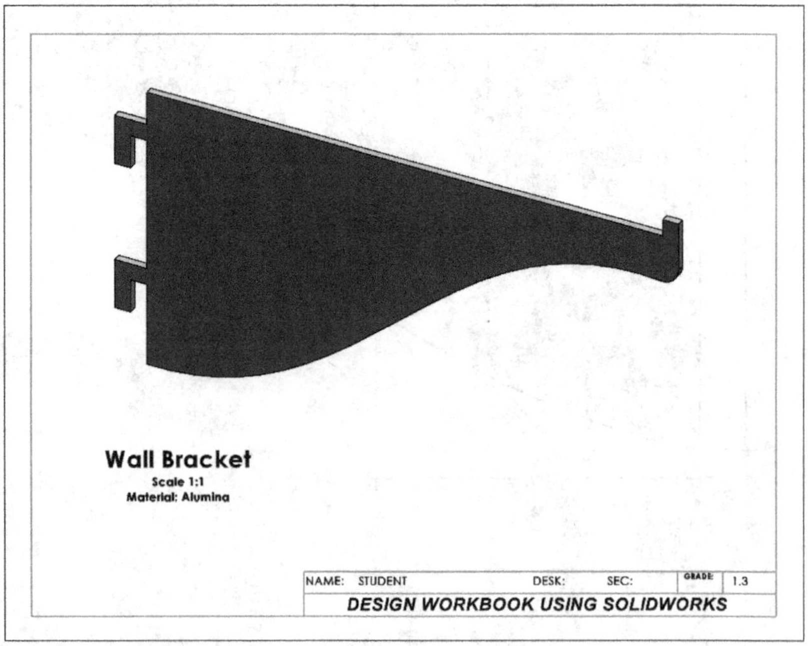

Exercise 1.4: MACHINE HANDLE

In the previous exercises, the parts were designed using English units (inches). International System (SI), or Metric units, is also common in engineering practice. In Exercise 1.4, the Machine Handle will be designed using millimeters as the base units. Since the machine handle is a cylindrical body, it will be created using a **Revolved Base** feature, by rotating a 2D sketch about an axis of revolution.

Start a new part using the **ANSI-METRIC** part template and save it as **MACHINE HANDLE** to continue. In the FeatureManager select the **Front** plane and add a new sketch. Add a horizontal **Centerline** starting at the origin going to the right. Use the **Line** and **CenterPoint Arc** tools to complete the profile, and dimension as indicated.
To fully define the sketch, add a **Coincident** relation between the arc's center and the bottom horizontal line.

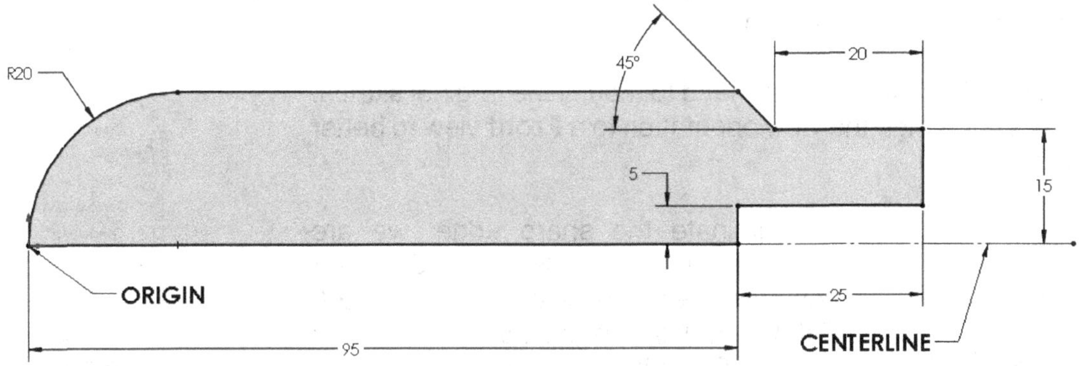

When the sketch is finished you are ready to create the revolved part. From the Features tab select the **Revolved Boss/Base**.
In the PropertyManager, if the sketch has a single **Centerline**, it will be automatically selected as the Axis of Revolution. If the sketch has multiple centerlines, you will need to select which one will be the axis of revolution.

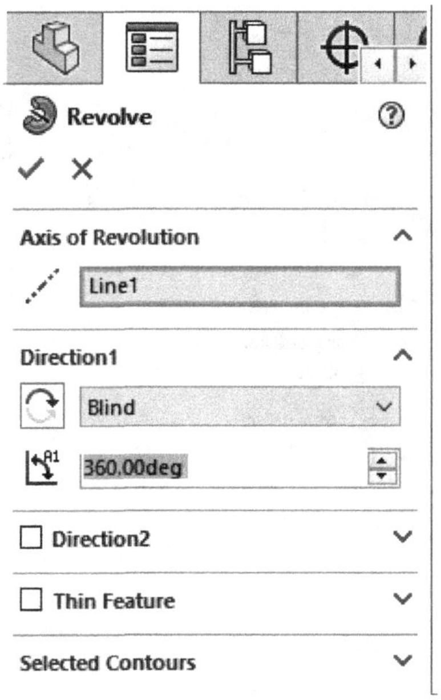

Now you need to define how many degrees to revolve the sketch about the axis of revolution. For this exercise use the default of **360°** to make a full revolution as shown and click OK to finish. Your part's **Isometric** view will look as follows.

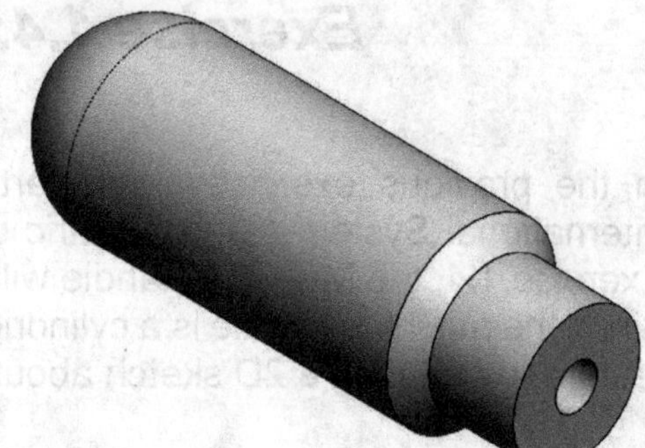

There are many different designs for a Machine Handle, and the one here is quite typical. After reviewing it, you notice that the sharp edge on the right side poses a safety risk. SOLIDWORKS allows you to easily make changes after creating the solid model. One way is to go back and edit the original sketch.

In the FeatureManager, SOLIDWORKS maintains a chronological list of the features used to model a part. Your first feature is "**Revolve1**," which is the operation you just completed. On the left of it you can see a right arrow. Click on this arrow to reveal the sketch used to create it, in this case "**Sketch1**."

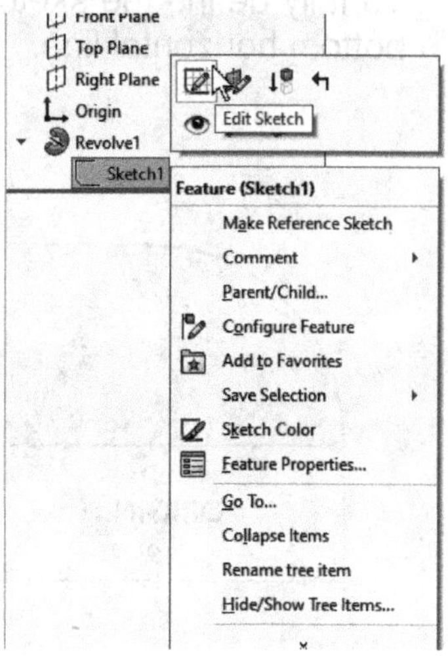

Selecting **Revolve1** or **Sketch1** with the left mouse button will show a context menu with common operations. If you right click instead, you will also reveal a context menu with expanded options as shown.

Select the **Edit Sketch** command to modify the original sketch. If needed, change the view orientation to a **Front** view to better see the sketch.

To eliminate the sharp edge, we are going to use the **Sketch Chamfer** command, which is nested in the **Sketch Fillet** icon. Click in the down arrow next to the **Sketch Fillet** and select the **Sketch Chamfer** command.

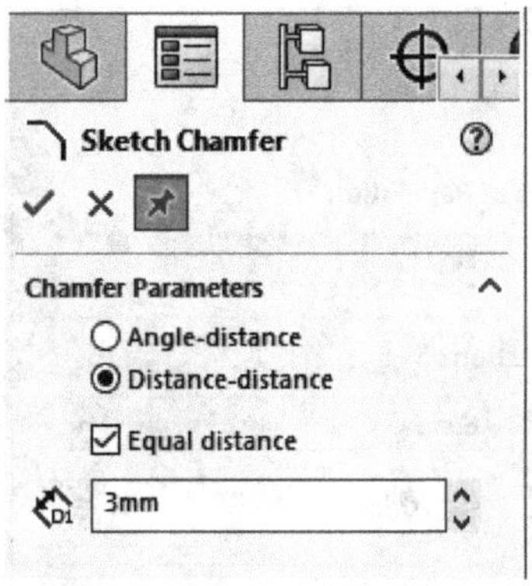

In the **Sketch Chamfer** PropertyManager select the Distance-Distance chamfer and turn on the Equal distance option and set the distance to 3mm.

Now click on the endpoint in the corner, or the two lines that form the sharp edge, and click OK to finish.

After the chamfer is added to the sketch, you need to rebuild the model to update the 3D model. Select the menu **Edit**, **Rebuild** or the Rebuild command from the main toolbar.

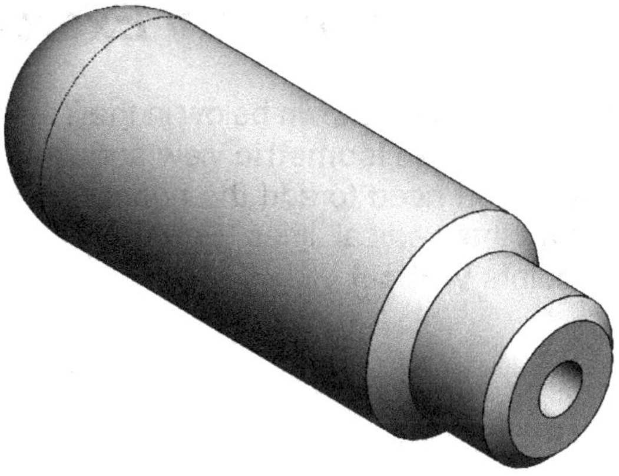

With this example you can see how easy it is to modify the sketch of a 3D solid model by simply editing its sketch to add a sketch chamfer, or any other change needed.

If you would like to change the color of your model, right click on the model name at the top of the FeatureManager tree and select the Appearances command from the context toolbar, to assign a different color to the model.

Save your part as **MACHINE HANDLE.sldprt**, make a new drawing using the **ANSI-METRIC** drawing template and add an **Isometric** view.

Make the **Isometric** view's scale 3:2 and add the missing annotations as shown, including a note with the Exercise number (**1.4**) in the upper right-hand box of the Title Block. Save your drawing to your personal folder as **MACHINE HANDLE.**

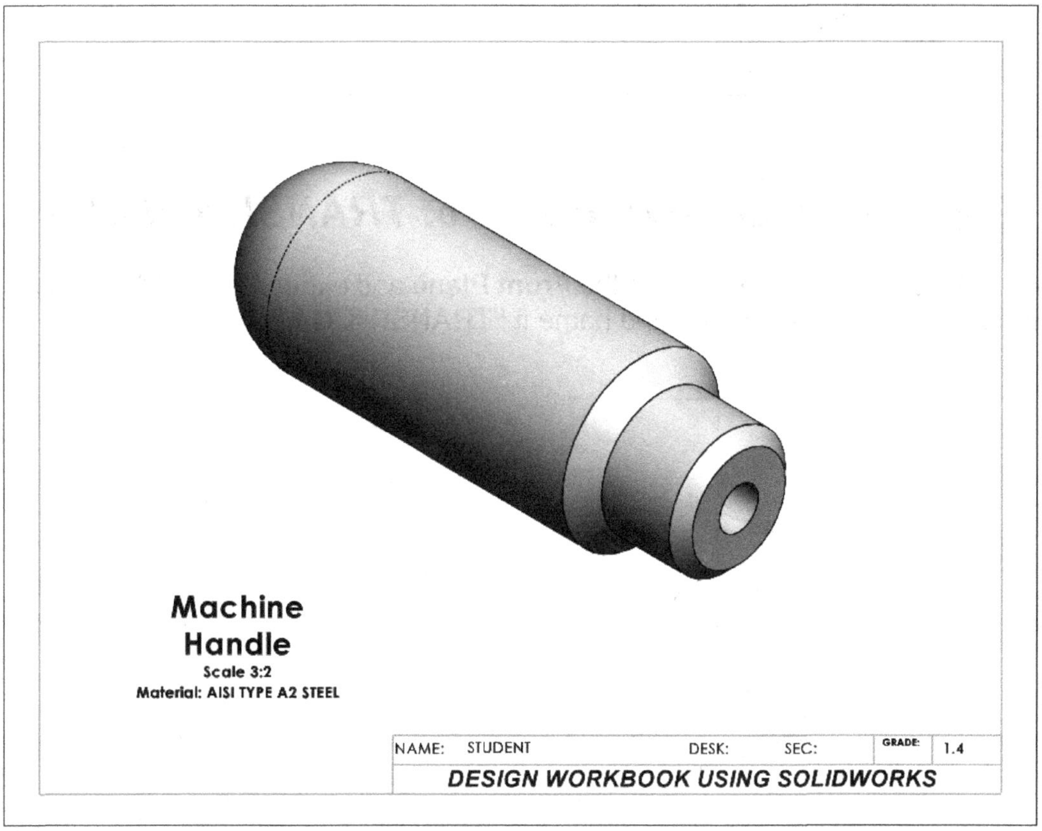

Supplementary Exercise 1-5: SLOTTED BASE

Make the sketch shown below in the **Top Plane** and extrude it **0.50** inches. Make a new drawing with an **Isometric** view and save it as **"SLOTTED BASE."** To fully define your sketch, you need to add the necessary geometric relations. Make the horizontal lines **Equal**, the vertical lines **Equal**, the angled lines **Equal**, and make the indicated endpoints **Vertical**.

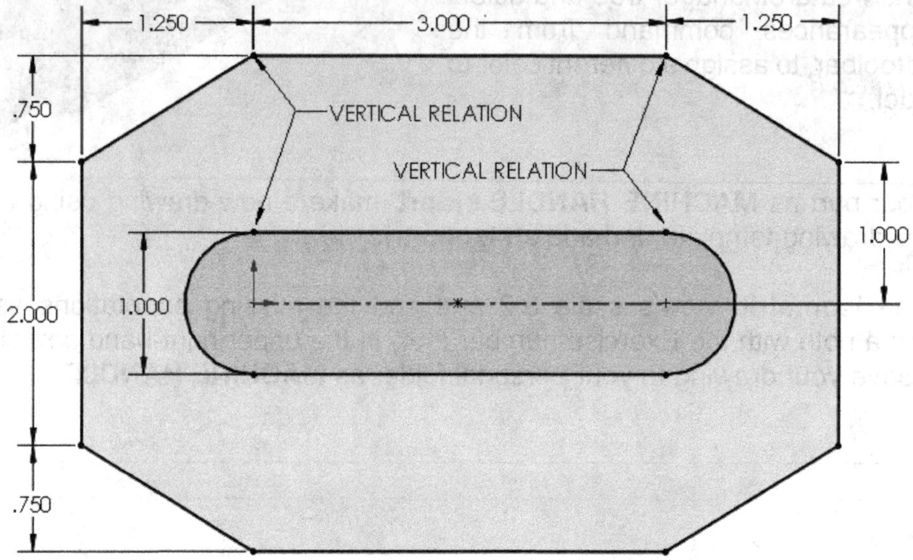

Supplementary Exercise 1-6: TRANSITION LINK

Make a sketch of the figure below in the **Front Plane** and extrude it **0.375"**. Make a new drawing with an **Isometric** view and name it **"TRANSITION LINK."**

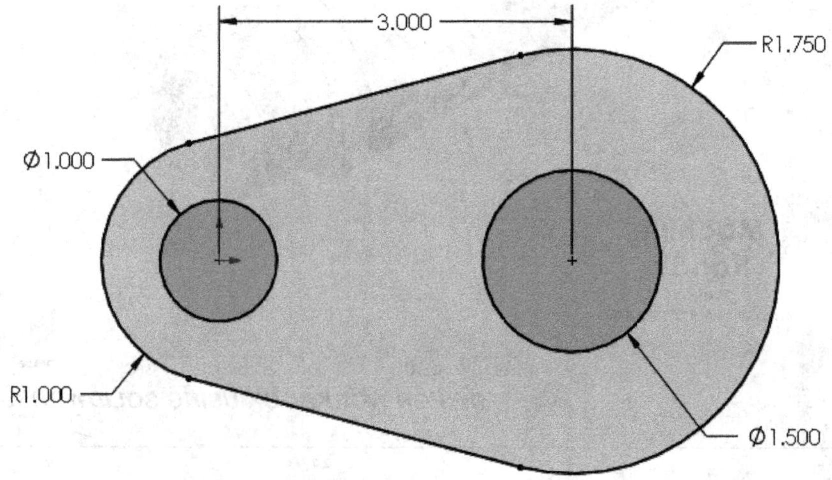

Supplementary Exercise 1-7: TEE BRACKET

Make a sketch of the figure below in the **Top Plane** using the **ANSI-METRIC** template and extrude it **40 mm** as indicated. Make a new drawing, add a **Trimetric** view in it and save it as **TEE BRACKET**.

TIP: Draw the outside perimeter and use the **Offset Entities** command from the Sketch tab to draw the inside profile. Select the perimeter and set the offset distance to 5mm. Use the "Reverse" option in the **Offset Entities** command.

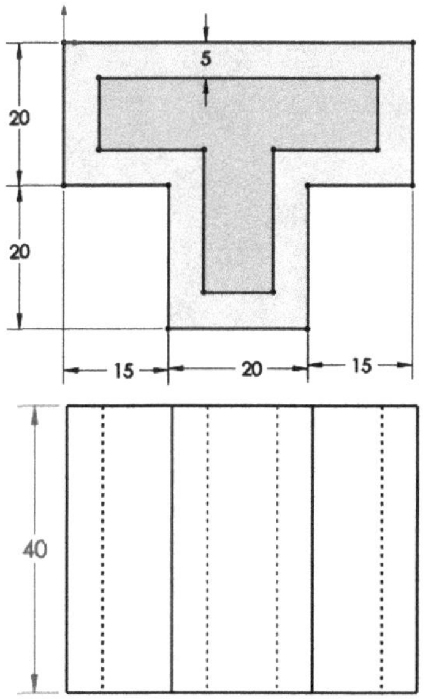

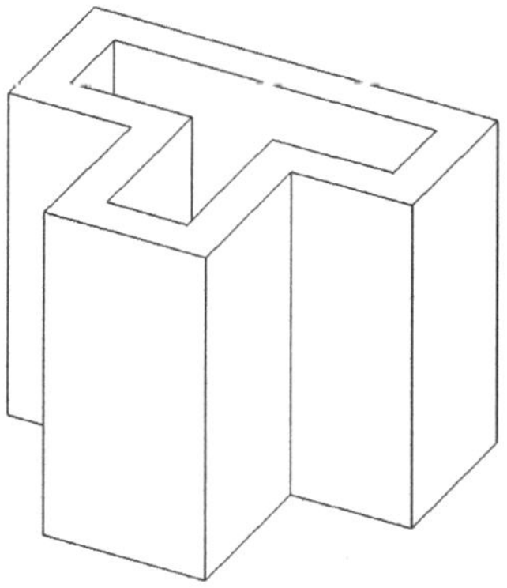

TRIMETRIC VIEW OF THE TEE BRACKET

Supplementary Exercise 1-8: CABLE SPOOL

Use the **ANSI-INCH** template, add a sketch in the **Front Plane** with the profile shown below and make a **Revolved Base**. Remember to add a centerline to revolve about. Make a new drawing with an Isometric view and save it as **CABLE SPOOL**.

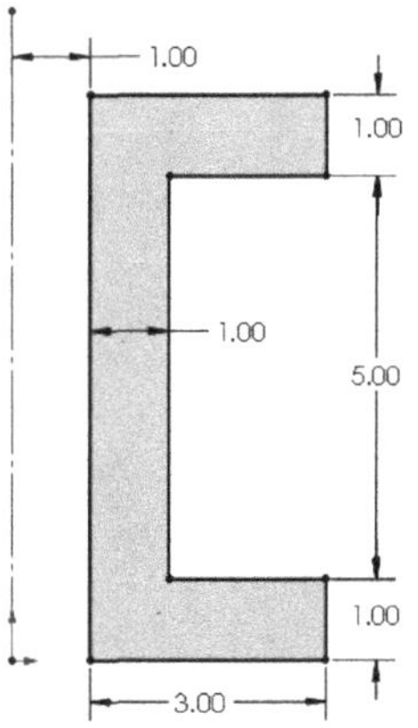

SKETCH PROFILE

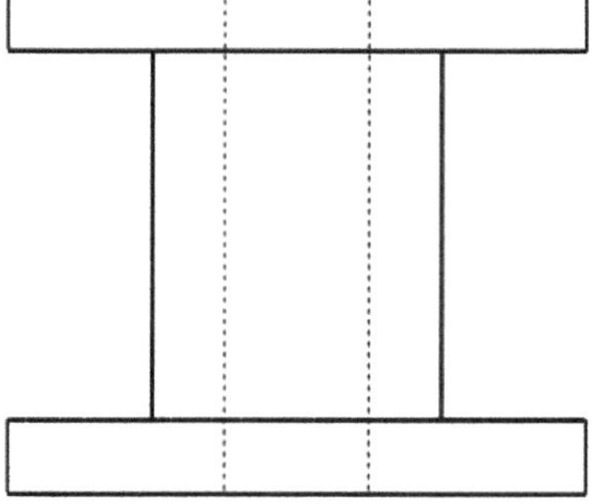

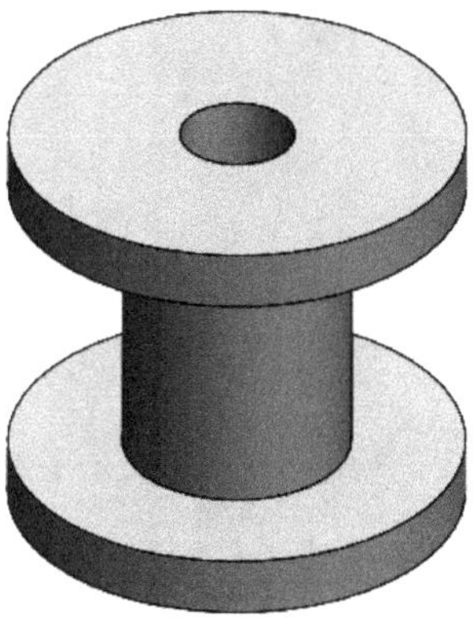

ISOMETRIC VIEW OF THE CABLE SPOOL

NOTES:

Design Workbook Lab 2
Advanced 2D Sketching

ADVANCED 2D SKETCHING

In the first chapter you learned some of the basic 2D sketch tools available in SOLIDWORKS. The first exercises reinforced how to draw a **Line**, **Circle**, **Rectangle**, **Arc**, **Polygon**, **Centerline**, and **Spline**. You also learned how to modify a 2D sketch using **Dimensions**, as well as the **Trim**, **Mirror**, **Fillet**, and **Chamfer** commands. In this lesson you will learn more advanced 2D sketch commands available in SOLIDWORKS.

SKETCH ENTITIES MENU

The complete list of sketch drawing tools available is listed in the menu **Tools, Sketch Entities,** as shown here, and most of them have additional options available to help you define your design's geometry, depending on your design intent. The commands more commonly used are included in the default sketch toolbar. The complete list of sketch entities available include:

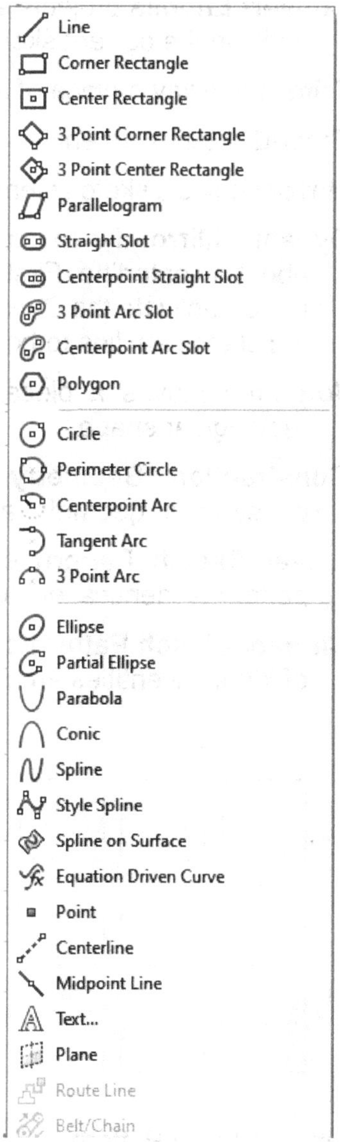

Line
Rectangle (multiple definition options)
Parallelogram
Slot (multiple definition options)
Polygon
Circle
Perimeter Circle
Centerpoint Arc
Tangent Arc
3 Point Arc
Ellipse (multiple definition options)
Partial Ellipse
Parabola
Spline
Spline on Surface
Point
Centerline
Text

While some of these 2D sketch entities are more commonly used than others, you will have a chance to use most of them while completing the exercises in this book.

SKETCH TOOLS MENU

All the 2D sketch editing tools are included in the menu **Tools, Sketch Tools**, and just like sketch entities, the most used are included in the default Sketch toolbar. The list of tools includes:

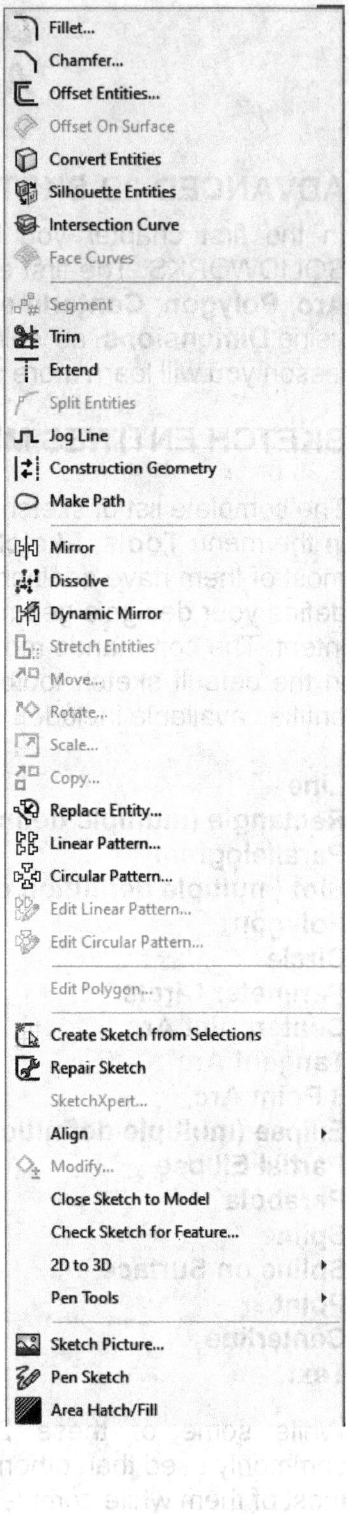

Fillet is used to round a corner with a radius.

Chamfer is used to cut a corner at an angle.

Offset Entities is used to create a copy of a sketch entity at an offset distance from the original.

Convert Entities projects a 3D feature's edge as a sketch entity in the current sketch.

Trim cuts away a piece of a sketch entity.

Extend extends an entity to meet another entity.

Mirror copies a sketch element about a selected mirror line.

Dynamic Mirror dynamically mirrors new sketch geometry about a centerline. First select the entity about which to mirror, activate the Dynamic Mirror command, and draw the sketch entities to be mirrored.

Jog Line moves a piece of the line up or down in a rectangular shape.

Construction Geometry converts sketch entities to construction geometry and vice versa.

Linear Sketch Pattern creates a linear or rectangular pattern of identical entities.

Circular Sketch Pattern creates a radial (or polar) pattern of identical entities around a center point.

Linear Sketch Pattern

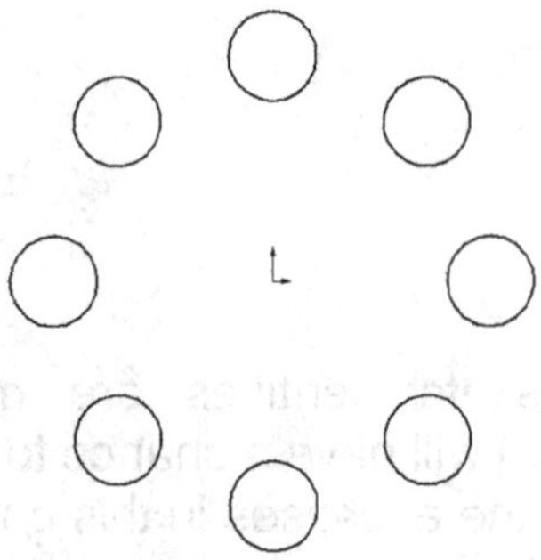

Circular Sketch Pattern

Exercise 2.1: METAL GRATE

In Exercise 2.1, you will design a Metal Grate. A grate usually has multiple identical slots, but instead of adding them individually, you can draw one slot, and use the **Rectangular Pattern** command to copy the rest. After the sketch is complete you can extrude it to build the base feature.

To make the grate, create a new part using the **ANSI-METRIC** template. Select the **Front** plane in the FeatureManager and then click on the **Sketch** command from the Sketch Tab. Since the grate is symmetrical about the X and Y directions, it is to your advantage to locate the part centered in the **Origin**. To do this, select the **Rectangle's** drop-down list and click on **Center Rectangle**.

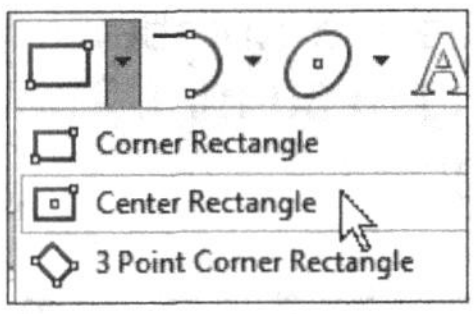

Draw the rectangle starting at the **origin** and dimension it **280 mm** wide by **195 mm** high. Next, select the **Rectangle** tool, draw the initial rectangular slot and dimension as shown next. Once the rectangles are dimensioned, your sketch will be fully defined (geometry will change from Blue to Black). Use the **Sketch Fillet** command to add a **3 mm** fillet to the corners of the small rectangle.

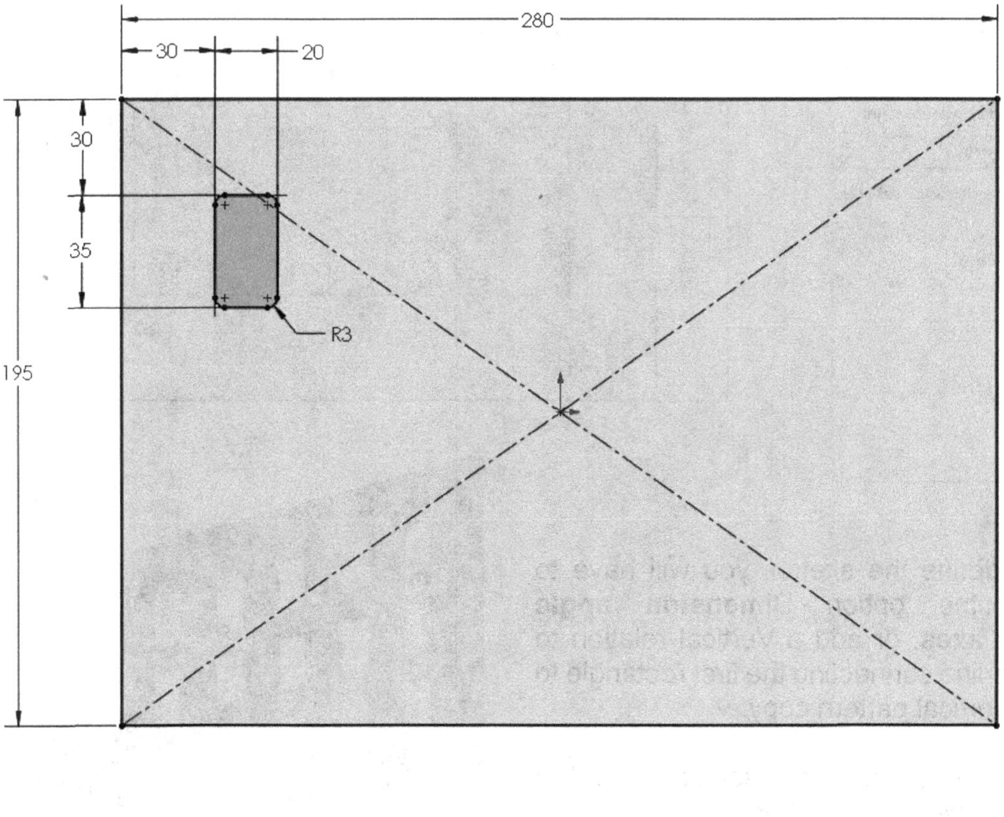

Now select the **Linear Sketch Pattern** command from the Sketch toolbar, or from the menu **Tools**, **Sketch Tools**, **Linear Pattern**. When the Linear Pattern PropertyManager is displayed, activate the Entities to Pattern selection box and select the lines and fillets of the small rectangle.

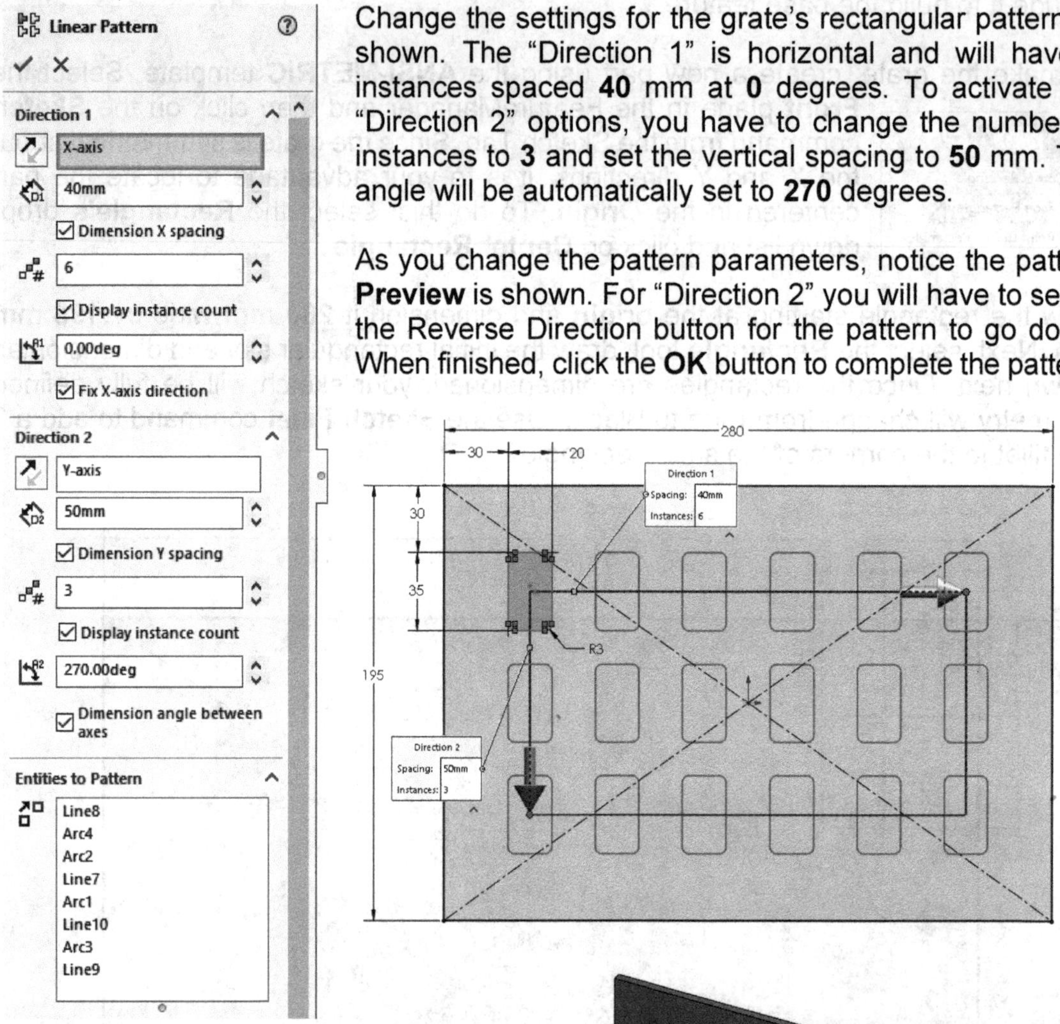

Change the settings for the grate's rectangular pattern as shown. The "Direction 1" is horizontal and will have **6** instances spaced **40** mm at **0** degrees. To activate the "Direction 2" options, you have to change the number of instances to **3** and set the vertical spacing to **50** mm. The angle will be automatically set to **270** degrees.

As you change the pattern parameters, notice the pattern **Preview** is shown. For "Direction 2" you will have to select the Reverse Direction button for the pattern to go down. When finished, click the **OK** button to complete the pattern.

To fully define the sketch, you will have to activate the option **Dimension angle between axes**, or add a Vertical relation to the centerline connecting the first rectangle to the first vertical pattern copy.

From the Features Tab select **Extruded Boss**, set the end condition as **Blind**, make the extrusion **5 mm** and click OK to finish.

The next step is to add a lip to the metal grate in order to provide a support when it is attached. Change to a **Front** view, select the front face of the Metal Grate and add a new Sketch.

To create the lip's sketch, you will leverage the previously made geometry. With the front face pre-selected, click in the **Convert Entities** command from the sketch tab. The outer edges of the face will be projected as sketch entities into the active sketch.

Notice the lines are all black; each projected entity is automatically added an **On Edge** geometric relation to the projected edges. This means that if the previous geometry is changed, the projected sketch entities will update accordingly.

To create the smaller internal rectangle, select the **Offset Entities** command. Click to select one of the previously projected entities, set the offset distance to **15 mm** and make sure the **Select Chain** option box is selected; this way all connected entities will be automatically selected.

If the offset preview is outside, select the **Reverse** option to create the offset inside and click **OK** to finish.

The offset command adds a single 15 mm dimension, letting you know the offset value. Also, each entity is added an **Offset** geometric relation to fully define the sketch. Add a **5mm** Fillet to the inside corners of the offset rectangle as shown.

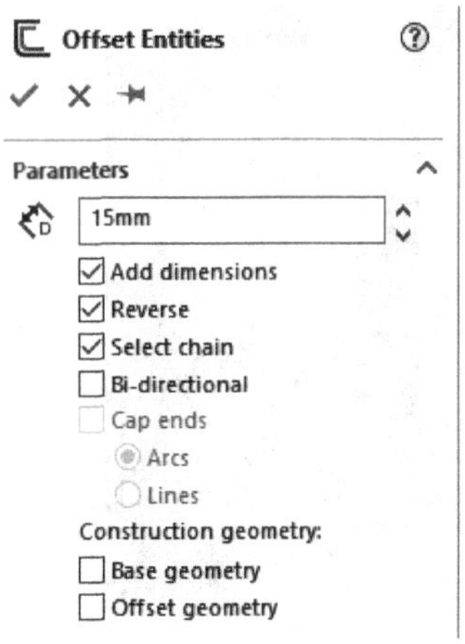

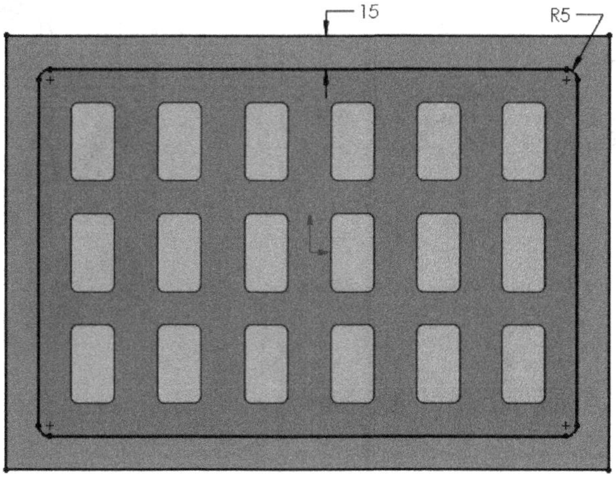

Before extruding the sketch, it is a good idea to change your view orientation to an **Isometric** view; this way it will be easier to see which direction the extrusion is going. Select the **Extruded Boss** command from the Features tab, set the end condition to **Blind** and change the extrusion depth to **5 mm.** Click **OK** to complete the boss.

Now you need to add the four attachment holes to the corners of the grate. Select the face of the last extrusion and add a new sketch on it. Draw an **8 mm** diameter circle in the upper left corner of the grate and dimension its center **9 mm** from the side and **9 mm** from the top edge. To better maintain the design intent, you'll use the **Sketch Mirror** command to create the other three circles. Starting in the origin, draw a vertical and a horizontal centerline as shown.

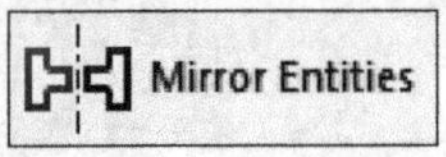

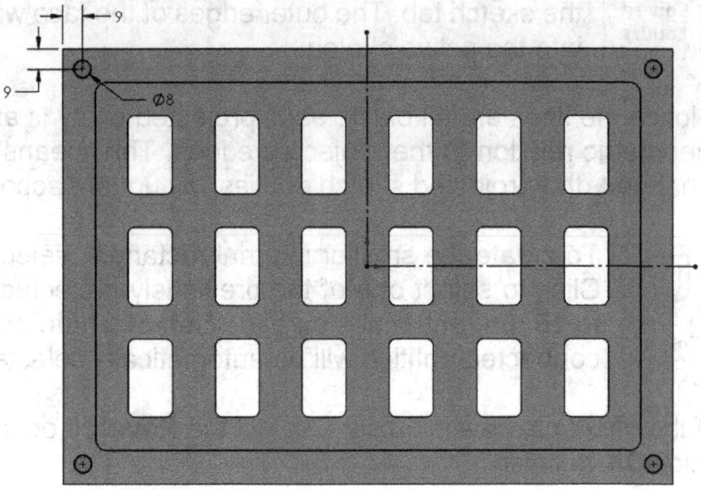

From the Sketch toolbar select the **Mirror Entities** command. In the "Entities to Mirror" selection box select the circle; in the "Mirror about" selection box select the vertical centerline and click OK. Next use the **Mirror Entities** again and this time select the two circles and mirror them about the horizontal centerline.

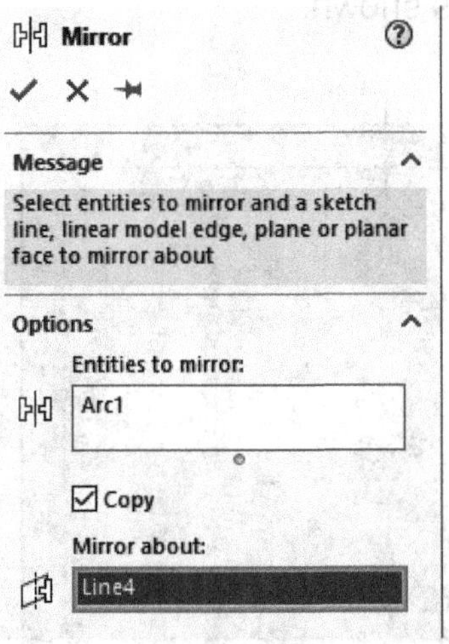

Change the model view's orientation to an Isometric. From the **Features** tab select **Extruded Cut**, set the End Condition to **Through all** and click OK to finish. The four corner attachment holes are now created on the grate.

The part is now complete, and you can view the lip feature more clearly by rotating the model.

If you'd like to, change the color of your model, and save your part as **METAL GRATE.sldprt**. Make a new drawing, add a **Trimetric** view, change the view's scale to 2:3 and complete the missing notes as before, including the exercise number.

Save your drawing as **METAL GRATE.slddrw** and print a copy to submit to your lab instructor.

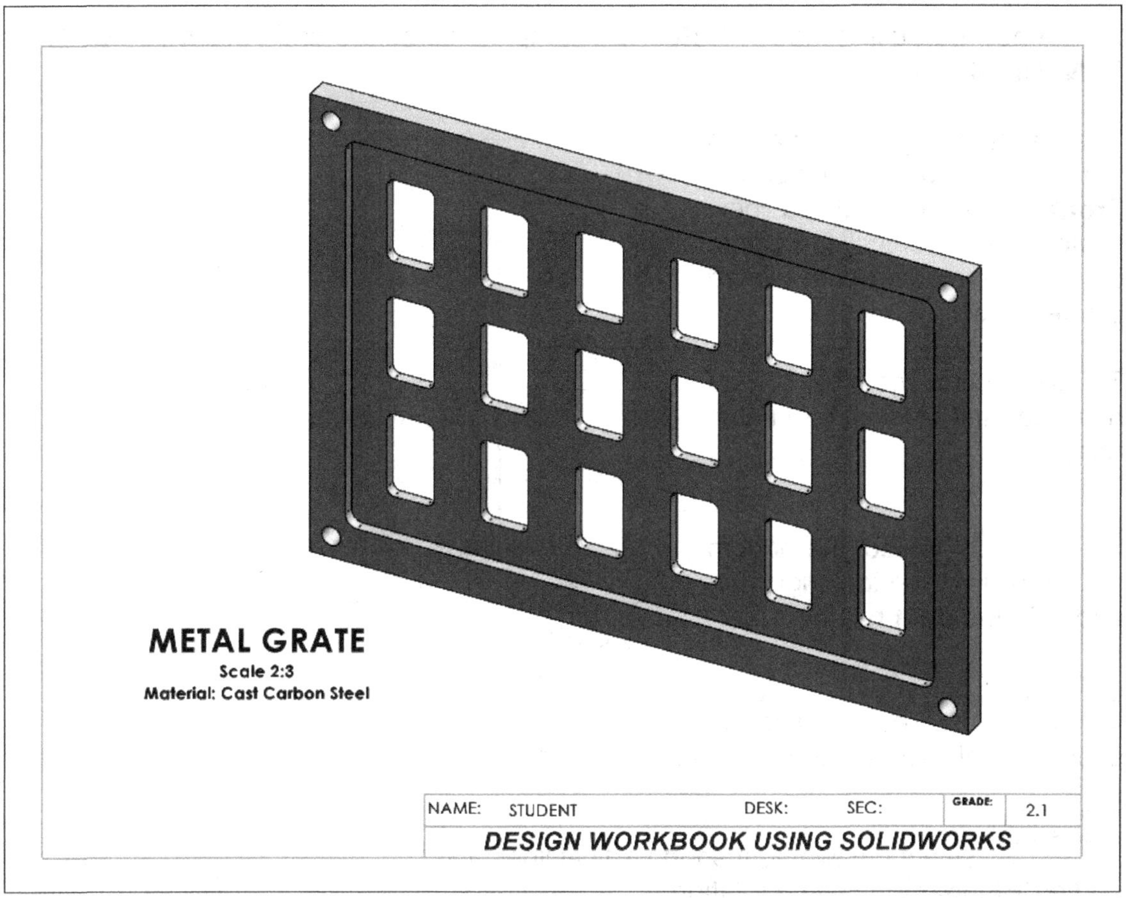

Exercise 2.2: TORQUE SENSOR

In Exercise 2.2 you will design a Torque Sensor casing. Since it is a cylindrical (axis-symmetrical) object, it will be created using a **Revolved Base** feature, and certain features will need to be made using a **Circular Pattern** to complete the part.

Start a new part using the **ANSI-INCHES** template as before and save it as **TORQUE SENSOR.sldprt**.

The first feature you will make is the revolved base. Select the **Front** plane and add a new sketch. Add a vertical centerline; draw and dimension the closed profile shown.

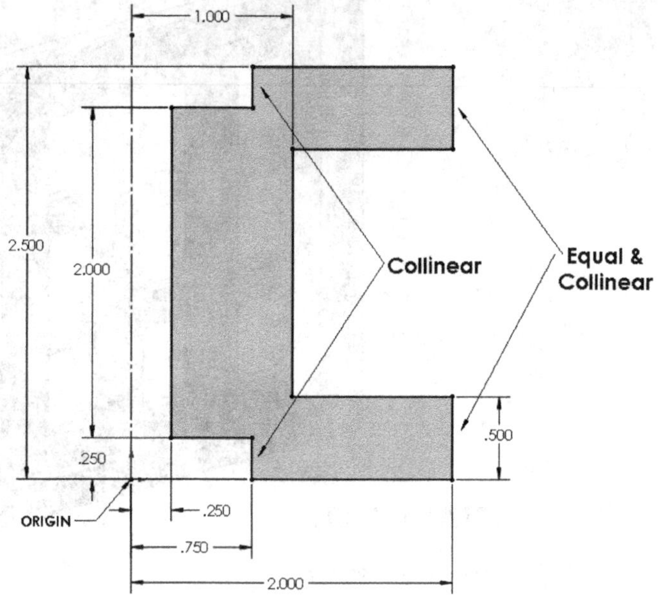

To fully define the sketch, you have to add a few geometric relations.

Select the bottom horizontal line and the origin and add a **Coincident** relation.

Select the two left short vertical lines and add a **Collinear** relation.

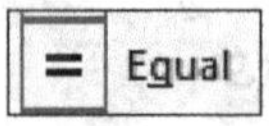

Select the two right short vertical lines and add a **Collinear** and **Equal** relation.

From the Features tab, select the **Revolved Boss** command. Make sure the centerline is selected as the axis of revolution, enter **360°** to make a full revolution and click OK to finish.

The resulting cylindrical part will have a **4.00"** outside diameter as shown.

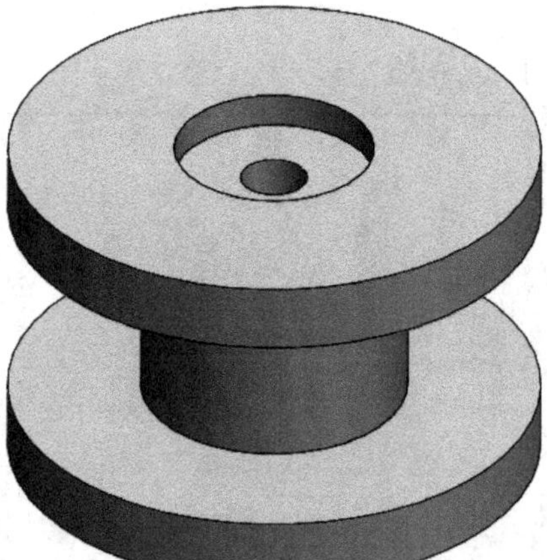

The next step is to add a circular pattern of holes at the top of the part. Change to a **Top** view orientation, select the top face of the part, and add a new sketch.

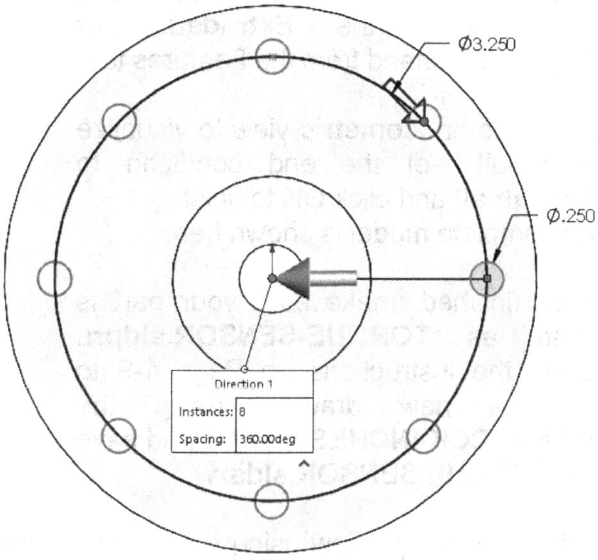

To locate the first hole, draw a **Circle** centered in the origin, and dimension it **3.25"**.

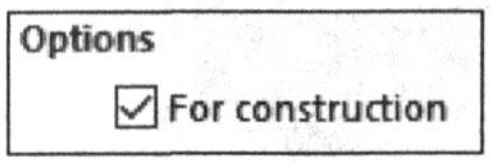

Select the circle and activate the "**For construction**" option in the PropertyManager to change the circle into construction geometry. Now draw the first circle of the hole pattern horizontal to the origin at the right quadrant of the construction circle. By doing this you will automatically capture a **Coincident** relation. Finally dimension it **0.25"** diameter as shown.

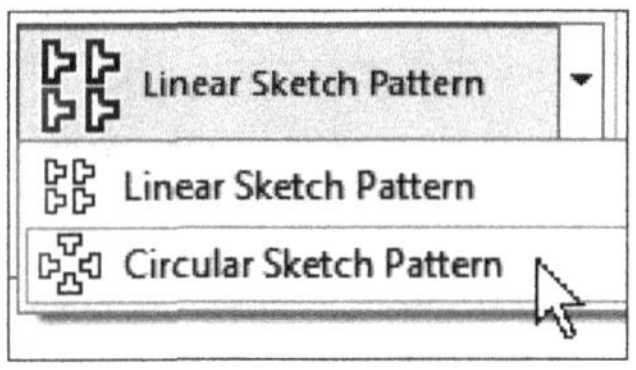

From the drop-down menu in the **Linear Sketch Pattern** select the **Circular Sketch Pattern** command.

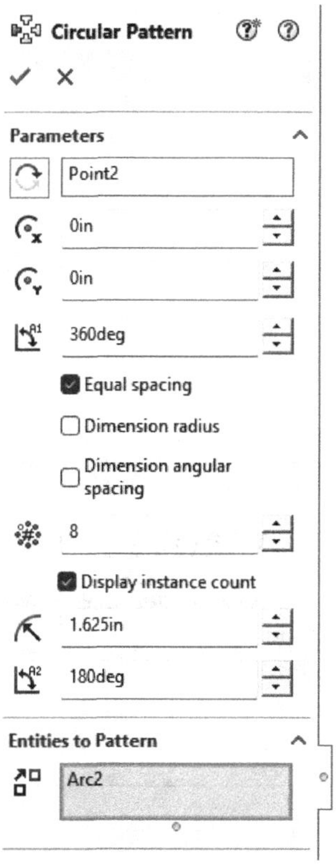

In the **Pattern Center** selection box select the sketch origin.

From the Circular Pattern options activate the "**Equal spacing**" option and set the angle to **360°.**

Change the number of instances to **eight (8)**, activate the "**Entities to Pattern**" selection box and select the circle to be patterned. When finished entering the options and selections click **OK** to continue.

NOTE: In this case, activating the "**Dimension radius**" option would cause the sketch to become Over defined, because the location of the circle has already been defined by the larger construction circle.

When your sketch is complete, select the **Extruded Cut** command from the Features tab.

Change to an **Isometric** view to visualize the result, set the end condition to **Through all** and click **OK** to finish.
The complete model is shown here.

When finished, make sure your part is saved as **TORQUE-SENSOR.sldprt**. Follow the instructions on Page 1-9 to make a new drawing using the **TITLEBLOCK-INCHES.drwdot** and save it as **TORQUE SENSOR.slddrw.**

Add an **Isometric** view using the **Shaded with Edges** display mode and change the view's scale to 1:1. Remember to add the missing annotations to the drawing including the exercise number and print a copy for your lab instructor.

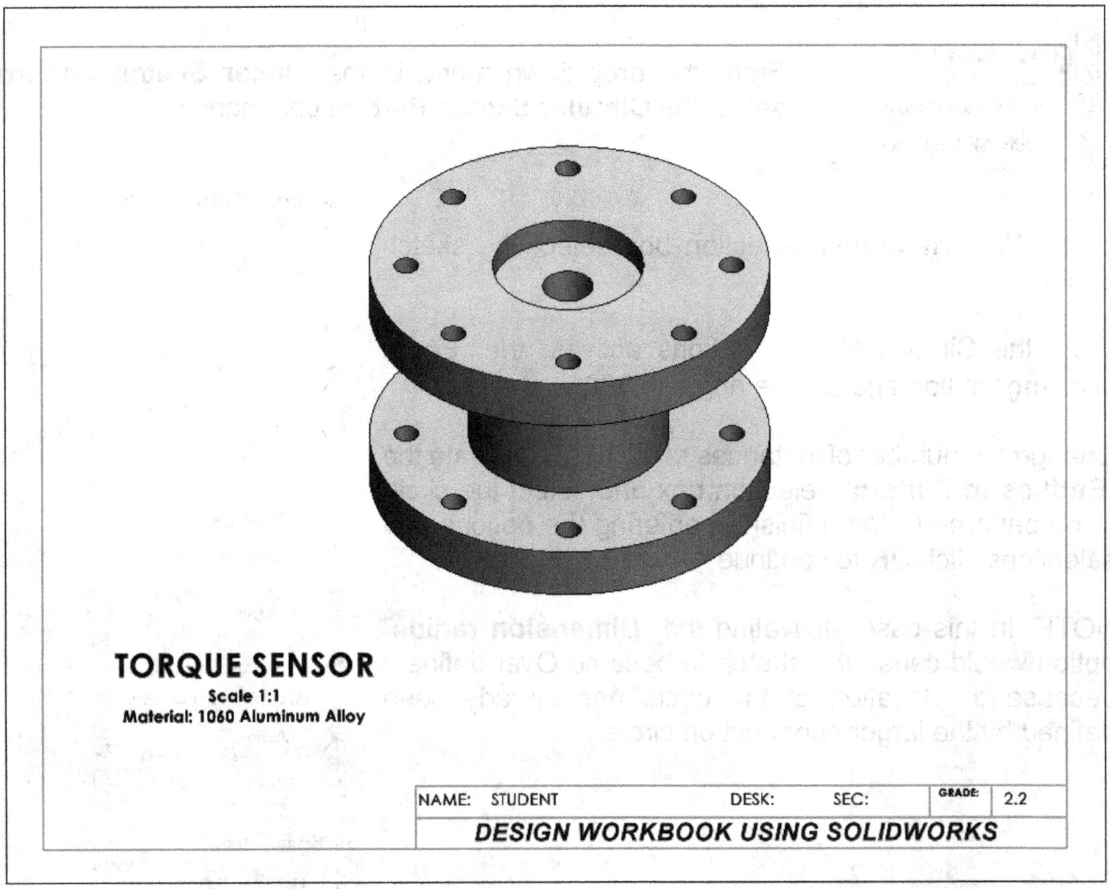

Exercise 2.3: SCALLOPED KNOB

In Exercise 2.3, you will design a Scalloped Knob that has some complicated geometry around its edges. This particular knob design will be a hexagon type. Since the hexagonal features are equally spaced around the center of the knob, you will use a circular pattern to maintain the design intent.

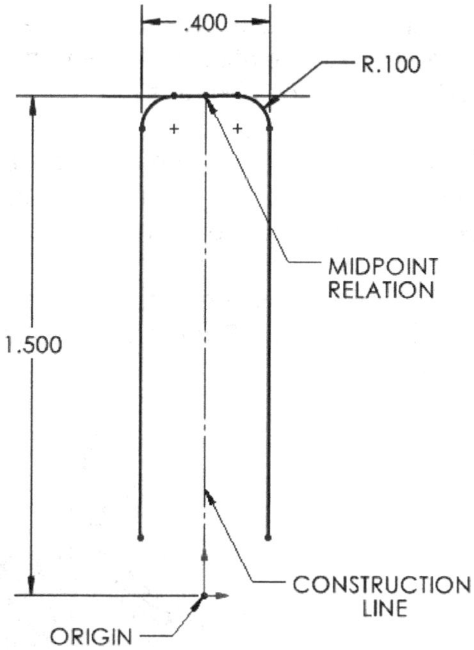

Start by making a new part using the **ANSI-INCHES** template and save your part as **SCALLOPED KNOB.sldprt**.

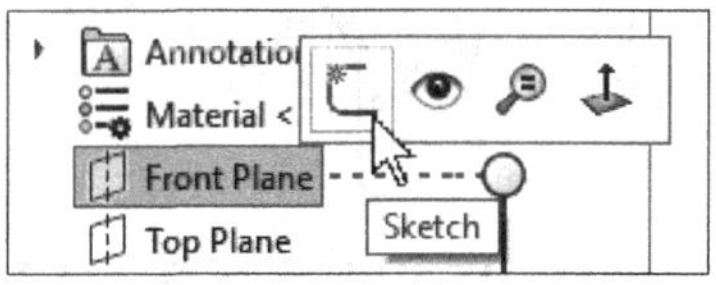

Select the **Front** plane in the FeatureManager and add a new sketch. You can select the **Sketch** icon from the pop-up toolbar immediately after selecting the plane. Using the **Line** and **Sketch Fillet** tools, draw and dimension the profile to fully define the sketch as shown. Make both vertical lines equal and add a **Centerline** from the **Origin** to the **Midpoint** of the top horizontal line. If the Centerline is not vertical, add a **Vertical** geometric relation to it.

From the **Linear Sketch Pattern** drop-down menu select the **Circular Sketch Pattern** command.

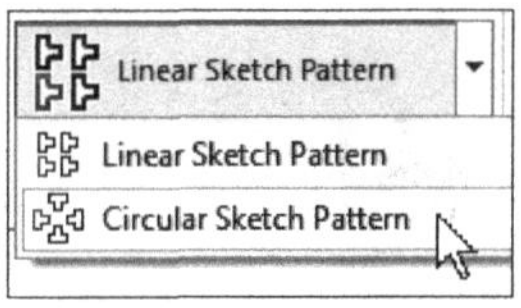

Activate the "**Pattern Center**" selection box and click in the Origin. Set the number of copies to **6,** activate the "**Equal Spacing**" option and set the Total Angle to **360°**.

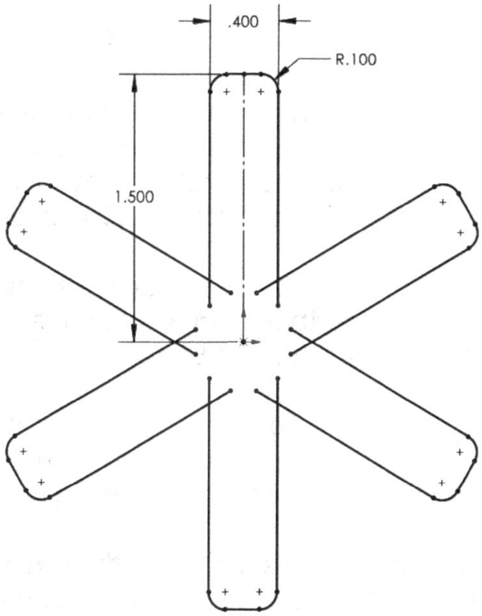

In the "**Entities to Pattern**" selection box add the three lines and the two fillets and click **OK** to continue. Adding the Centerline to the circular pattern is not necessary.

After the circular pattern is complete you will notice that some of the lines may overlap. The next step is to add the sketch fillets. By adding a fillet between the intersecting lines, a fillet will be created, and the lines will be trimmed at the same time.

Select the **Sketch Fillet** command and set the radius to **0.45"** in the PropertyManager. Now select two intersecting lines to preview the fillet. Select the rest of the intersecting pairs of lines and click OK to finish.

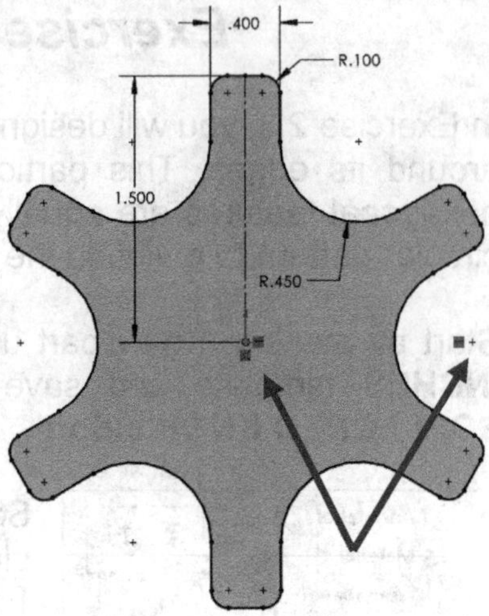

A single **0.45"** radius dimension will be added. Remember that all fillets created at the same time have an **Equal** relation to the dimensioned fillet. Your sketch should look like this.

If needed, to fully define your sketch, add a **Horizontal** relation between the **Origin** and the center of the **Sketch Fillet** on either side.

From the Features tab select **Extruded Boss**. Extrude the sketch **0.375"** using the **Blind** end condition and click **OK** to finish.

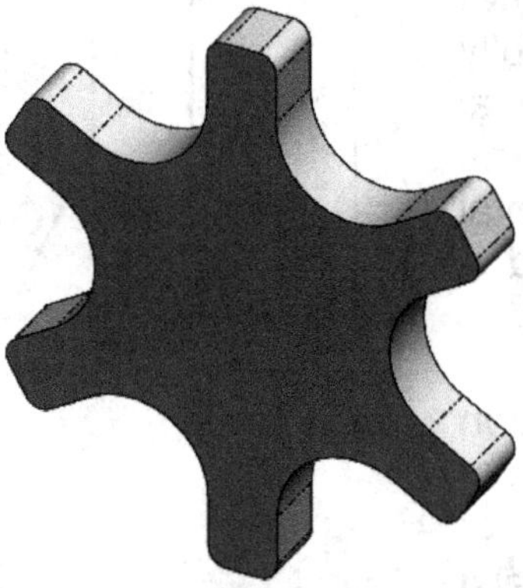

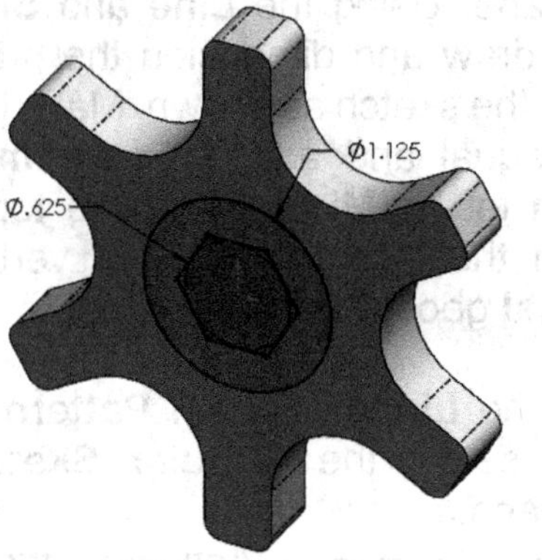

To finish the part, you will now add the attachment base. Change to a **Front** view orientation, select the front face, and add a new sketch. Draw a **Circle** centered at the origin and dimension it **1.125"** in diameter.

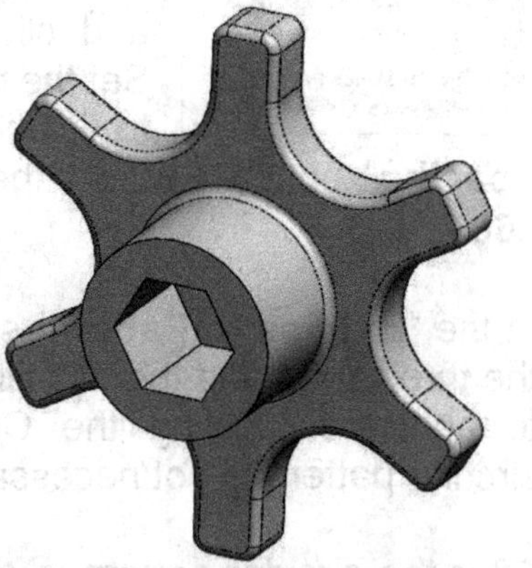

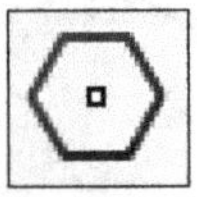

Now draw a **Hexagon** at the origin. Check **Inscribed Circle** option and dimension the diameter to **0.625"**.

To fully define the sketch, select the top (or bottom) line and add a **Horizontal** relation; this will prevent the hexagon from rotating.

From the Features tab select the **Extrude** command. Change to an **Isometric** view for better visibility and extrude the sketch **0.75"** away from the part using the **Blind** end condition.

To finish the knob, you need to add a few fillets. From the Features tab select the **Fillet** command and set the radius to **0.05"**. To remove the sharp edges, you can:

- Select the front and back edges of the first extrusion and the edge connecting the first extrusion with the second one.

Or

- Select the front and back faces of the first extrusion and click **OK** to finish.

Use the Appearances command to change the color of the part to your liking and save your model as **Scalloped Knob.sldprt**.

Make a new drawing using the **TITLEBLOCK-INCHES** template, add a **Trimetric** view using the **Shaded with Edges** display style and a scale of 2:1. Add the missing notes and save your drawing as **SCALLOPED KNOB.slddrw.**

Print a copy to submit to your lab instructor. The finished drawing is shown next.

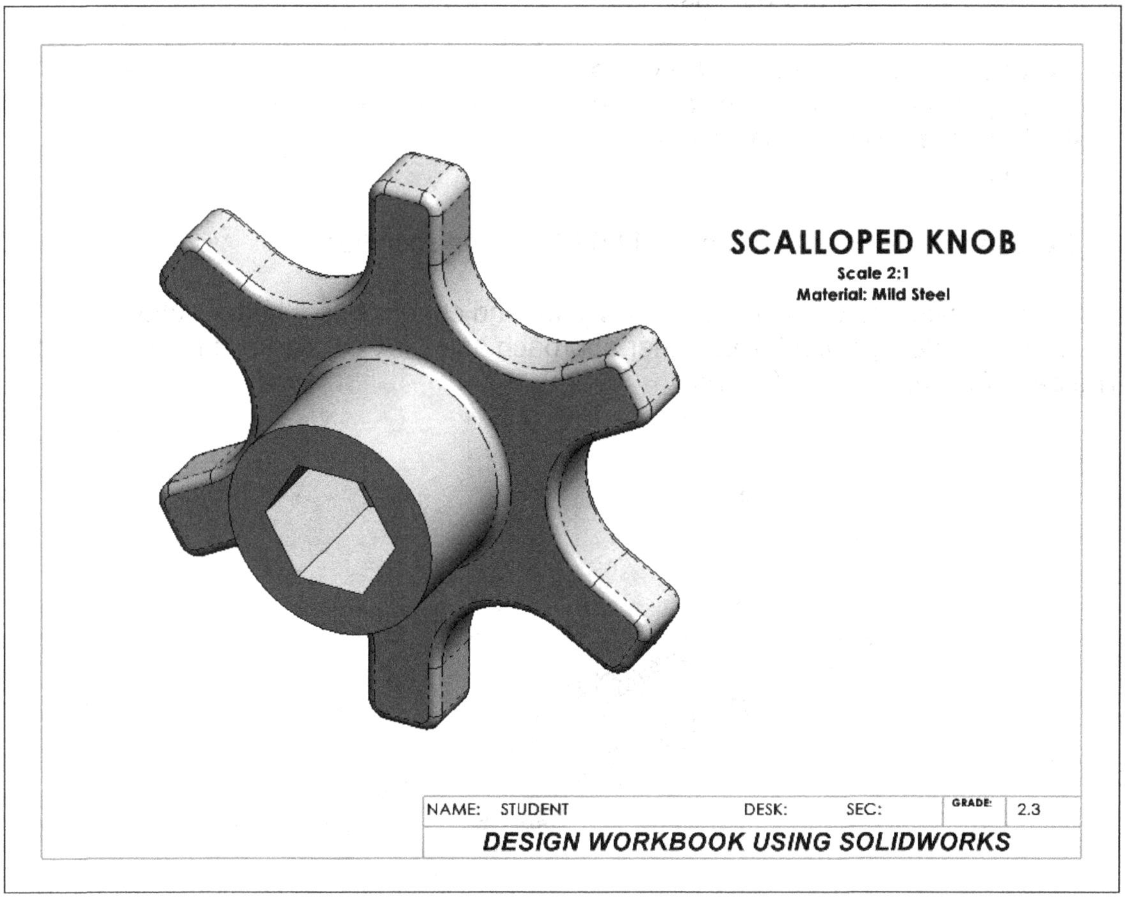

Exercise 2.4: Linear Step Plate

In Exercise 2.4, you will design a Linear Step Plate used for linear motion control in machinery. There are a lot of holes on this part, and you will find the linear pattern and mirror commands to be very useful in this case. Start a new part using the **ANSI-INCHES** template and save it as **LINEAR STEP PLATE.sldprt**.

Select the **Right** Plane and add a new **Sketch**. Your part will be automatically rotated to a Right view. Draw a vertical centerline through the Origin, and with the centerline selected, go to the menu **Tools, Sketch Tools**, and select **Dynamic Mirror**. Draw the right half of the profile, and the sketch on one side of the centerline will be automatically duplicated on the other side. Use the **Smart Dimension** command to fully define the sketch.

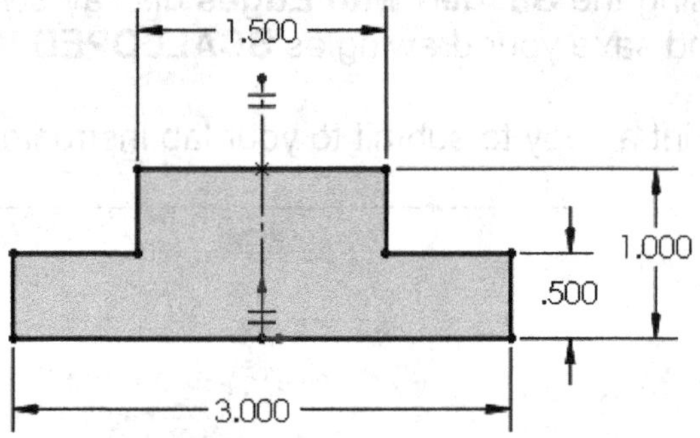

From the Features tab select the **Extruded Boss** command. To make the base feature symmetrical about the sketch plane you have two options:

- Extrude the sketch **4.20"** in **Direction 1** using a **Blind** end condition, and **4.20"** in **Direction 2** using a **Blind** end condition.

or

- Extrude the sketch **8.4"** using the **Mid Plane** end condition.

If the part is symmetrical, using the **Mid Plane** end condition is a better option to maintain your design intent. Use either approach and click **OK** to finish the extrusion. The base feature looks as follows.

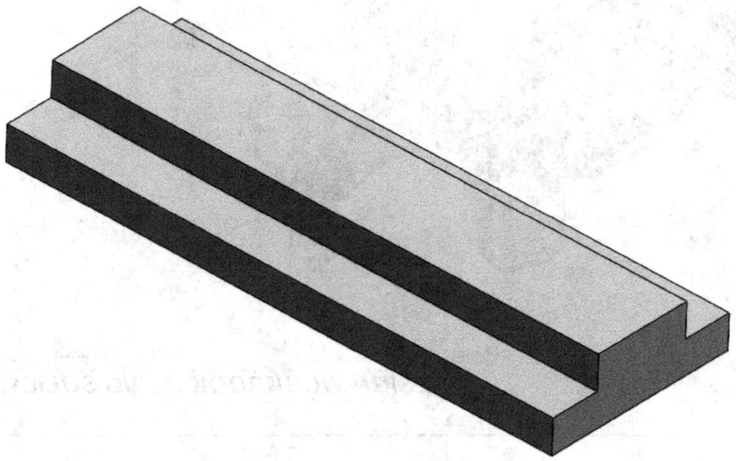

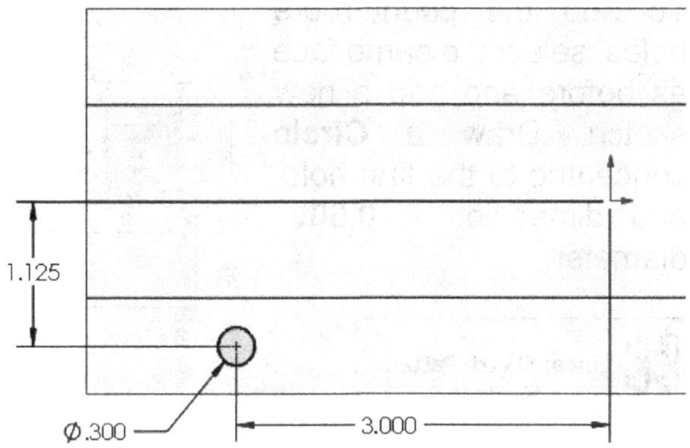

The next step is to add the holes to the part. Change to a **Top** view orientation and add a new sketch in the face of the lower step. Draw a small circle and dimension as shown.

Now create a Linear Pattern. Select the **Linear Sketch Pattern** from the Sketch tab. In the **Direction 1** options set the number of copies to **6** with a spacing of **1.200"** and activate the "**Dimension x spacing**" and "**Fix X-axis direction**" options. The last options are used to fully define the sketch and prevent it from moving.

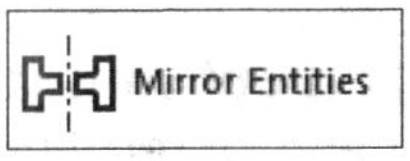

Now that the circles on one side are complete, you need to add the six circles on the other side. To copy the linear pattern, you will use the **Mirror** command. First draw a **Horizontal** centerline starting at the Origin and then select the **Mirror Entities** command. In the Mirror options add the six circles from the linear pattern in the "**Entities to mirror**" selection box. Activate the "**Mirror about:**" selection box and select the horizontal centerline. When you see the mirrored circles preview, click **OK** to finish.

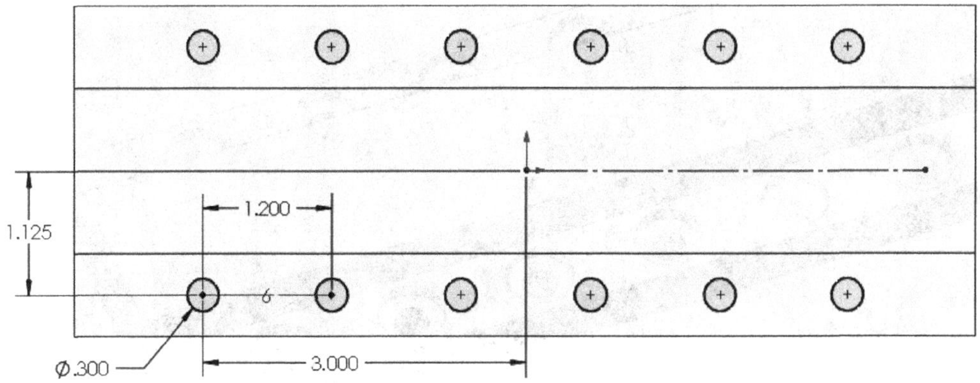

From the Features tab select **Extruded Cut** command. Use the **Through All** option and click **OK** to finish. Now that you have made the small holes for the part, you need to add the larger diameter holes to create the counterbores.

NOTE: SOLIDWORKS has a **Hole Wizard** command that can easily create many different types of holes for your designs, including counterbore holes, which can then be patterned and mirrored, simplifying the process significantly. The **Hole Wizard** will be covered in a later lesson; in this lesson the focus is on sketch patterns.

To add the counterbore holes, select the same face as before and add a new sketch. Draw a **Circle** concentric to the first hole, and dimension it **0.600"** diameter.

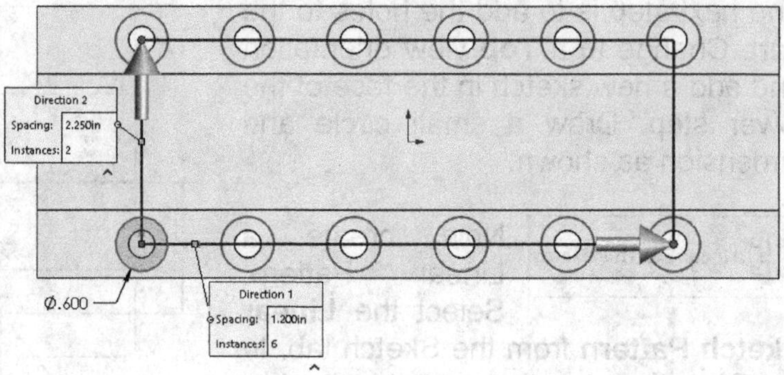

Now add a **Linear Pattern** as before, but this time you will make the pattern in two directions. In **Direction 1** make six **(6)** copies spaced **1.200"**. In **Direction 2** make two **(2)** copies spaced **2.250"**, turn on the "**Dimension Y spacing**" and "**Dimension angle between axes**" to fully define the pattern. Click **OK** to finish the pattern.

From the Features tab select the **Extruded Cut** command. Use the **Blind** option to a depth of **0.125"** into the material and click **OK** to finish. Now that you have bored the counterbores into the plate's two steps, your part will look like the following image.

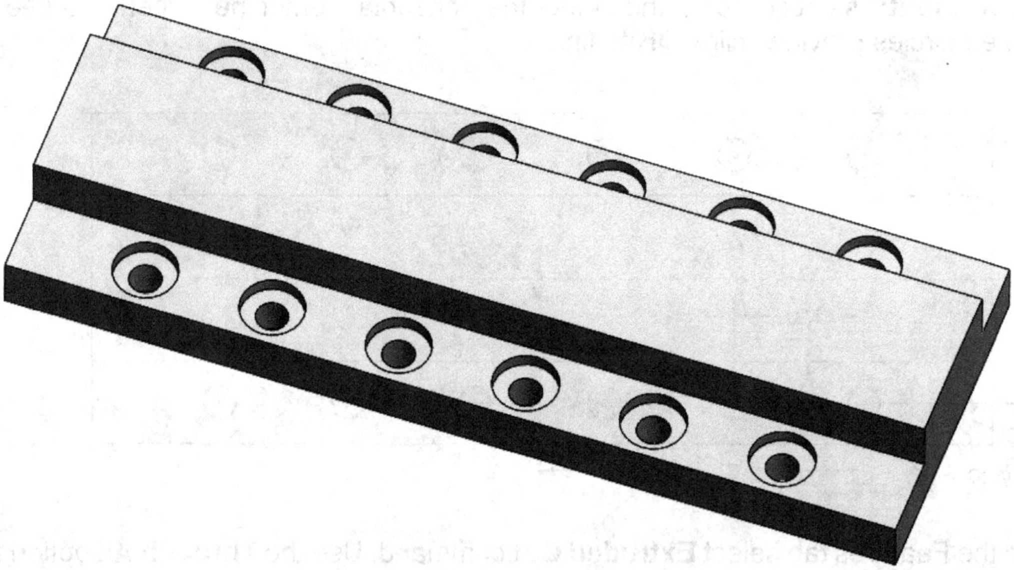

NOTE: If you make a mistake or need to modify a 3D feature's definition, you can click on the feature to be modified in the FeatureManager and select **Edit Feature** from the pop-up context menu. If you need to modify the 2D sketch, select Edit Sketch.

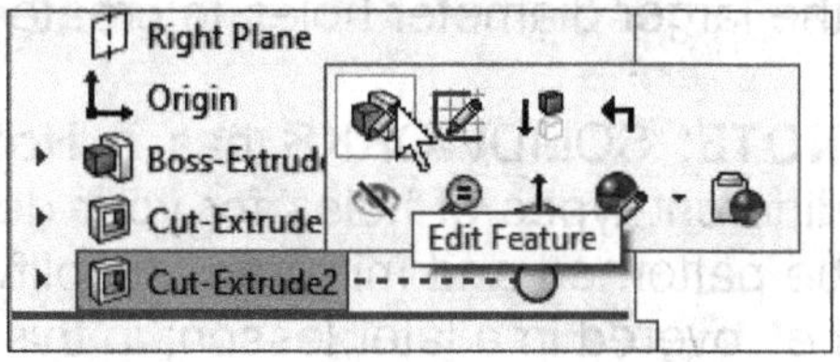

The next step for your design is to create four holes on the top of the plate. Change to a **Top** view orientation and add a new Sketch on the top face. Draw the first **Circle** on the left, and dimension it as indicated.

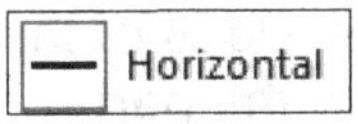

To maintain the design intent, select the center of the circle and the origin, and add a **Horizontal** relation to fully define the circle's location.

Use a **Sketch Linear Pattern** to get a second circle **1.20"** to the right of the first circle. Draw a vertical **Centerline** starting at the Origin, and finally use the **Mirror Entities** command to copy the two circles. The completed sketch is shown.

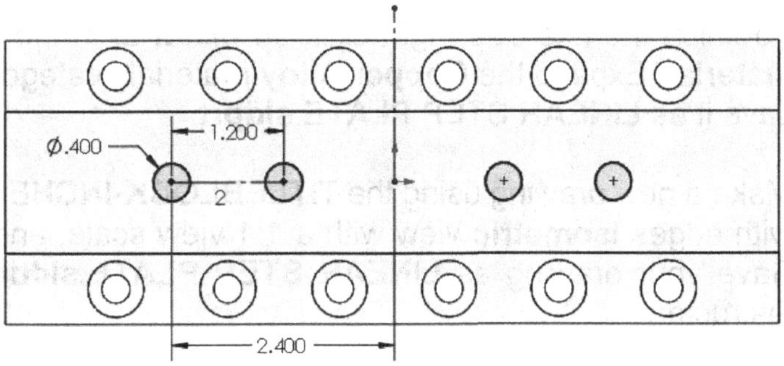

Change to an **Isometric** view for better visibility. From the Features tab select **Extruded Cut**, use the **Through All** end condition and click **OK** to finish. You have completed the small holes all the way through the thick part of the plate.

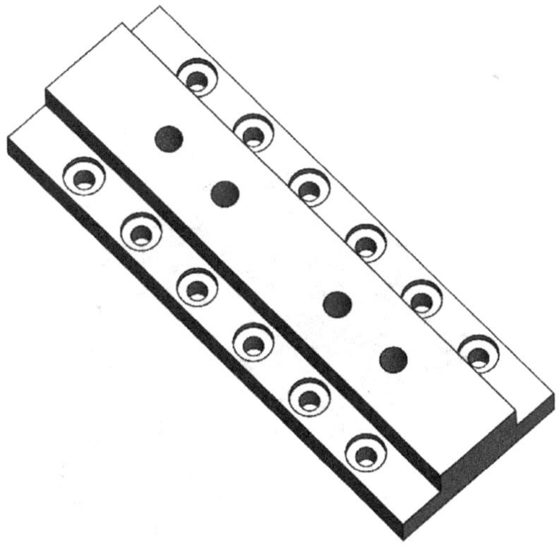

Now you need to add two counterbore slots. Change to a **Top** view and add a new sketch on the top face of the part. Select the **Slot** command from the sketch tab and use the **Straight Slot** option. Start the slot concentric to the left circle of the top face and end it concentric to the hole immediately to the right. This will make the slot concentric with the two circles. Dimension the slot's height to **0.80"**. Using the same approach, draw a second slot on the holes to the right of the origin, or use the **Sketch Mirror** command.

NOTE: Using the slot's "**Add dimensions**" option will add width and height dimensions when drawing the slot.

Since we defined the slot's width using the holes' centers, the width dimension is redundant and it will be colored grey, meaning it is a reference dimension. In other words, it cannot control the slot's size.

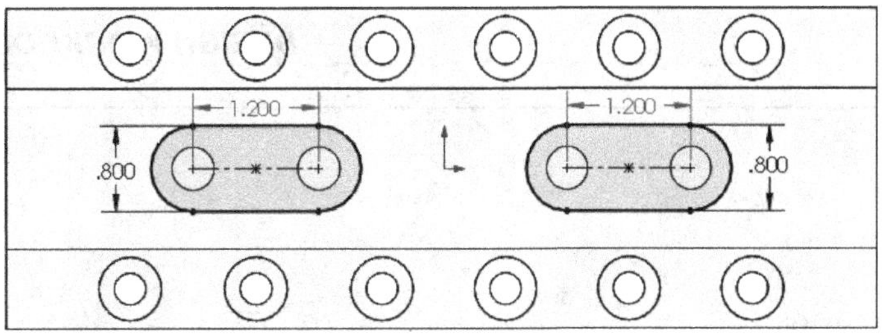

To make the counterbores select the **Cut Extrude** command using a **Blind** end condition of **0.250".** The final step is to add a chamfer to the indicated model edges. In the Features tab click on the **Fillet** command's pull-down menu and select **Chamfer**. Set the chamfer value to **.125" x 45 deg**, select the six short edges at the sides of the step plate and the two long edges of the top surface. Click **OK** to complete the part.

To finish the exercise, right click on **Material** in the FeatureManager and select **Edit Material.** Expand the **Copper Alloy** materials category, assign **Brass** to your part and save it as **LINEAR STEP PLATE.sldprt**.

Make a new drawing using the **TITLEBLOCK-INCHES.drwdot** template, add a shaded with edges **Isometric** view with a 1:1 view scale, and add the necessary annotations. Save your drawing as **LINEAR STEP PLATE.slddrw** and print a copy for your lab instructor.

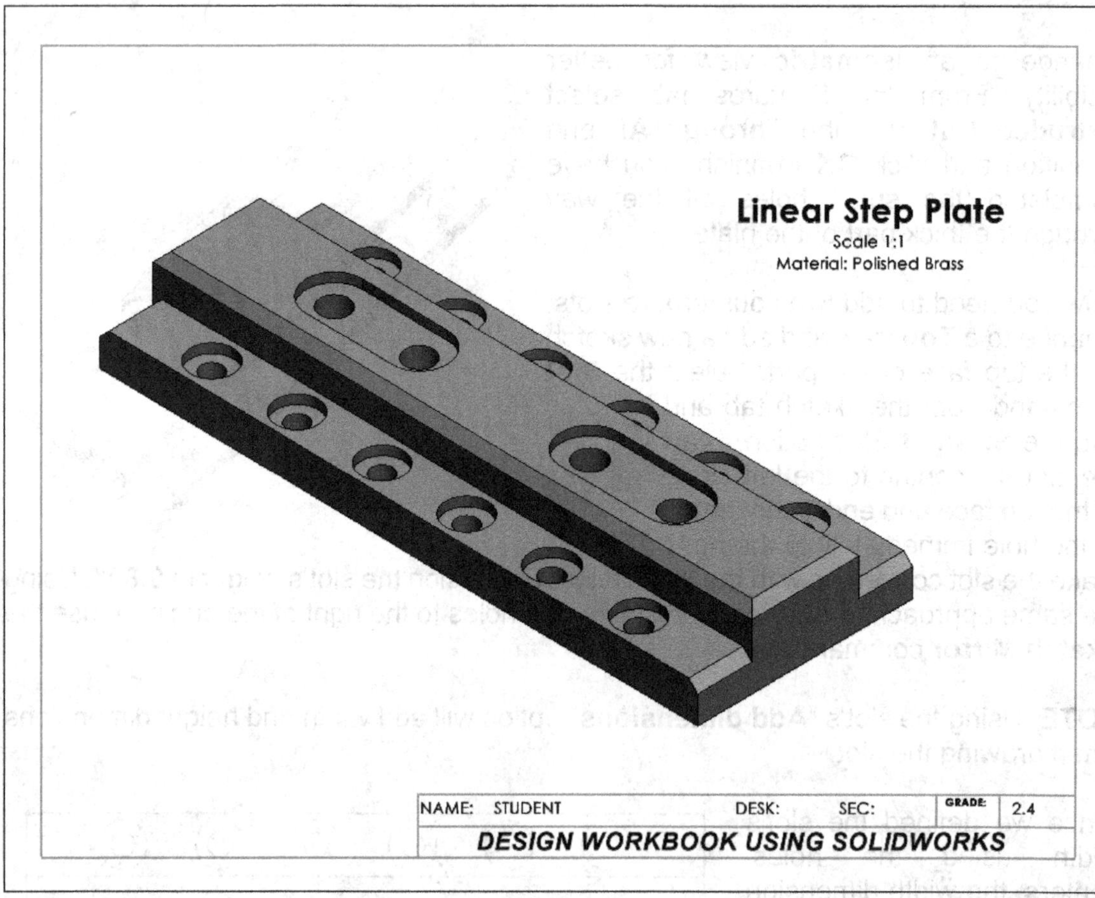

Supplementary Exercise 2-5: FLANGE

Build the following model using the **Revolve** and **Circular Pattern** commands learned in this lesson.

Add the sketch for the **Revolve** feature in the **Top Plane** and dimension it using the next drawing as a guide. The units of measure are in Inches. Make a drawing with an Isometric view and save it as **"FLANGE."**

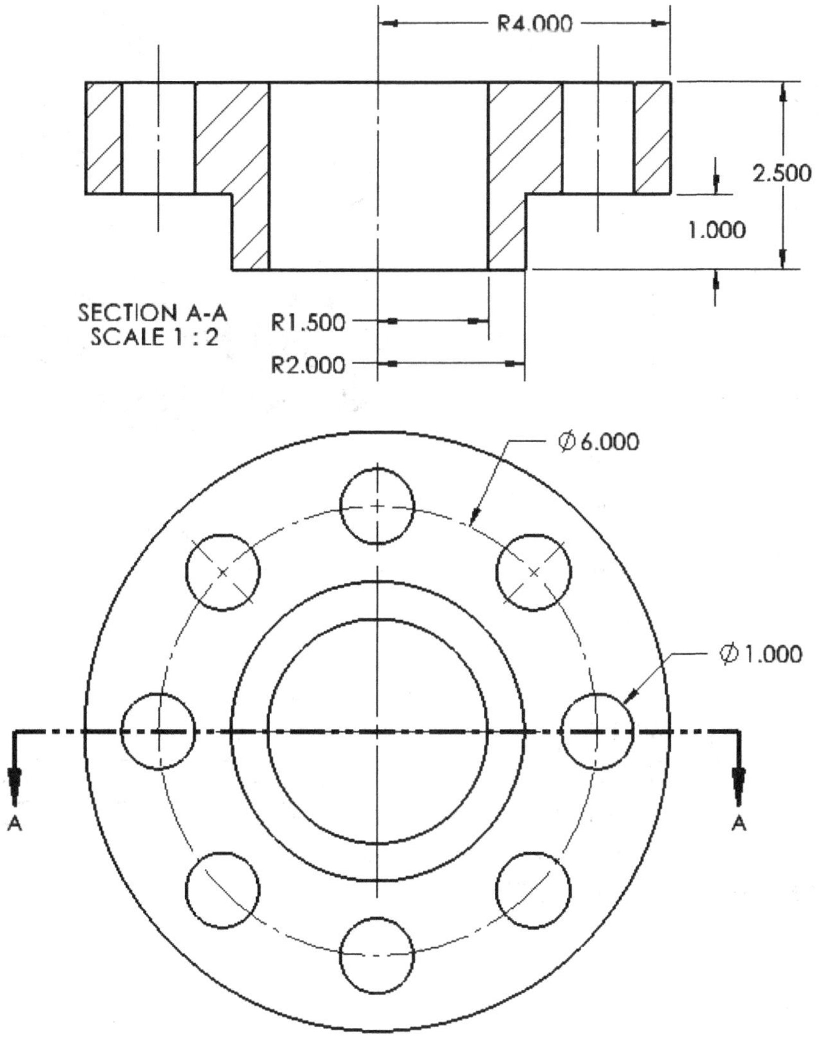

Supplementary Exercise 2-6: STEEL VISE BASE

Make a model of the figure below using the commands learned so far. All holes and slots are through. Make a drawing with an **Isometric** view and save it as **STEEL VISE BASE**.

TIP: Use the **Dynamic Mirror** command for the base feature.

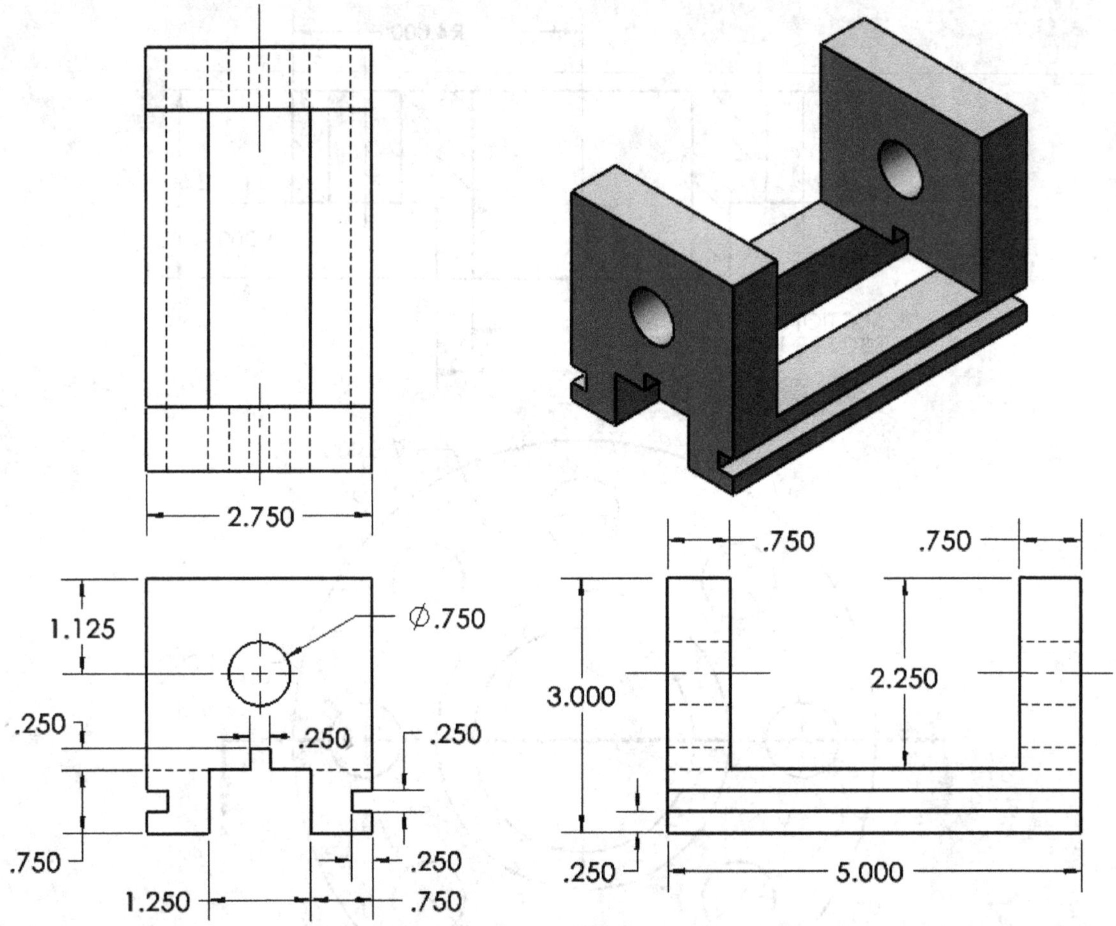

Design Workbook Lab 3
3D Modeling Part I

In the first two Labs you created some solid parts using simple 2D sketches. In these parts, most of the geometric information was defined in a 2D sketch, and then it was either extruded or revolved to make the 3D model. While many simpler components can be designed using the tools learned so far, SOLIDWORKS includes advanced tools to create more complex designs. In this Lab you will learn how to use some of these tools to create and modify your 3D models.

SKETCH RELATIONS

So far you have learned how to create and dimension basic sketch geometry and how to add basic sketch relations. In this lab you will learn more about adding geometric relations to improve your **Design Intent**, in other words, when your design's dimensions are changed, the 3D model updates predictably. These geometric relations can be added to current sketch entities, 3D model features, or sketch geometry in a different sketch.

In short, the design intent is the plan you make so that your parts update *predictably* when you make changes. For example, if your design is symmetrical, when you make a dimensional change it should remain symmetrical. Under the **Display/Delete Relations** button, you can also access the **Add Relation** command as shown here. The following list includes most of the sketch relations available in a 2D sketch.

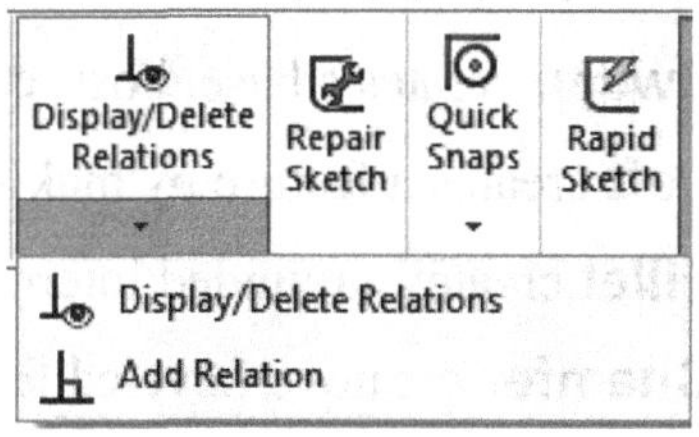

Horizontal makes one or more lines horizontal, or two or more endpoints horizontal.

Vertical makes one or more lines vertical, or two or more endpoints vertical.

Collinear makes two or more lines lie along the same infinite line.

Co-radial makes two or more arcs share the same center point and radius.

Perpendicular makes two lines perpendicular to each other.

Parallel makes two or more lines parallel to each other.

Tangent makes an arc, ellipse, or spline, and a line, edge, or arc tangent to each other.

Concentric makes two or more arcs, or a point and an arc to share the same center.

Midpoint makes an endpoint to coincide with the midpoint of another line or curve.

Intersection makes two lines and one point to remain at the intersection of the lines.

Coincident makes a point and a line, arc, or ellipse to lie on the line, arc, or ellipse.

Equal makes two or more lines or two or more arcs have the same length or radii.

Symmetric makes a centerline and two points, lines, arcs, or any geometric sketch element to be equal and symmetric about the selected centerline.

Fix makes an entity's size and location fixed. Should only be used as a temporary solution, and not a substitute to properly defining sketch geometry.

THE FEATURES TOOLBAR

The Features toolbar shown next includes most of the commands available in SOLIDWORKS to create and modify 3D features. Below are descriptions of these features.

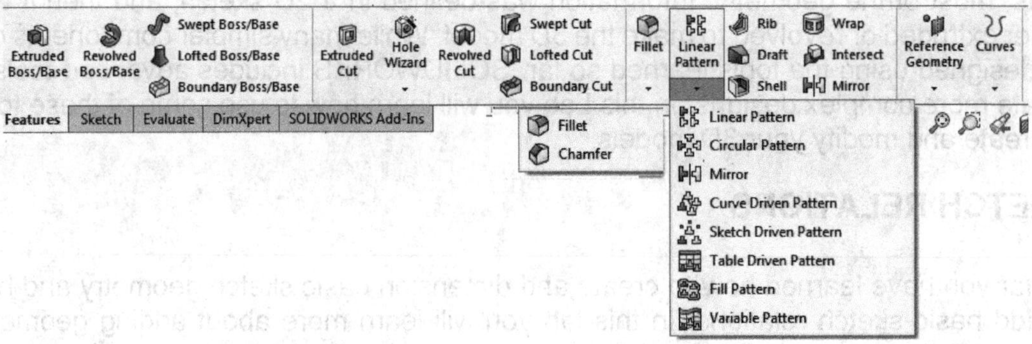

Extrude Boss/Base creates a base or boss by extruding a sketch in a linear direction.

Revolve Boss/Base creates a base or boss by revolving a sketch around a centerline.

Extruded Cut removes material from a solid body by linearly extruding a sketch through it.

Revolved Cut removes material from a solid body by revolving a sketch around a centerline.

Sweep creates a base, boss, cut, or face by moving a profile along a designated path.

Loft creates a feature by making a transition between multiple profiles.

Fillet creates a rounded internal or external face on the part by selecting an edge or face.

Chamfer creates a beveled feature on the selected face, edge or vertex.

Rib adds material of a specified thickness determined by a contour in an existing part.

Shell hollows out the part, optionally removing the selected faces.

Draft tapers faces of a part using a specified angle.

Wrap wraps a closed sketch contour onto a cylindrical face.

Hole Wizard allows you to quickly add different types of standard holes to your part.

Linear Pattern creates multiple instances of selected 3D features along one or two linear directions.

Circular Pattern creates multiple instances of one or more 3D features uniformly around an axis.

Mirror Feature creates a mirror copy of one or more 3D features about a flat face or plane.

Reference Geometry creates reference geometry like planes, axis, coordinate systems, and points.

Curves create different types of curves including spiral and helix.

Exercise 3.1: Clevis Mounting Bracket

In Exercise 3.1, you will design a Clevis Mounting Bracket. In this case the design intent requires that the part be symmetrical, and the hole in the clevis remains concentric to the outer circular edge even if the design dimensions change and should also go through the width of the part, regardless of its size. In order to maintain the design intent conditions, you will need to use different sketch relations, features and end conditions in your part.

Start a new part using the **ANSI-INCHES.prtdot** template and save it as **CLEVIS MOUNTING BRACKET**. In order to maintain the part's design intent, first you will create the base, then the side wall, a 3D mirror of the side wall, and finally the base cutouts.

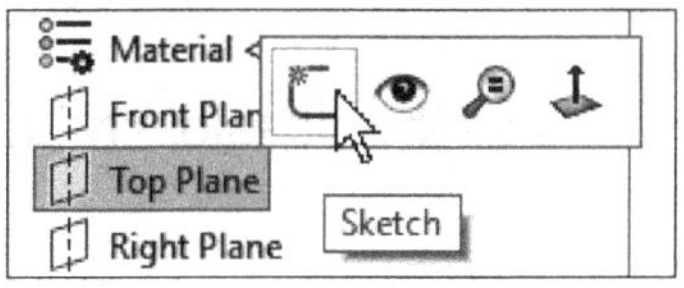

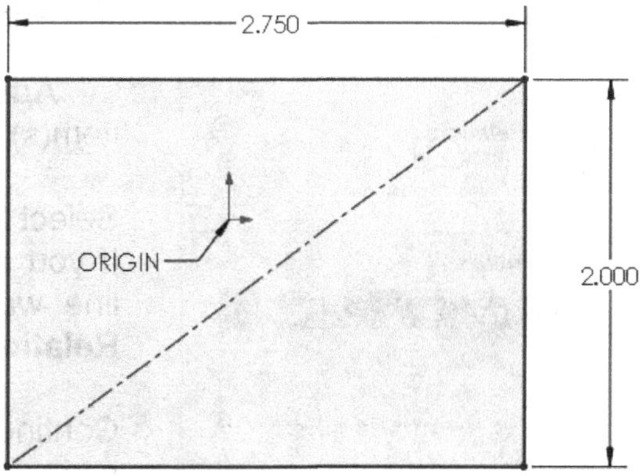

To start, select the **Top** plane in the FeatureManager and add a new sketch. The model's orientation is changed to a Top view. Draw a rectangle with a centerline across the diagonal and dimension it as shown. Do not start the rectangle at the origin. Open the **Add Relations** command from the Display/Delete Relations drop-down icon. In the "**Selected Entities**" box add the Centerline and the Origin.

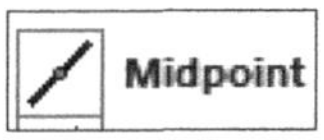

From the available relations click on **Midpoint**. By adding this relation, the centerline, and therefore the rectangle, will be centered about the origin, making the base feature symmetrical about the **Front** and **Right** planes. Make a **0.500" Extruded Boss** and click **OK**.

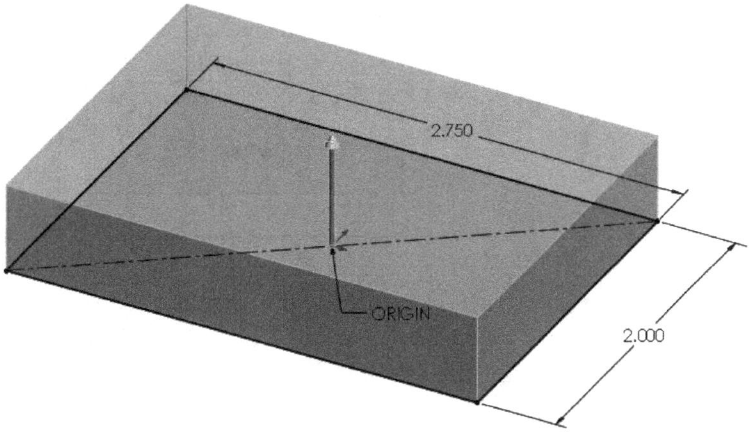

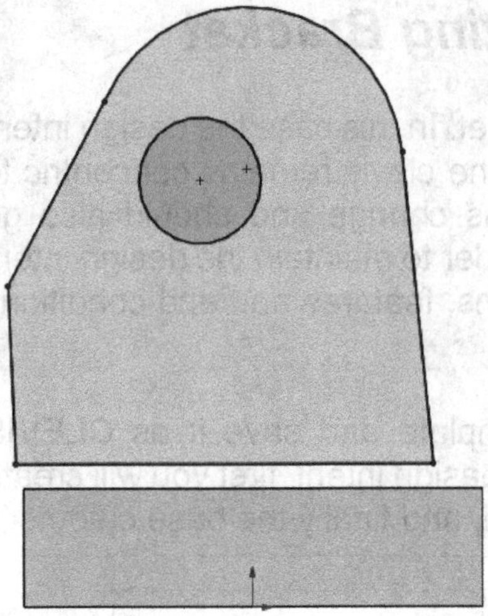

For the next step change to a **Right** view orientation. Select the right face of the base feature and add a new sketch. In order to practice manually adding sketch relations, draw a sketch profile approximately as shown above the base feature.

Open the Sketch Relations command. The different sections of the **Add Relations** command are:

"**Selected Entities**" displays the currently selected entity(ies);

"**Existing Relations**" displays existing relations of the selected item(s);

"**Add Relations**" lists all the possible relations the item(s) can have.

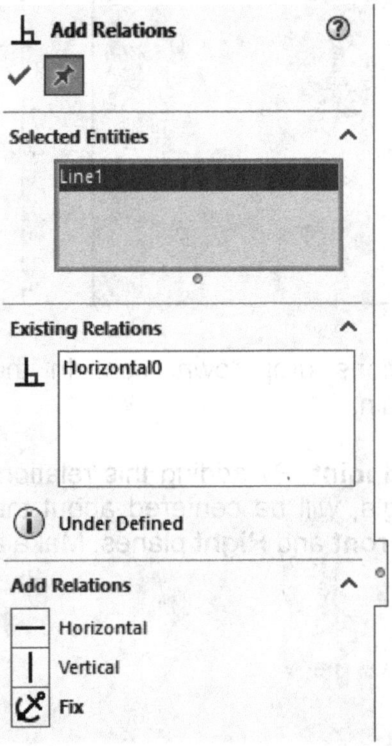

Select the bottom line and add a **Horizontal** relation. If you automatically captured this relation when the line was drawn, it will be listed in the "**Existing Relations**" box when you select it.

Continue adding the sketch relations needed. In each case select the corresponding entities to add the relations indicated.

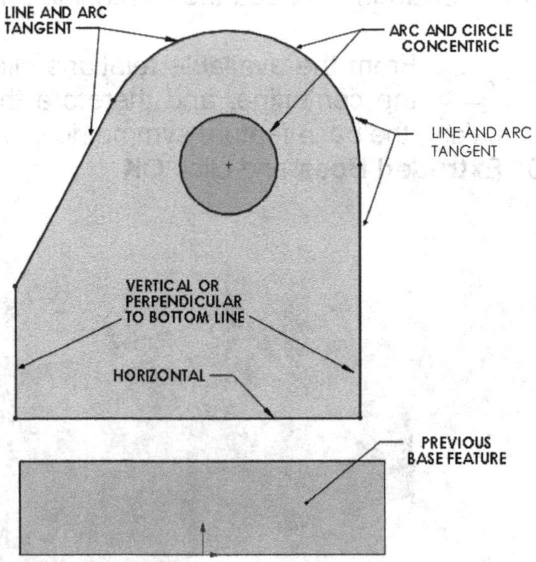

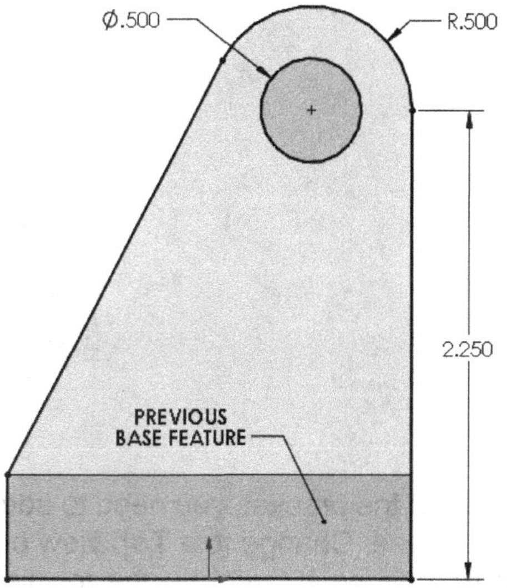

To finish the sketch, add the needed relations between the lower endpoints to the base, and the dimensions indicated next.

At this point, your sketch will be fully defined, and all geometry should be black. To verify the proper geometric relations have been added, double click on the **2.25"** dimension, and temporarily change it between **1.00"** and **3.00"**.

Notice the tangent relations are maintained as the dimension is changed, and the sketch updates predictably based on your design intent.

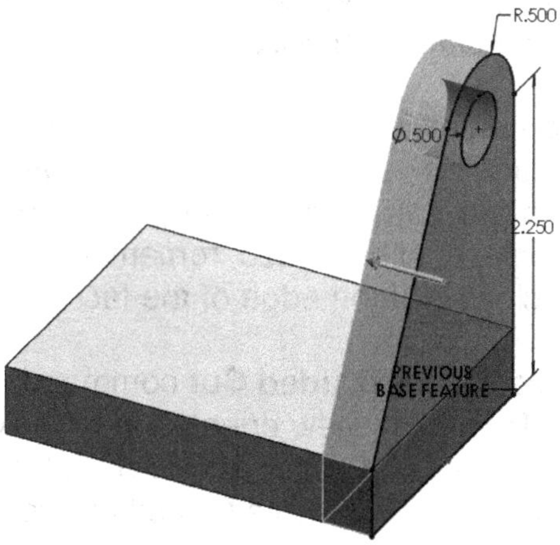

Select the **Extruded Boss** command and change to an **Isometric** view orientation to preview the extrusion. Use the Reverse Direction button to make the new boss go *into* the existing base feature. Make the extrusion **0.375"** deep and click **OK**.

Now you need to create a new extrusion equal to the last one on the other side of the bracket. Instead of adding a new sketch and extruding it, you will use the **Mirror** feature command which will create a symmetrical 3D feature.

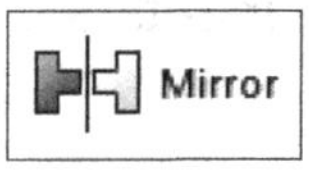

To use the **Mirror** command (and almost any command), you can pre-select the mirror plane and the feature(s) to mirror, *or* you can select the command, and make the selections after.

In the **Mirror** command you need to select the mirror plane, which in your case will be the **Right** plane of the part.

If you remember the first feature was centered about the origin, making it symmetric about the **Front** and **Right** planes, this is almost always a good idea if you anticipate your part will be symmetrical, or have a certain level of symmetry.

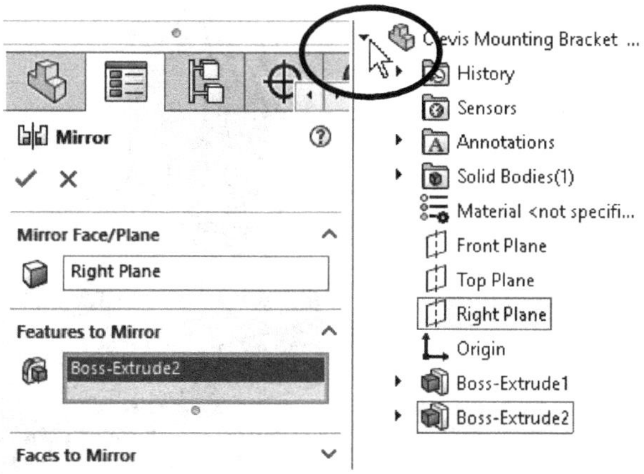

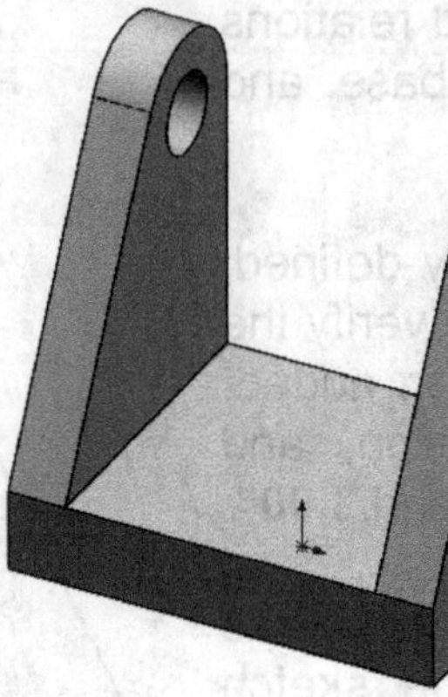

To finish the bracket, you need to add a cutout in the base. Change to a **Top** view orientation and add a new sketch in the top face of the base feature. Draw the sketch profile shown using the **Line**, **Circle** and **Arc** tools, and dimension as indicated.

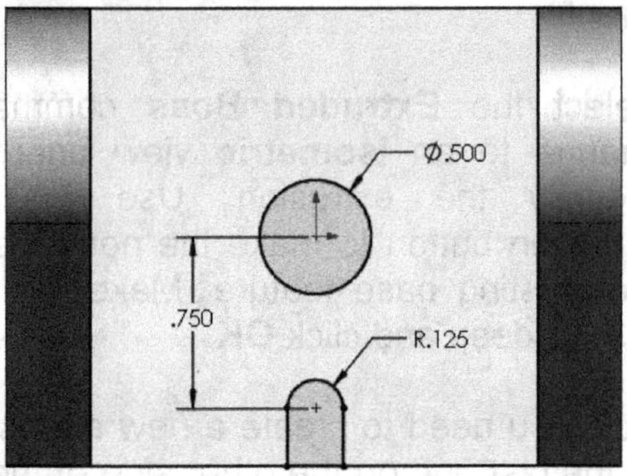

To center the slot in the part, add a **Vertical** geometric relation between the arc's center and the part's origin. Keep in mind the slot's profile must be closed; remember to add a line to close it at the edge of the face.

Select the **Extruded Cut** command. Use the **Through All** end condition and change to an **Isometric** view orientation to make sure the cut is going down. Click **OK** to finish. Use the **Fillet** command to round the inside edges where the base feature meets the vertical faces using a **0.125"** radius.

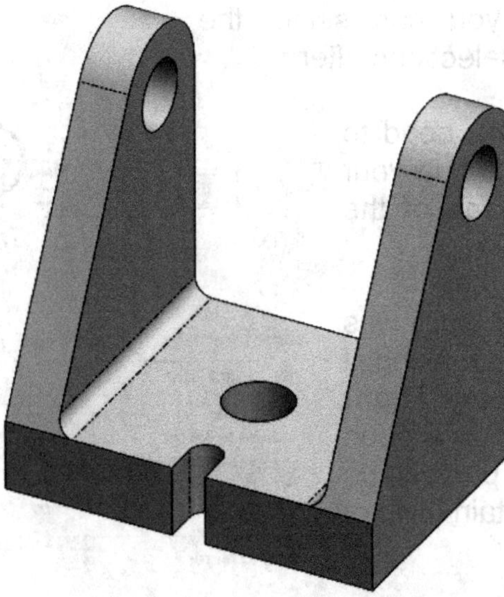

Your Clevis Mounting Bracket is now complete. To define the part's material right click on **Material** in the FeatureManager, select **Edit Material** from the context menu, expand the **Copper Alloy** category and select **Leaded Commercial Bronze**.

If you haven't already, save your part as **CLEVIS MOUNTING BRACKET.sldprt**. Make a new drawing and add an **Isometric** view using the Shaded with Edges display mode.

Add the missing notes and save your drawing as **CLEVIS MOUNTING BRACKET.slddrw**. The finished drawing is shown next.

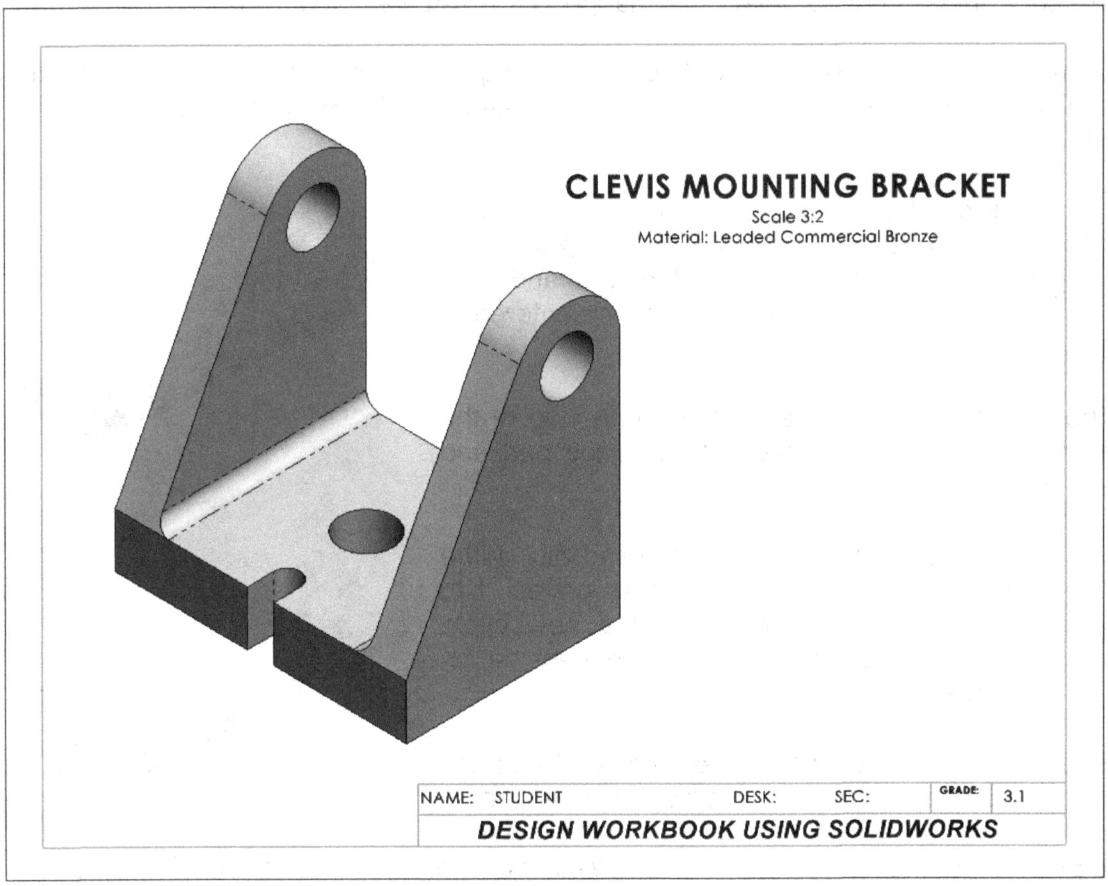

Exercise 3.2: MANIFOLD

In this Exercise you will design a Manifold. A manifold's purpose is to distribute air (or any fluid) to multiple locations through its ports. Since the manifold's ports are all similar, you will be able to design one of them and use the **Pattern** and **Mirror** features to replicate them.

Start a new part using the **ANSI-METRIC** template and save it to your personal folder as **Manifold.sldprt**. Select the **Right** plane in the FeatureManager and add a new sketch. Your part will automatically change to a **Right** view orientation.

Draw two concentric **Circles** centered at the Origin, dimension the larger circle **60** mm in diameter and the smaller circle **45** mm in diameter. These two dimensions will fully define your sketch.

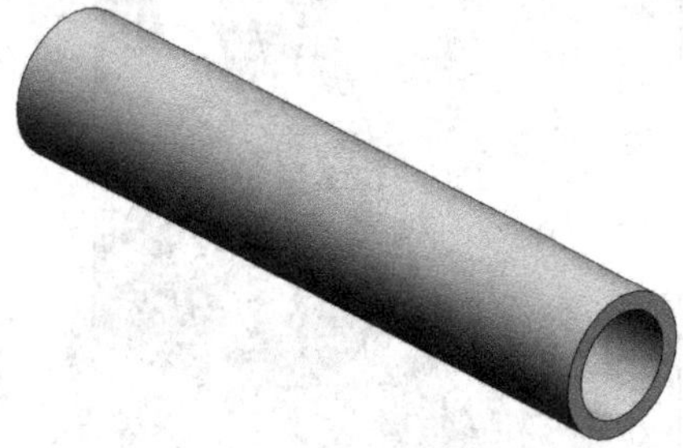

From the Features tab select the **Extruded Boss/Base** command, select the **Mid Plane** end condition, make it **300 mm** and click **OK**. You now have a long flute centered at the origin, as shown.

The next step is to add a collar on one side of the manifold. Select the right-side face of the part and add a new sketch.

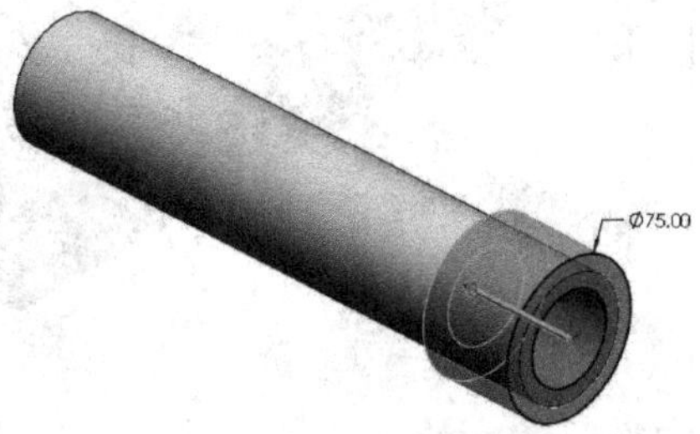

Select the inner circular edge and click in the **Convert Entities** command to project it to the new sketch. Now draw a new circle centered at the origin and dimension its diameter **75 mm**.

Select the **Extruded Boss/Base** command and use the **Blind** end condition. Reverse the direction to extrude it on top of the base feature, set the distance to **50 mm** and click **OK** to finish. You now have a small collar in one end of the manifold.

To duplicate the collar on the other side of the manifold you will use the **Mirror** command. In the previous exercise you learned how to use the **Mirror** command *and* display the FeatureManager at the same time to make the necessary selections. In this case you will use a pre-selection approach.

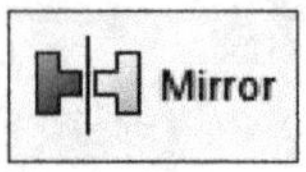

Hold down the **Ctrl** key and pre-select the **Right** plane and the previously made **Boss-Extrude2** feature. Notice both features are highlighted in the screen. From the Features tab select the **Mirror** command and click **OK** to finish. You now have a collar on each end of the manifold.

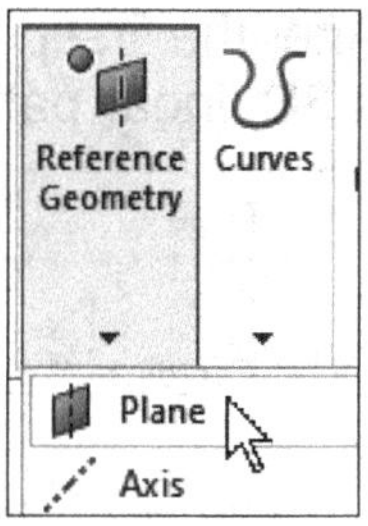

The next step is to add the first port in the manifold. To do this you will need to create a reference plane to draw a new sketch above the part, since there are no planes or flat faces to add a sketch to. If needed, change to an **Isometric** view orientation for clarity. From the Features tab select the **Reference Geometry** command and click on **Plane**, or from the menu **Insert, Reference Geometry, Plane**. In the **First Reference** selection box add the **Top** plane from the FeatureManager, (remember you can expand it while the command is active) set the distance to **42.5 mm** above the **Top** plane and click **OK** to finish. If needed, activate the **Flip offset** option to reverse the side to create the new plane.

Now change to a **Top** view orientation, select **Plane1** and add a new sketch. Draw a circle and dimension it as indicated. Using the **Add Relations** command, select the circle's center (*not the perimeter*) and the part's origin, and add a **Horizontal** relation to fully define the sketch.

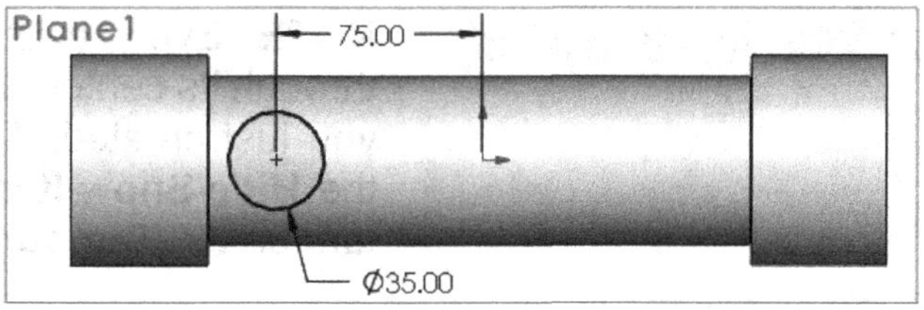

Select the **Extruded Boss/Base** command in the Features tab and change to an **Isometric** view for visibility. Set the **Direction 1** end condition to **Up to Surface**, select the long outer face of the manifold body (it will be highlighted in pink), and click **OK** to finish. Now you have a new boss extruded from **Plane1** down to the selected face.

Now add a new sketch at the top of either the new boss or **Plane1**, draw a circle concentric with the previous boss and dimension its diameter of **17.5 mm**.

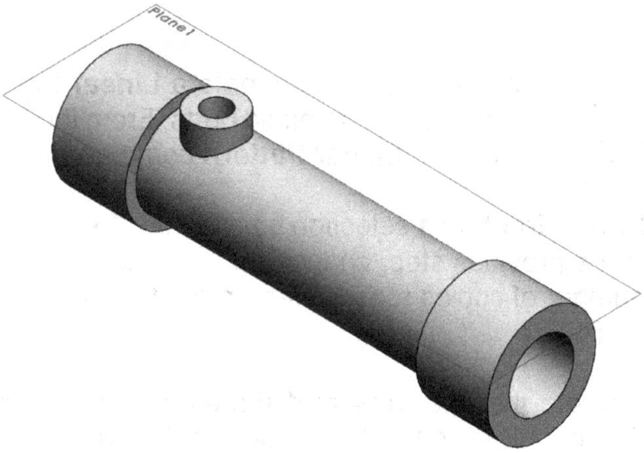

If needed, return to an Isometric view and select the **Extruded Cut** command from the Features Tab. Set the Direction 1 end condition to **Up to Surface**, select the cylindrical face inside of the manifold and click **OK** to finish. You have now made a hole going from Plane1 to inside the manifold.

To better see the hole, use the **Rotate View** command to rotate your model in the screen. Try to view the manifold through the center hole of the part. Later in this exercise you will learn how to use a temporary section view to inspect the inside features of a part.

After the first port is complete you will create the rest using a 3D linear pattern. To create a 3D linear pattern, you need a linear edge or an axis to use as a direction reference.

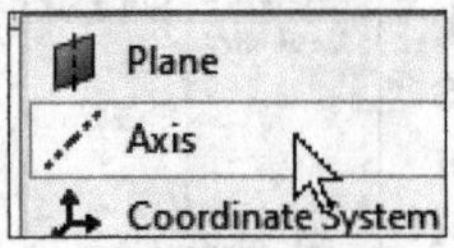

Since the manifold does not have a linear edge to be used as reference, you need to add a reference **Axis**. From the Features tab select **Reference Geometry** and click on **Axis** from the drop-down menu, or use the menu **Insert, Reference Geometry, Axis**.

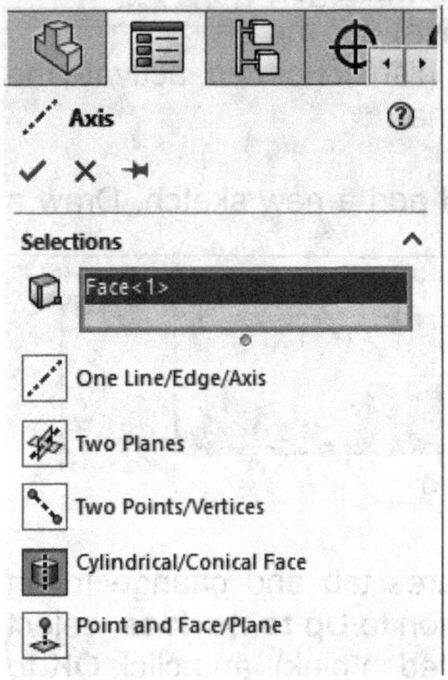

To define an auxiliary **Axis**, you can use a cylindrical face, a line, an edge, an axis, two points, or a plane and a point. In this exercise you will select the outer cylindrical face of the manifold and click **OK** to finish. Notice the preview of the **Axis** after selecting the face.

NOTE: Every cylindrical surface has an axis running through its center, and they can be used just as the **Axis** you just created. To view the **Temporary Axis,** select the **Hide/Show Items icon** from the **View** toolbar, and turn on the **Temporary Axis** button.

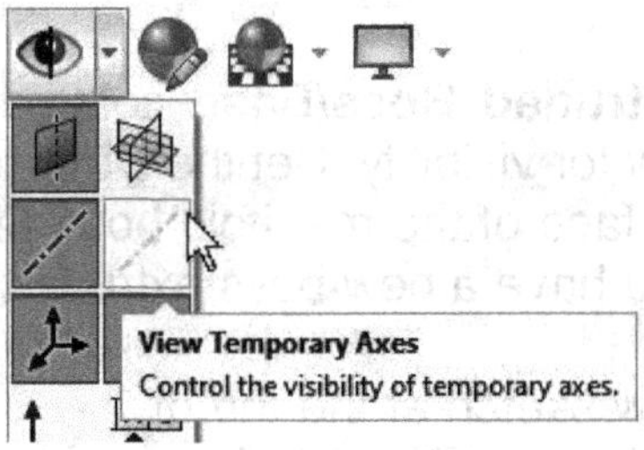

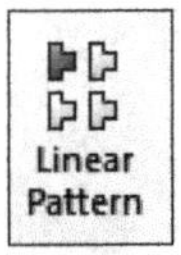

The next step is to create a **Linear Pattern** to add the rest of the manifold ports. From the Features tab select the **Linear Pattern** command.

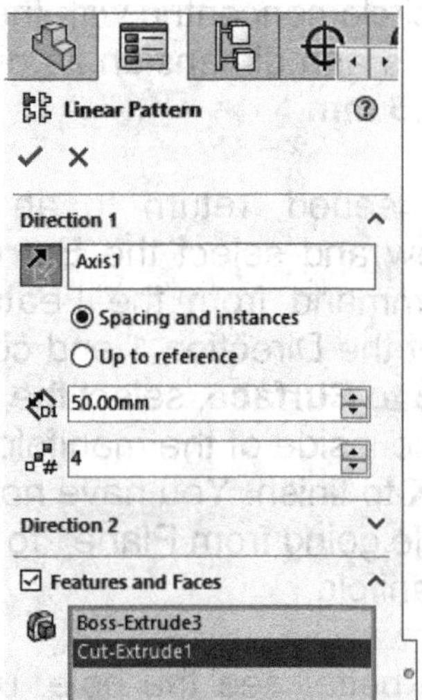

In the **Direction 1** selection box add the **Axis1** feature made in the previous step, change the spacing to **50 mm** and the number of copies to **four** (remember this value includes the original).

Activate the **Features and Faces** selection box and from the graphics area or the fly-out FeatureManager select the **Boss-Extrude3** and **Cut-Extrude1**. Notice the preview on the screen and click **OK** to finish. The manifold with the four ports is shown next.

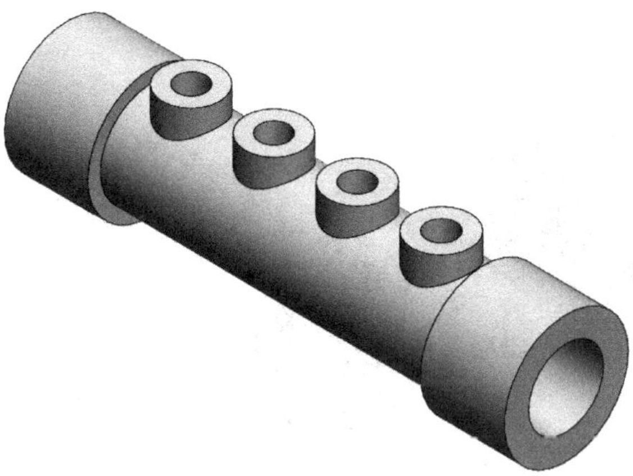

Now you need to add two additional ports at the bottom of the manifold. Since they are identical to the top ports, you can use a **Mirror**. From the Features tab select the **Mirror** command, expand the fly-out FeatureManager and select the **Top** plane in the **Mirror Face/Plane**. In the **Features to Mirror** selection box add the **Boss-Extrude3 and Cut-Extrude1**. Click OK to finish.

To add the last port, make a new **Mirror** feature using the **Right** plane and the previous **Mirror2** feature (the bottom port). After you complete this operation you will have a **Mirror3** feature as shown.

*<u>** In SOLIDWORKS 2026 SP 0.0 there is a software error where **Mirror3** fails to build. If this error still exists, change the Boss-Extrude3's (Page 3-9) end condition from **Up to Surface**, to **Blind**, 20mm.**</u>*

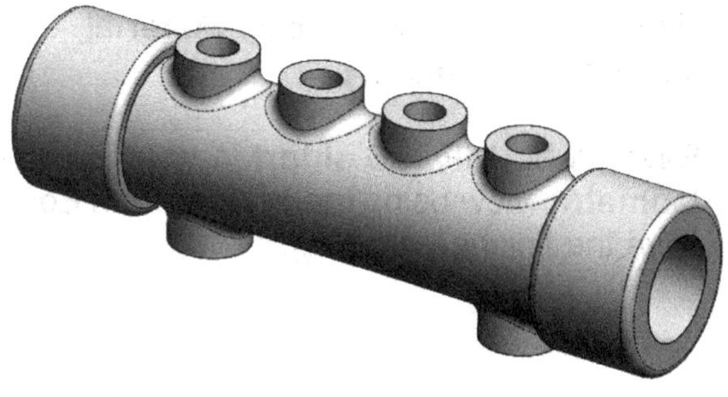

Now you must add a fillet to the edges where the ports meet the manifold's main body. Select the **Fillet** command, set the radius to **6 mm** and select the edge where the six ports meet the main body. Click **OK** to continue.

The final step is to add a **4 mm** fillet to the sharp edges of the two collars. The finished model will look like this:

At this time, you can create a temporary section view of the model to inspect the internal features of the manifold. From the **View** toolbar select the **Section View** command, or from the menu **View, Display, Section View**.

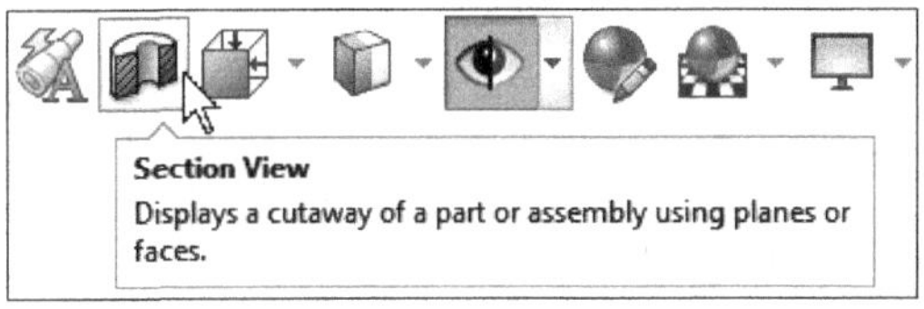

In the **Section View** options the Front plane is the default selection with a distance of 0 mm. By changing the distance, the section view will be moved parallel to the selected reference.

Optionally you can click and drag the arrow and rotation rings to redefine and rotate the section plane as indicated in the following image. After reviewing the section view cancel the **Section View** command to continue.

NOTE: If you click OK to close the **Section View** command, the model will remain sectioned. To exit the section view mode, click on the **Section View** command again to turn it off.

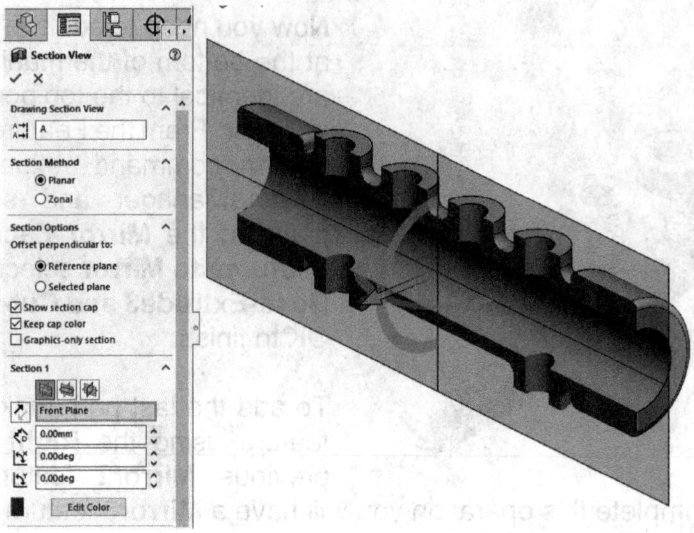

The last thing to do is to define the manifold's material. In the FeatureManager right click in **Material**, select **Edit Material**, expand the **Other Metals** section and set the material to **Titanium**.

Save as **MANIFOLD.sldprt** in your personal folder and make a new drawing with an **Isometric** view using the Shaded with Edges display mode. Add the missing notes and save the finished drawing.

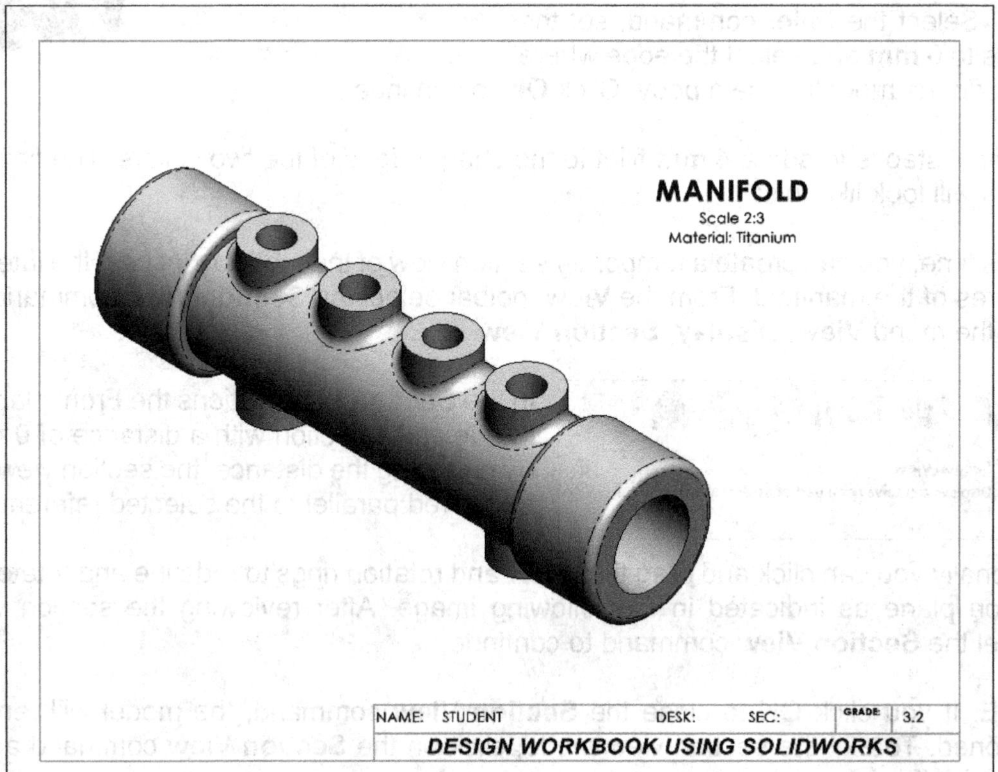

Exercise 3.3: HAND WHEEL

In this Exercise you will design a Hand Wheel with an elliptical cross-section, and five spokes. In this exercise you will learn how to make a Circular Pattern, as well as a couple other commands.

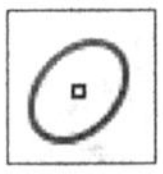

Start a new part using the **ANSI-INCHES** template and save it as **HAND WHEEL.sldprt**. Add a new sketch in the **Front** plane and draw a vertical centerline starting at the origin. From the Sketch tab select the **Ellipse** command and draw it as shown.

To draw the ellipse, click to locate the center, click to locate the minor axis horizontally to the left (or right) or the center, and finally click to locate the major axis vertically above (or below) the center.

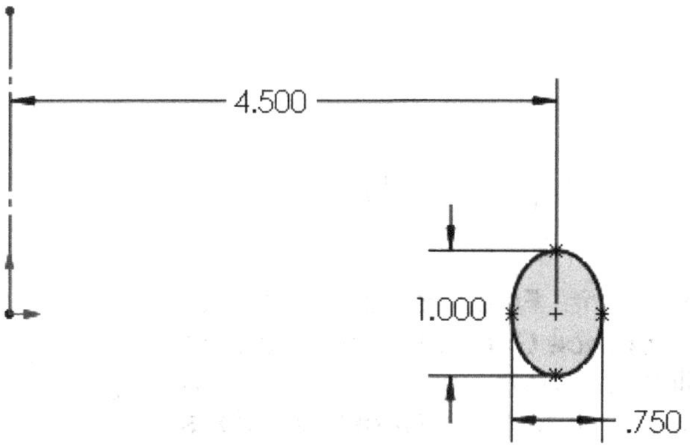

After adding the dimensions indicated you need to add a **Horizontal** relation between the horizontal axis reference points (or **Vertical** between the vertical axis reference points), and finally a **Horizontal** relation between the ellipse's center and the origin.

From the Features tab select the **Revolve Boss/Base** command. In the PropertyManager set the centerline is selected, and the default angle of **360°** is set. Click **OK** to finish the first feature.

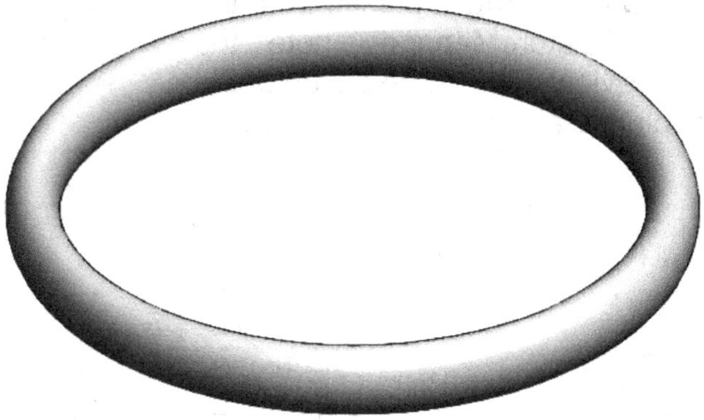

The next step is to add one spoke going from the origin out to the wheel. To make this spoke, which also has an elliptical cross section, change to a **Front** view orientation, select the **Front** plane in the FeatureManager and add a new sketch.

Next select the **Ellipse** command and start its center in the origin, dimension it and add a **Horizontal** relation between the major axis endpoints as shown.

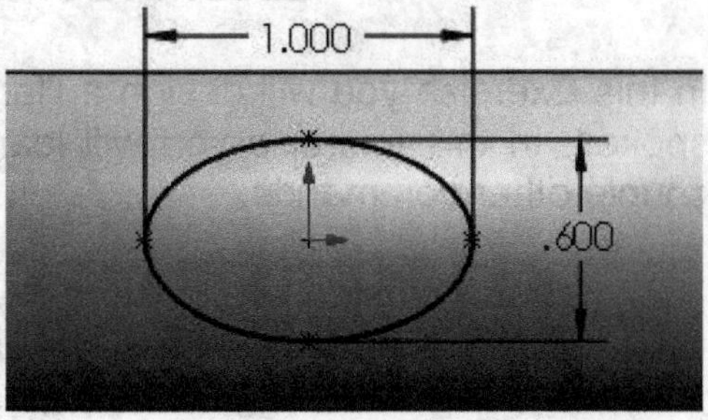

Select the **Extrude Boss/Base** command, set the **Direction 1** end condition to **Up to Next** and reverse the direction.

This end condition will make the extrusion until the next model's face. Change to an **Isometric** view orientation for better visualization and click **OK** to finish.

With the first spoke finished you can make a **Circular Pattern** to copy the rest of the spokes. In order to make the pattern, first you need to have a reference to use as the direction axis.

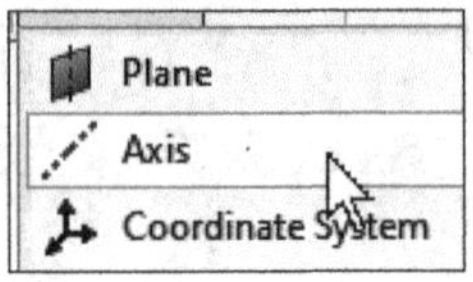

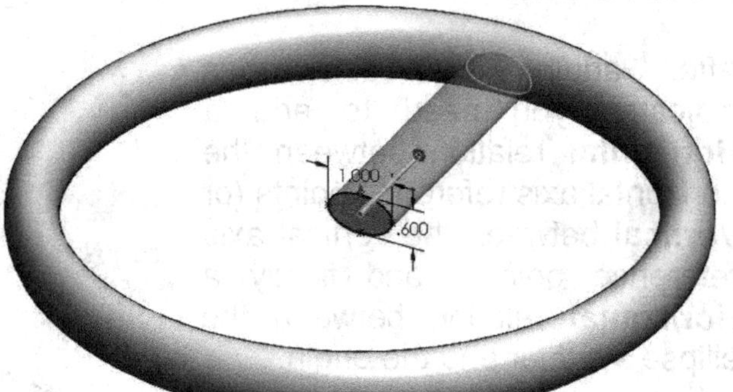

From the Features tab select the **Reference Geometry** command and click on **Axis**, or from the menu **Insert, Reference Geometry, Axis**.

In the **Axis** PropertyManager select the **Front** and **Right** planes from the fly-out FeatureManager, note the axis preview at the intersection of the two planes and click **OK** to finish.

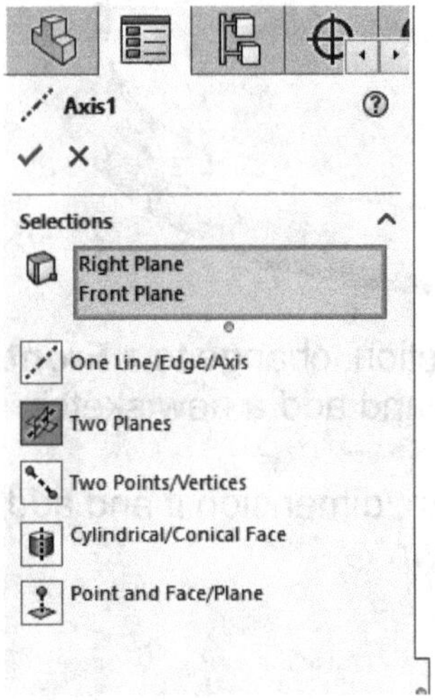

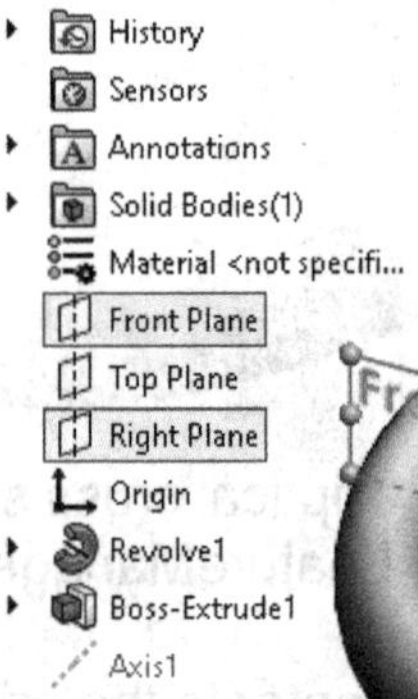

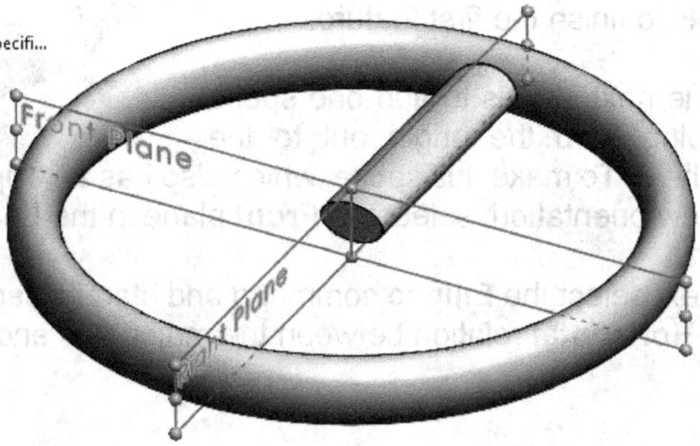

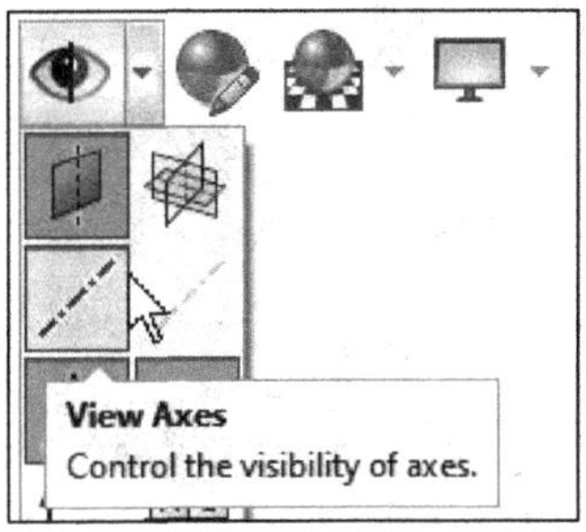

If the new **Axis1** feature is not visible on the screen, select the **Hide/Show Items** drop-down command in the **View** toolbar, and turn on the **Axis** features.

Now that you have the necessary reference, from the Features tab select the **Linear Pattern** drop-down and click on **Circular Pattern**, or from the menu **Insert**, **Pattern/Mirror, Circular Pattern**.

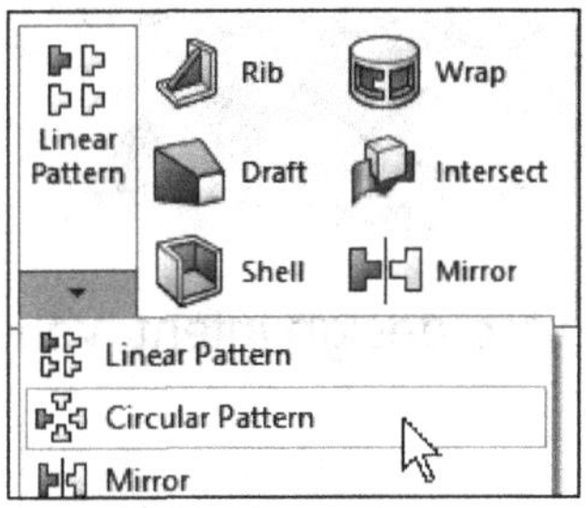

In the **Circular Pattern** command, add the **Axis1** feature in the **Direction 1** selection box from the fly-out FeatureManager, and add the spoke in the **Features and Faces** selection box. Set the number of copies to **5** and click **OK** to finish.

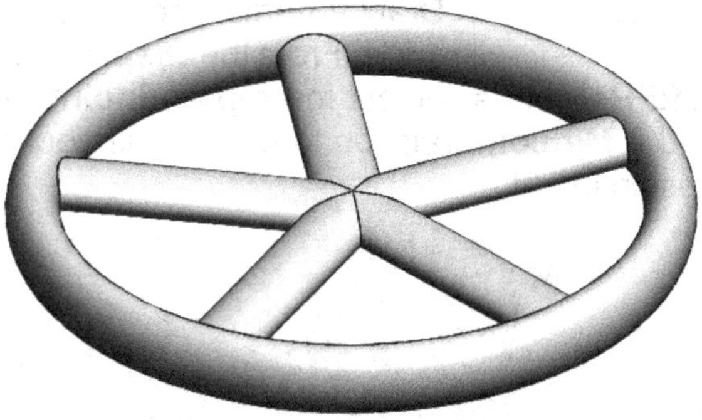

The complete circular pattern should look like this. In the **Hide/Show** command turn off the **View Axis** option.

The next step is to add the hub of the Hand Wheel. Select the **Top** plane in the FeatureManager and add a new sketch on it.

Change to a **Top** view orientation for better visibility and draw a circle centered at the origin. Dimension its diameter **2.000"** to fully define it.

Change to an **Isometric** view and select the **Extrude Boss/Base** command. Set the **Direction 1** end condition to **Mid Plane**, change the extrusion's depth to **1.500"** and click **OK** to finish. Notice the extrusion was made in both directions, symmetrical about the **Top** plane.

Now you need to add a hole and keyway through the hub to fit a shaft and key. Change to a **Top** view for visibility, select the top face of the hub and add a new sketch. Draw a circle at the origin, and a rectangle as shown, and use the **Trim** tool to make the closed profile.

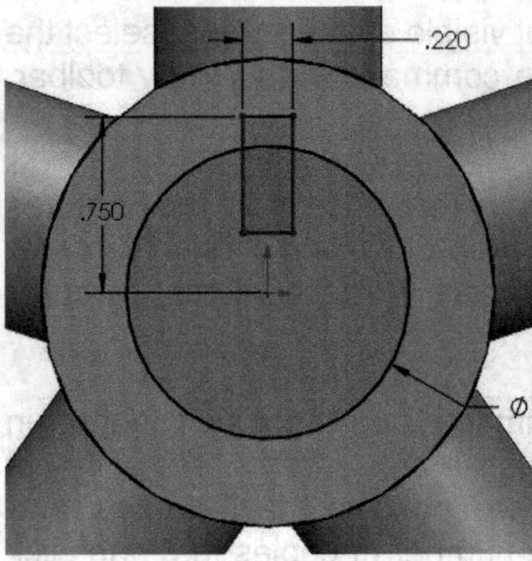

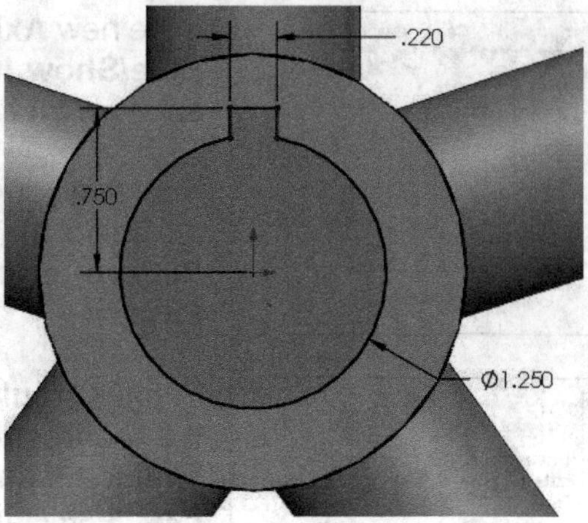

Since the keyway cut must be centered in the shaft, to maintain the design intent, you will use a **Midpoint** geometric relation to keep it centered.

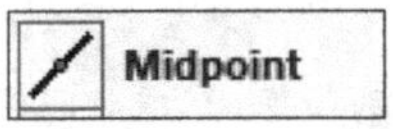

Draw a Vertical centerline starting at the origin going up almost to the circle. From the **Display/Delete Relations** drop down command select **Add Relations**. Select the **top endpoint** of the centerline and the horizontal line and add a **Midpoint** relation to fully define your sketch as shown, correctly maintaining the design intent.

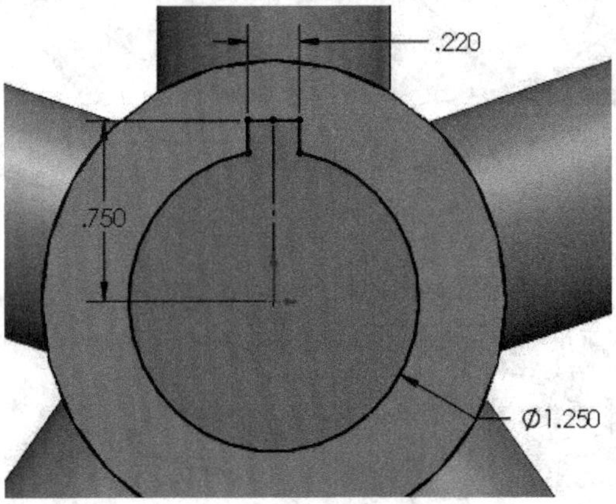

With the sketch profile complete, select the **Extruded Cut** command, set the **Direction 1** end condition to **Through All** and click **OK** to finish.

Now you must add a fillet at the intersections of the spokes and the hub, and the spokes and the hand wheel.

Select the **Fillet** command from the Features tab, or the menu **Insert, Features, Fillet**. Set the radius to **0.125"** and select the faces of the **5** spokes. By selecting the faces, you will add a fillet to all 10 edges at the same time.

The finished Hand Wheel model will look like this.

Select the part as **HAND WHEEL.sldprt** in your personal folder. Make a new drawing, add a shaded Isometric view as in previous exercises and add the missing notes. The finished drawing is shown next. Finally print a copy for your lab instructor. Save your drawing as **HAND WHEEL.slddrw**.

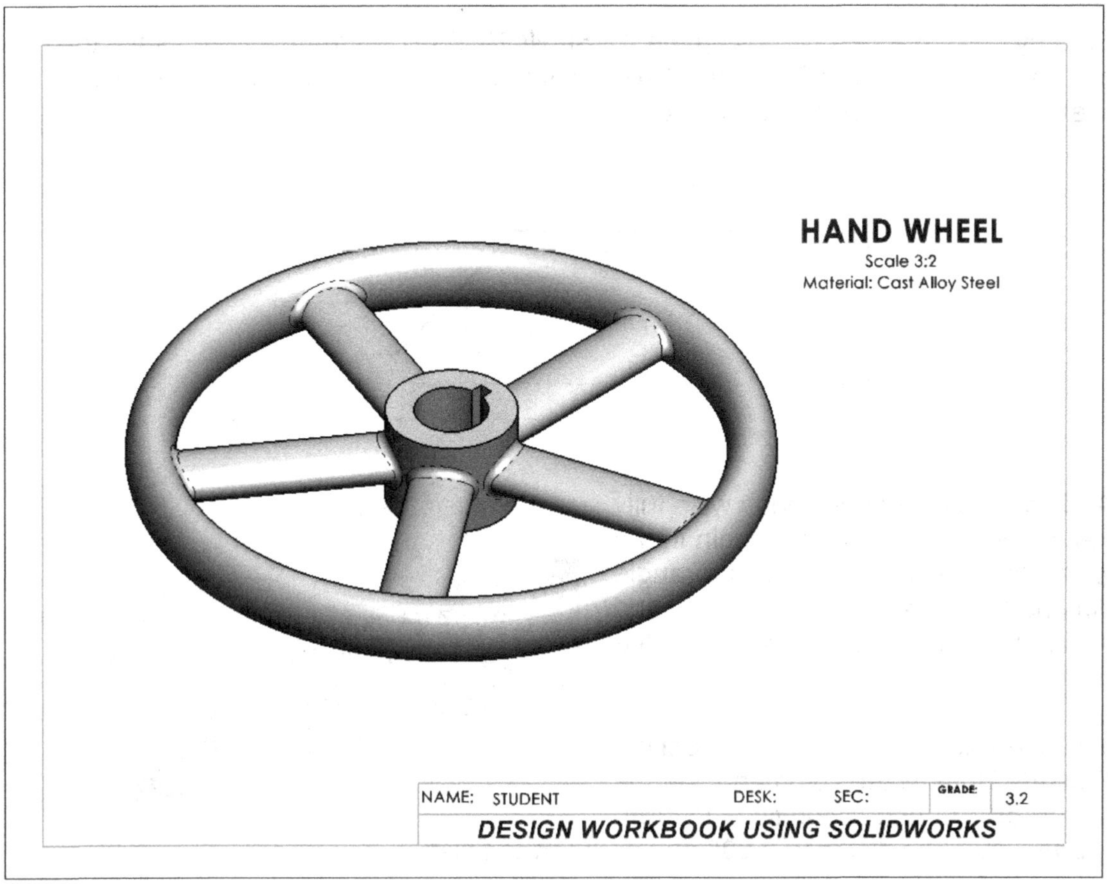

Exercise 3.4: TOE CLAMP

In this exercise you will practice previously learned concepts, and practice working on non-orthogonal planes while designing the Toe Clamp part.

Start by making a new part using the **ANSI-INCHES.prtdot** template and save the new part as **TOE CLAMP**. Select the **Front** plane in the FeatureManager and add a new sketch. Draw the profile shown next.

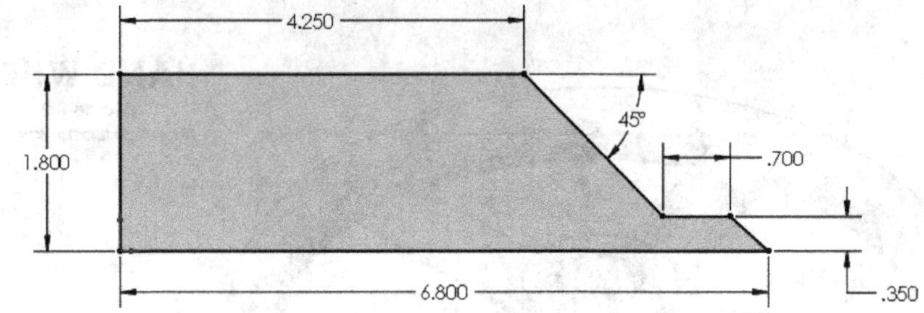

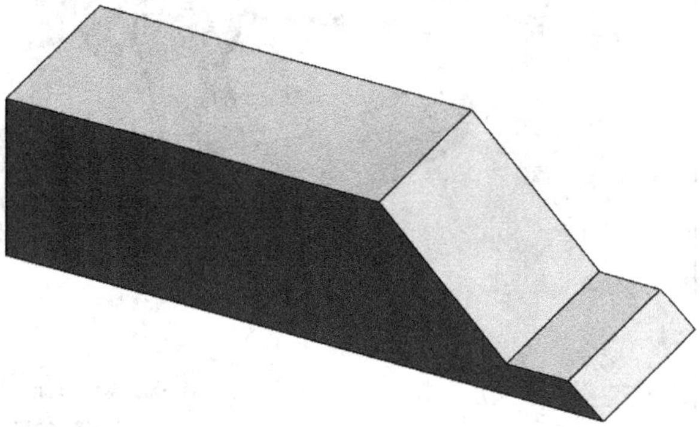

Adding these dimensions will fully define the sketch, with the origin in the lower left corner. To add the angle dimension, select the horizontal and inclined lines, and locate the dimension.

When the sketch is complete, select the **Extrude Boss/Base** command, set the **Direction 1** end condition to **Mid Plane**, make the distance **2.00"** and click **OK** to finish.

Now you will cut the counterbore and hole on the inclined face. To make it easier to work on this face you want to orient the part parallel to the screen.

To get this orientation select the face and click in the **Normal to** command in the View Orientation box, or in the context menu after the face is selected.

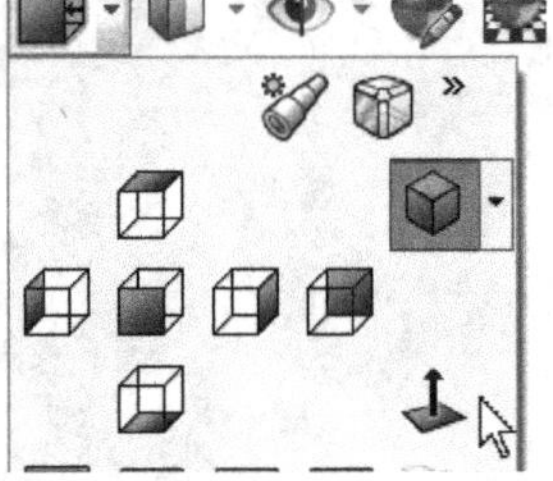

Now it will be easier to add the counterbore and hole on the inclined face. After the part is reoriented, select the inclined face, and add a new sketch.

Draw a circle and dimension it as shown next. To fully define the sketch, add a **Vertical** geometric relation between the center of the circle and the part's origin.

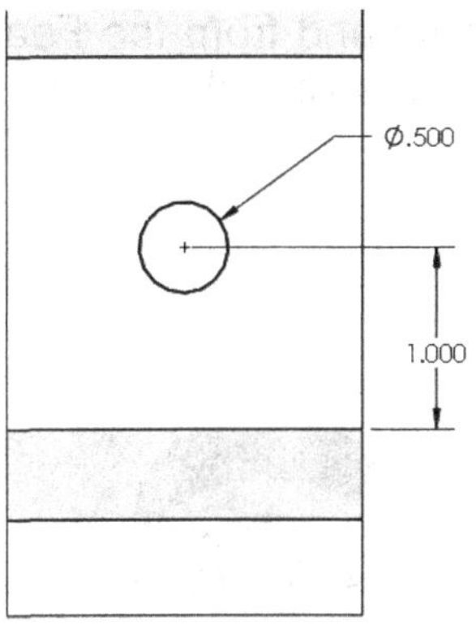

Select the **Extruded Cut** command, set the end condition to **Through all** and click **OK** to finish. The hole will be made perpendicular to the sketch face.

After adding the first hole, select the same angled face, add a new sketch and draw a circle concentric with the previous hole, and dimension its diameter **0.80"**.

Select the **Extruded Cut** command and make a cut **0.25"** deep using the **Blind** end condition to make the counterbore.

You can now make the V-cut to add stress relief for the Toe Clamp. This cut can be added to any of the horizontal faces, but for this exercise we'll use the horizontal flat face between the angled faces.

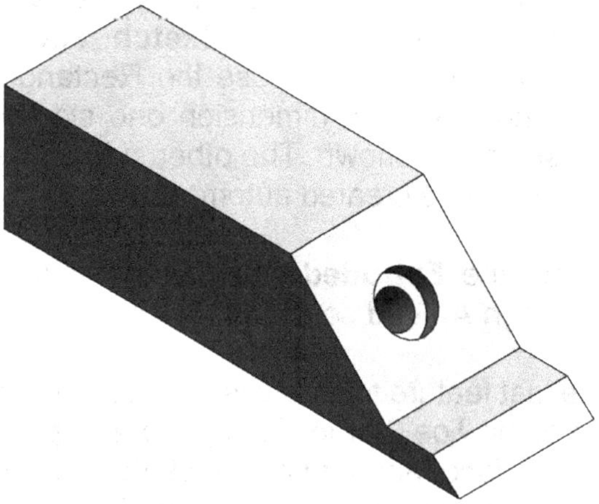

Change to a **Top** view orientation and add a new sketch on the flat face. To locate the cut centered in the part, draw a horizontal centerline starting at the middle of the edge, and then draw the triangle approximately as shown.

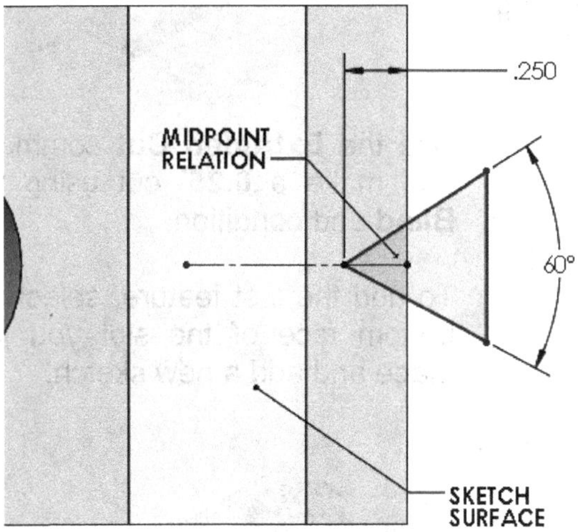

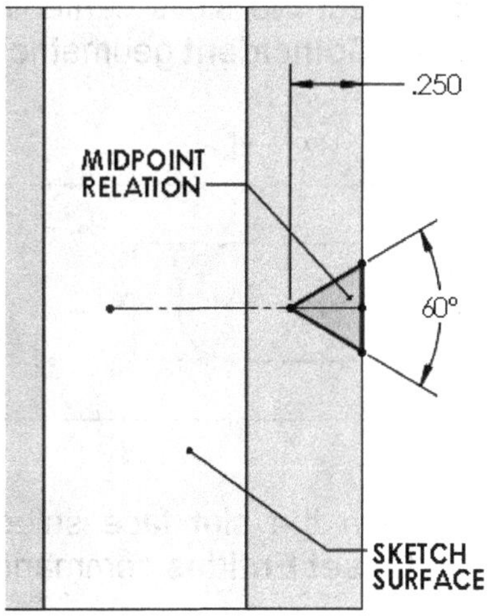

The easiest option to fully define the sketch and maintain the design intent is to add a **Midpoint** relation between the vertical line of the triangle and the endpoint of the centerline you added first.

After you add the relation, the sketch is fully defined, and regardless of which dimension is changed, the V-Cut will remain at the center of the part and symmetrical.

Now you can make the V-Cut. Select the **Extruded Cut** command from the Features tab and use the **Through all** end condition. Make sure the cut goes into the part through the beveled end, and then click **OK** to finish.

The next step is to add the side slots. Change to a **Right** view orientation and add a new sketch on the **Right** plane. The design intent requires the slots to be symmetrical about the part's center. To accomplish this, you will use the **Dynamic Mirror** command. Draw a vertical centerline starting at the origin, select it and go to the menu **Tools, Sketch Tools, Dynamic Mirror**. Now use the Rectangle tool and draw and dimension one side of the sketch as shown. The other side of the sketch will be created automatically.

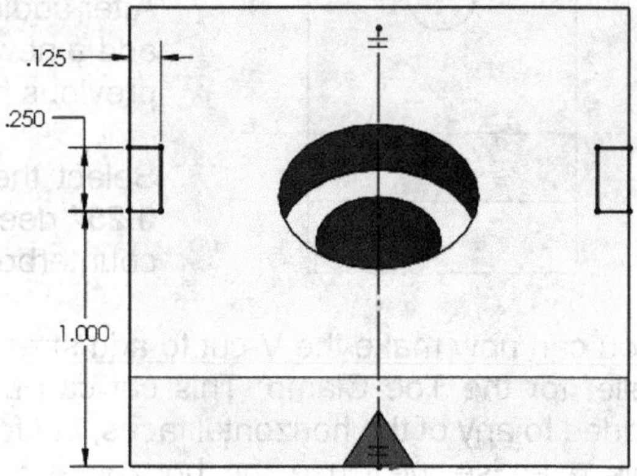

Select the **Extruded Cut** command and use the **Through All** end condition.

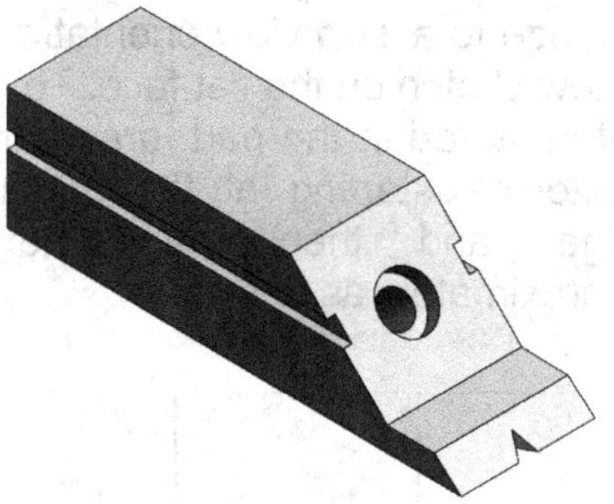

The last feature to create is the large counter slot on top of the Toe Clamp. Change to a **Top** view, select the top face, and add a new sketch. Using the **Slot** tool draw and dimension a slot. To fully define the sketch, select the slot's centerline and the origin, and add a **Coincident** geometric relation.

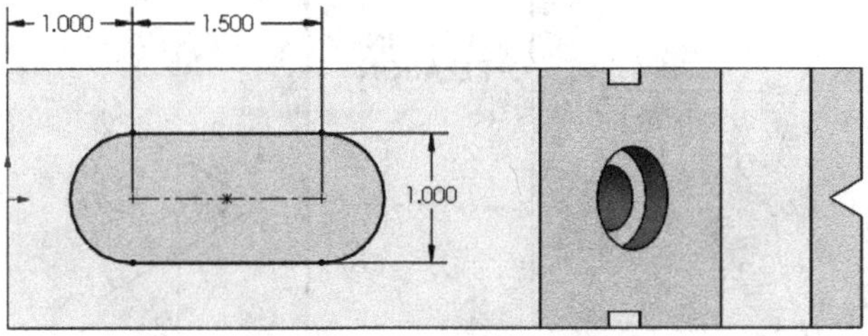

Use the **Extruded Cut** command and make a **0.25**" cut using the **Blind** end condition.

To add the last feature, select the bottom face of the slot you just made and add a new sketch.

With the slot face selected, click on the **Offset Entities** command.

Use the **Reverse Direction** option, set the distance to **0.20"** to make a smaller slot and click **OK**. Select the **Extruded Cut** command, set the end condition to **Through all** and click **OK** to finish.

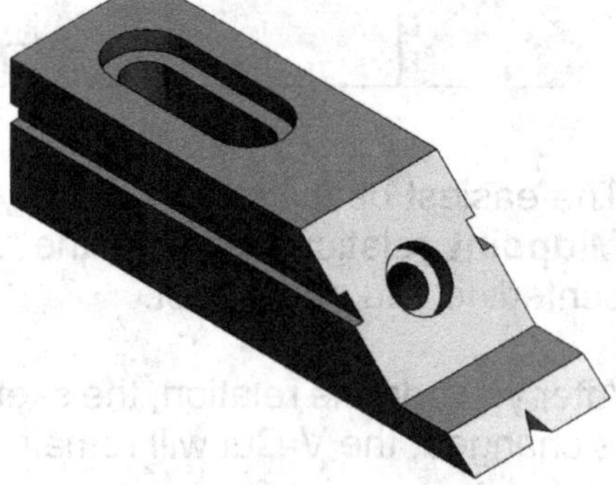

The last thing you need to do is to define the part's material. In the FeatureManager, right click on **Edit Material**, expand the **Steel** materials category, and assign **AISI Type A2 Tool Steel** to the **TOE CLAMP**.

Save your part as **TOE CLAMP**, make a new drawing, add a shaded isometric view and add the missing annotations. Save your drawing as **TOE CLAMP.slddrw**. Print a copy to submit to your lab instructor.

Supplementary Exercise 3-5: CONVEYOR RAMP GUIDE

Build a solid model of the figure below. Make a drawing with an **Isometric** view and name it **"CONVEYOR RAMP GUIDE." The dimensions indicated are in Inches.**

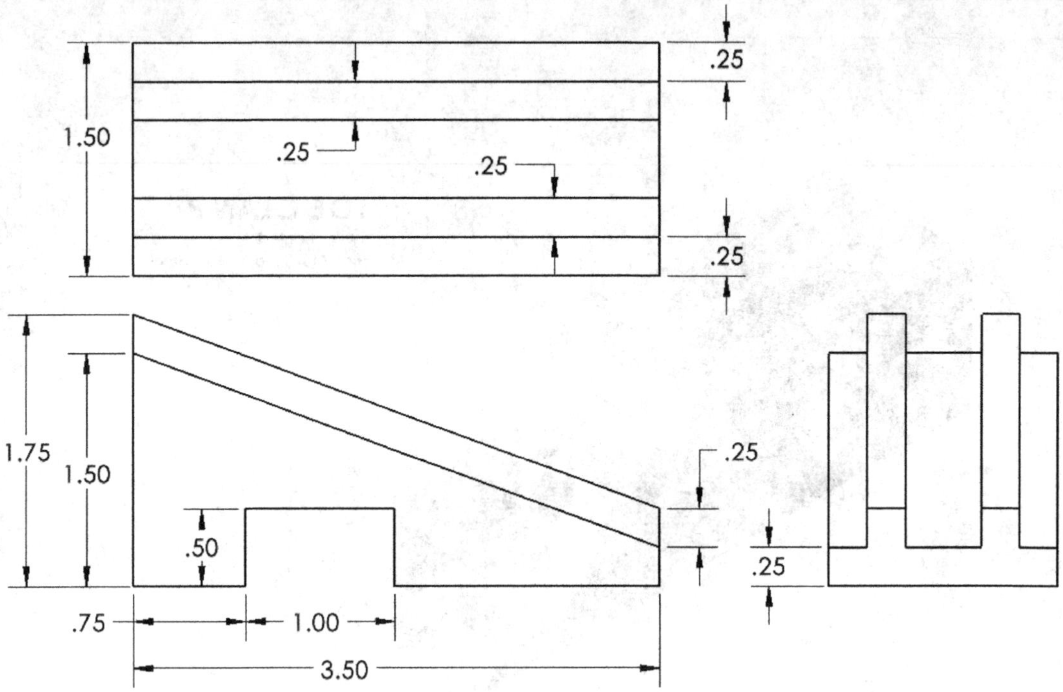

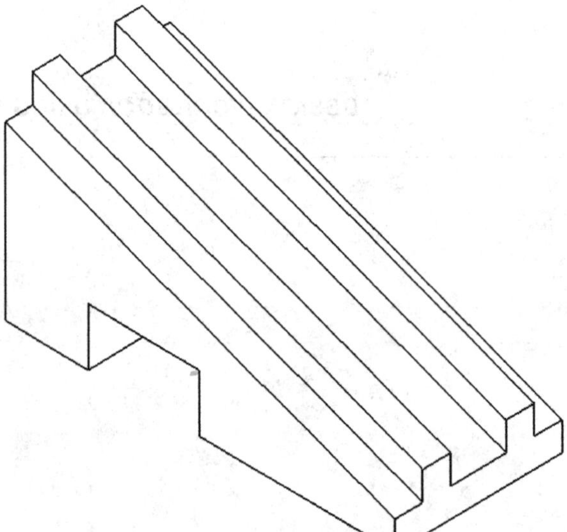

ISOMETRIC VIEW

Supplementary Exercise 3-6: DOUBLE SHAFT HANGER

Using the knowledge acquired so far, make a part of the figure below, make a drawing with an Isometric view and save it as **DOUBLE SHAFT HANGER**.

HINT: The part is symmetrical; make the first rectangle centered at the origin in the **Top** plane.

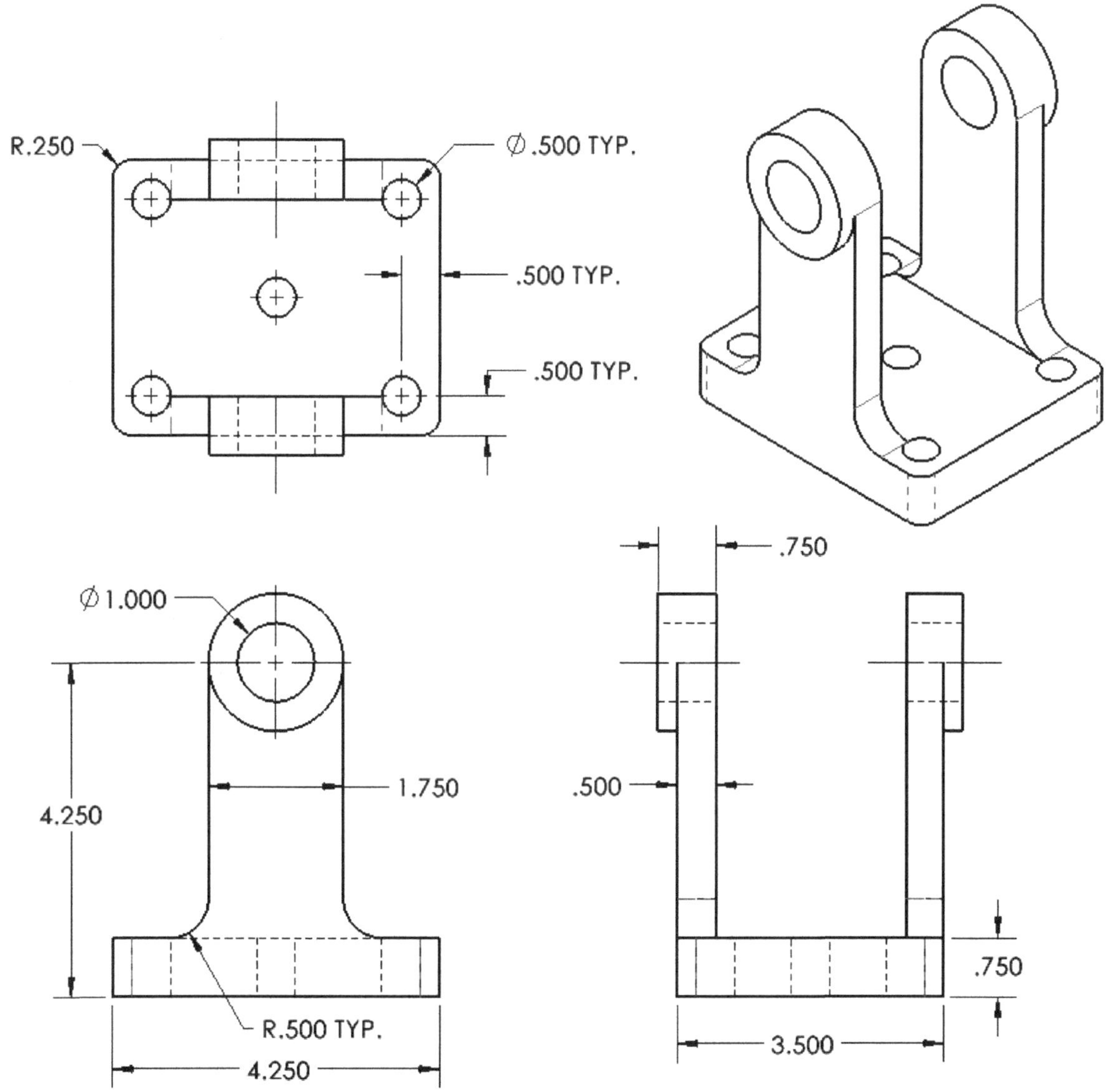

NOTES:

Design Workbook Lab 4
3D Modeling Part II

Introduction to Lab 4

In this lab you will learn to use advanced features to make 3D models that would be difficult or impossible to make with the prismatic shapes used previously. These are the commands you will be using:

Draft - Creates a taper to the selected model faces by a specified angle, using either a neutral plane or a parting line. **NOTE:** You can also apply a draft angle as a part of an extruded base, boss, or cut feature.

Offset Plane - You can create planes in parts or assemblies. You can use planes to sketch, to create a section view of a model, for a neutral plane in a draft feature, and so on.

Offset - You can create sketch curves offset from one or more selected sketch entities, edges, loops, faces, curves, set of edges, or set of curves by a specified distance. The selected sketch entity can be changed to construction geometry. The offset entities can be bi-directional.

Convert Entities - You can create one or more curves in a sketch by projecting an edge, loop, face, curve, external sketch contour, set of edges, or set of sketch curves onto the sketch plane. You can convert sketched entities into construction geometry to use in creating model geometry.

Face/Edge Fillet – A Fillet/Round creates a rounded internal or external face on the part. You can fillet all edges of a face, selected sets of faces, selected edges, or edge loops.

Shell - The shell hollows out the part, leaving open the selected faces, and creates a thin-walled feature on the remaining faces.

Loft - Creates a 3D feature by making transitions between different profiles. A loft can be a base, boss, cut, or surface.

Dome - Creates a 3D feature that begins with the shape of the selected face and lofts up to a rounded face at a specified height.

Sweep - Creates a base, boss, cut, or surface by sweeping a profile (section) along a path, according to these rules:

- The profile must be closed for a solid sweep feature; the profile may be open or closed for a surface feature.
- The path may be open or closed.
- The path may be a set of sketched curves contained in one sketch, a curve, or a set of model edges.
- The start point of the path must lie on the plane of the profile.
- The section, the path, or the resulting solid cannot be self-intersecting.

Exercise 4.1: Drawer Tray

Start a new part based on the **ANSI-INCHES** template and save it as **DRAWER TRAY.sldprt**. Select the **Top plane** in the FeatureManager and a new sketch. Your part will be rotated to a Top view automatically.

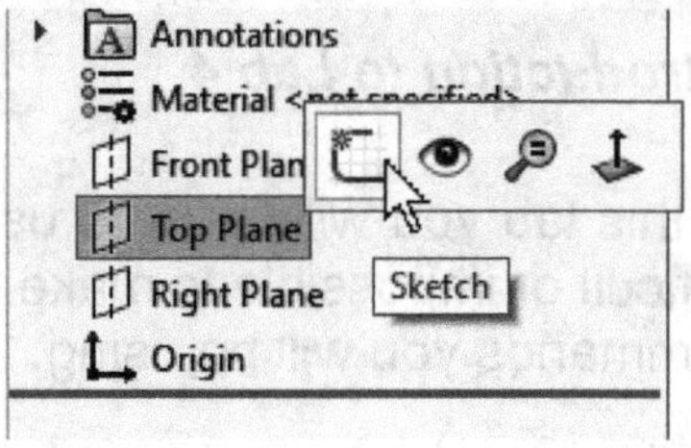

Use the commands learned so far to draw the profile shown next. Add a sketch Fillet in the corners to finish. Make sure the sketch origin is in the lower left corner.

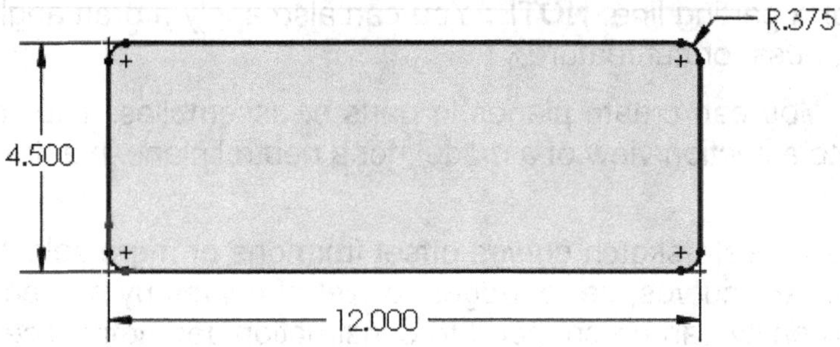

From the Features tab select the **Extruded Boss** command and make the extrusion **3.000"** deep using the **Blind** end condition.

Activate the **Reverse Direction** option to make the extrusion go down. Turn On the **Draft Angle** option and set the angle to **5°**, and click **OK** to finish. The first feature looks as shown here.

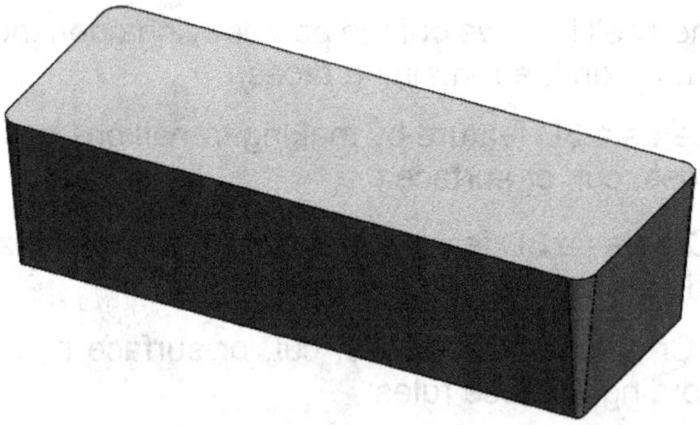

SOLIDWORKS allows you to rename features in the FeatureManager to make it easier to identify them later. To rename the **Boss-Extrude1** feature, slow double-click on it and rename it "**Tray Body.**"

The next step is to add fillets to the bottom of the part. To rotate the view, use the **Middle Mouse Button** and drag the mouse, or use the arrow keys until you see the bottom face. From the Features tab select the **Fillet** command, change the radius to **0.375"**, select the bottom face of the "**Tray Body**" feature and click **OK** to finish. By selecting the face, all its edges will be rounded.

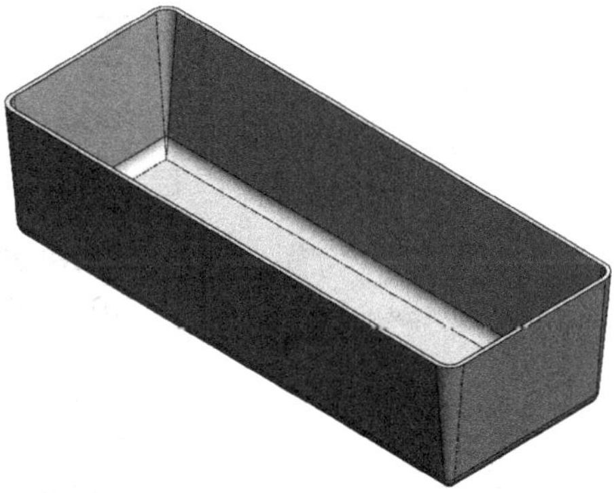

To create a hollow part change to an **Isometric** view orientation, select the **Shell** command from the Features tab, in the Parameters section set the wall thickness to **0.10"** and select the top face of the part.

This is the face that will be removed from the model, and every other face will have the specified thickness. Click OK to shell the part.

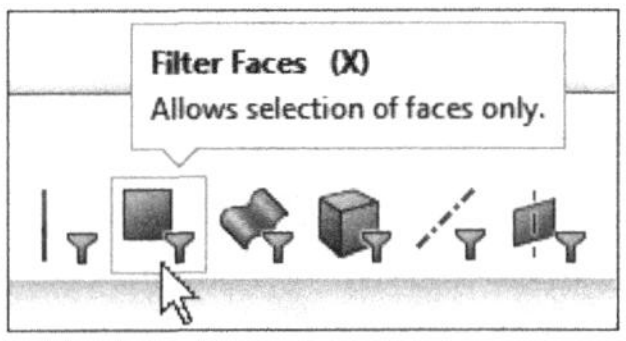

To reinforce the tray, you will add a rim around the top edge. To do this you have to add a new sketch in the top thin face of the part. To make the selection of this face easier, you can use the **Face Selection Filter**. Press the **F5** key (default shortcut) to display the selection filter toolbar and activate the **Faces** filter. Optionally you can press the **X** key (default shortcut) to activate/deactivate the **Face Selection Filter**.

As long as the selection filter is active you can only select faces. Your mouse cursor will have a small funnel to indicate a selection filter is active.

NOTE: If you don't use the selection filter you have to zoom in close to select the face.

After selecting the top thin face add a new sketch in it, and immediately turn off the selection filter. If you don't turn it off, you will not be able to select other entities.

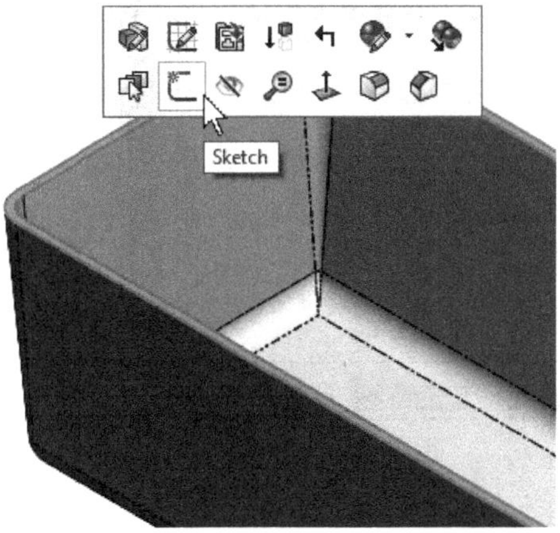

With the top face selected, click on the **Convert Entities** command to project the outer edges of the selected face as new sketch entities.

Now select the **Offset Entities** command and select one of the previously projected sketch lines. Make sure the **Select Chain** option is activated to select the entire profile. If the preview is shown outside, turn on the **Reverse** option to add the offset inside.

Set the distance to **0.10"** and click **OK** to finish.

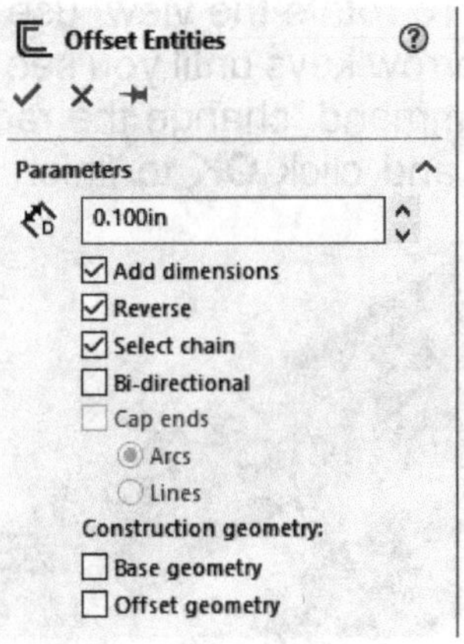

Select the **Extruded Boss** command and make a **Blind** extrusion going down. Set the distance to **0.625"** and click **OK** to create the vertical rim around the upper edge of the tray.

The next step is to add three equally spaced dividers inside the tray. They will be located **0.25"** below the top face of the tray.

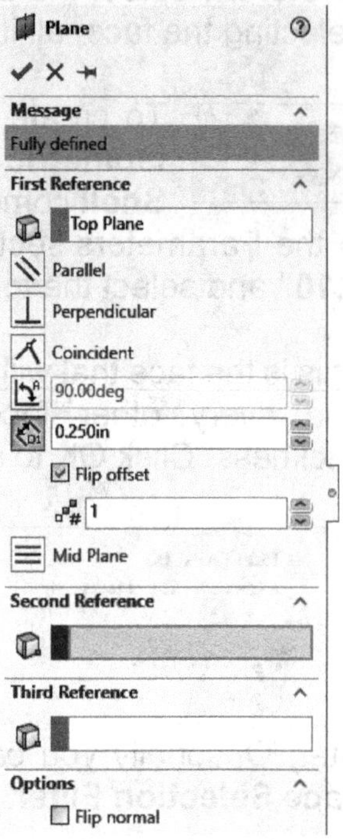

From the Features tab, click in the **Reference Geometry** command and select the **Plane** command.

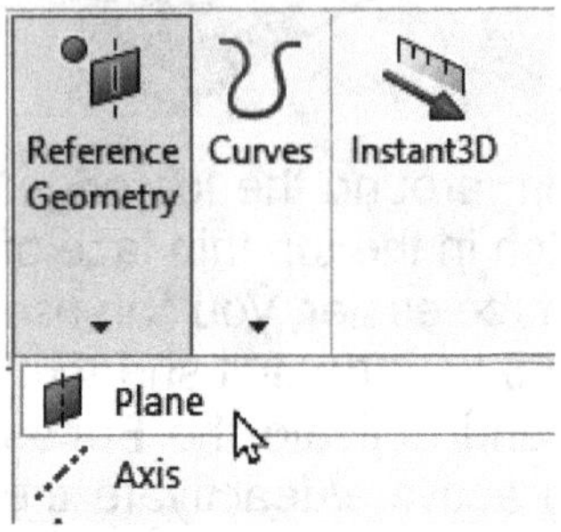

From the fly out FeatureManager select the **Top** plane, set the distance to **0.25"** and use the **Flip Offset** option to add the plane below the top face and click **OK** to finish.

Now slow double click to rename the new plane in the FeatureManager and change its name to **Dividers Plane**.

To create the divisions, you will use the **Rib** command. This command uses a sketch with a **single open profile**, or **multiple open profiles**, and extrudes it in the designated direction, defining the width and direction of the extrusion, and optionally adding a draft.

Change to a Top view and add a new sketch in the Dividers Plane just created. Draw a single line as shown.

From the Features tab select the **Rib** command. In the Parameters section set the **Thickness** to Both Sides, set the **Rib Thickness** to **0.10”** and, if needed, change the **Extrusion Direction** to Normal to Sketch. Make sure the arrow is pointing down into the part. Turn on the **Draft** option, set the draft angle to **2°** and make sure the **Draft outward** option is checked.

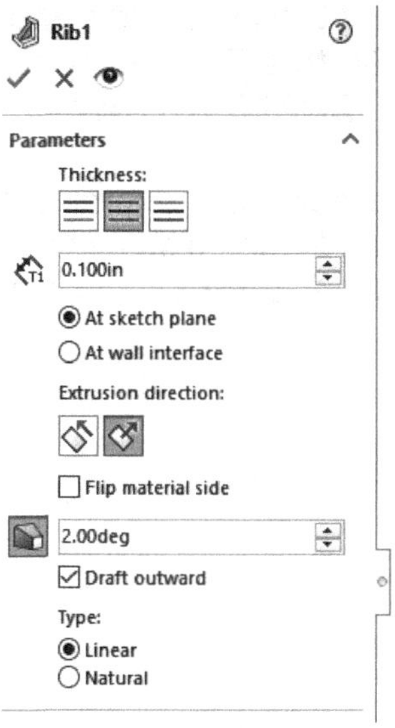

Click **OK** to finish; the completed rib is seen next.

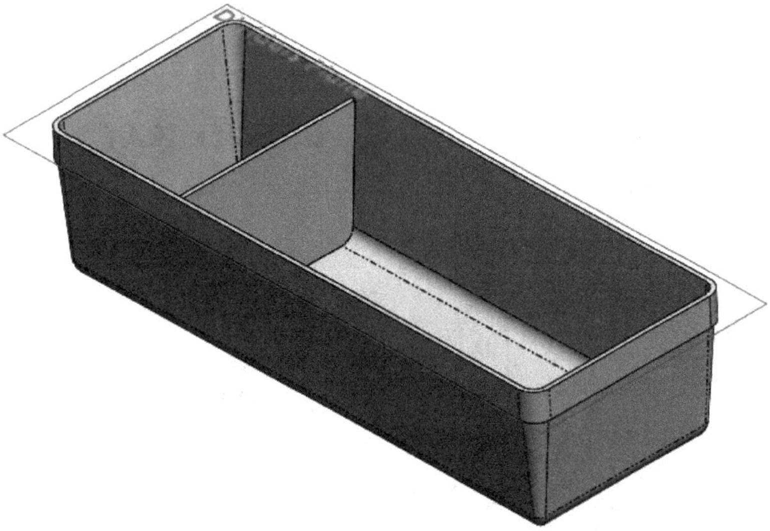

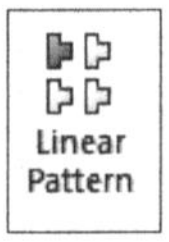

To add the rest of the ribs, use the **Linear Pattern** command, select the long edge of the tray for the **Direction 1**, set the number of copies to **3** and make the spacing equal to **3.00”**.

In the Features and Faces selection box add the Rib1 feature and click **OK** to finish. In the **Hide/Show** Items command turn off the Plane’s visibility.

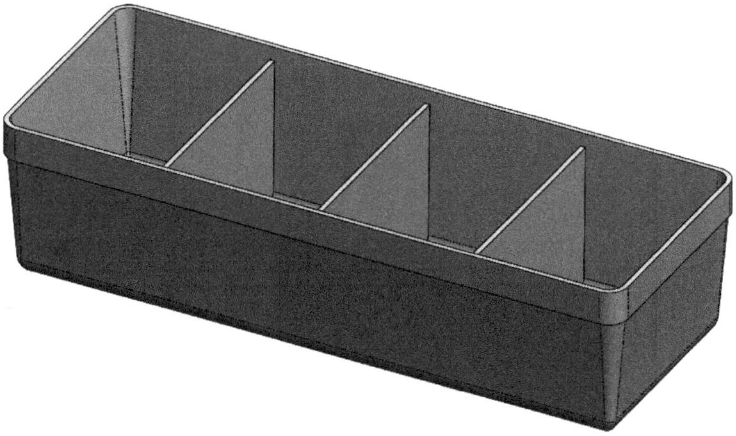

Right click on **Material** in the FeatureManager. Select **Edit Material**, expand the **Plastics** Tab and apply the **PC High Viscosity** material for the part. Save your model as **DRAWER TRAY**.

To finish this exercise, make a new drawing, add an Isometric view using the Shaded with Edges display mode and add the missing notes. Save your drawing as **DRAWER TRAY.slddrw** and print a copy for your instructor.

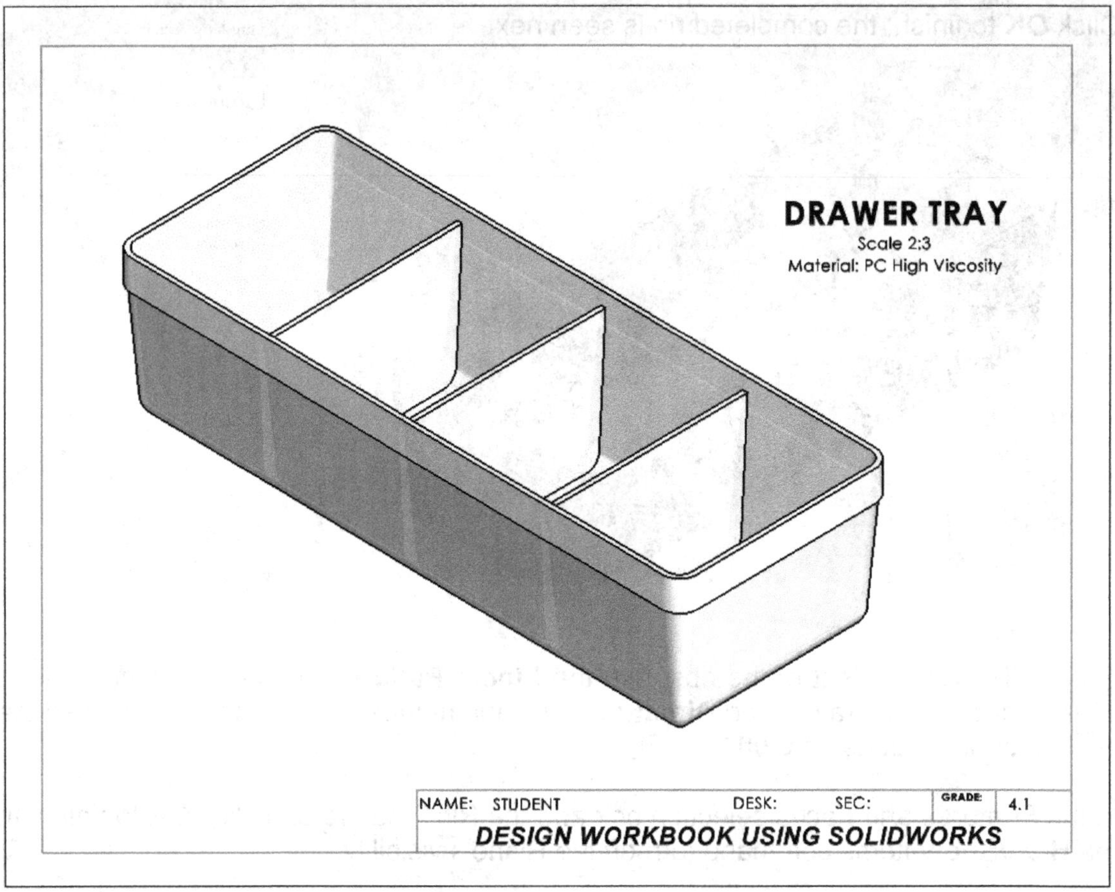

Exercise 4.2: Tap-Light Dome

Start a new part using the **ANSI-INCHES** template and save it as **TAP-LIGHT DOME**. Add a new sketch in the **Top** plane and draw a circle centered in the origin with a diameter of **4.70"**. After the sketch is complete make an **Extruded Boss** command **0.375"** thick.

In the FeatureManager rename the **Boss-Extrude1** feature as "**Dome Base.**"

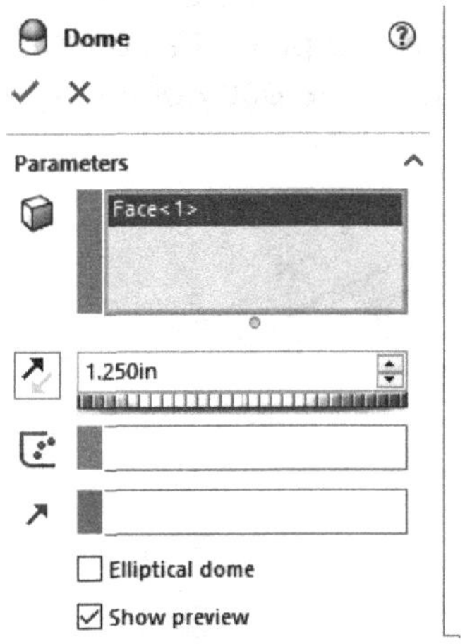

Now you need to add the dome on top of the disk. Select the menu **Insert, Features, Dome**. In the **Faces to Dome** selection list add the top face of the base feature, set the **Dome Height** to **1.25"** and click **OK**.

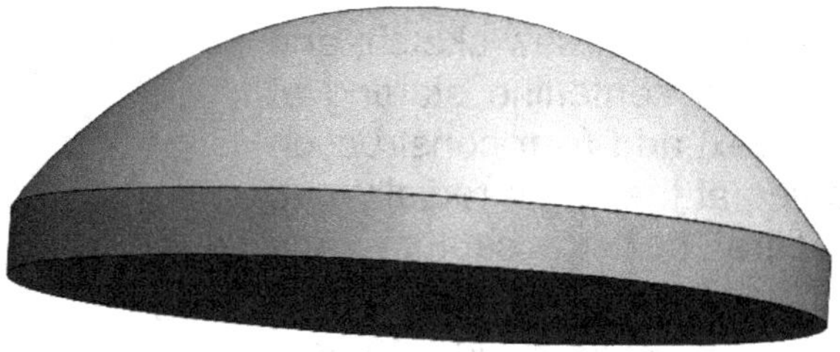

The next step is to make the part hollow. From the Features tab select the **Shell** command, rotate the part, and select the bottom flat face to remove it. In the Parameters section, set the shell thickness to **0.075"** and click **OK** to finish.

Now you need to add a lip at the bottom of the part. Change to a **Bottom** view orientation, select the thin face at the bottom of the part and add a new sketch. If needed, use the **Face** selection filter (Shortcut **X**) to make selection easier.

With the bottom face selected, click in the **Convert Entities** command to project the outside edge of the face.

Now click on the **Offset Entities** command, set the offset distance to **0.20"** and select the previously projected outside edge. Make sure the offset is added outside the part and click **OK** to finish.

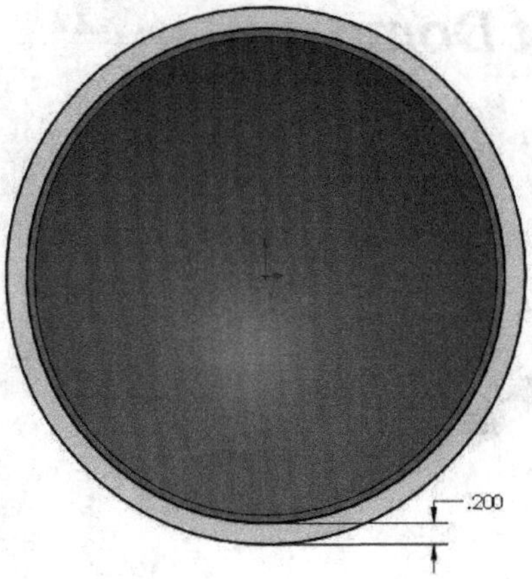

Select the **Extruded Boss** command and make a **0.075"** thick boss going up. Change to an **Isometric** view to visualize the extrusion's direction and click **OK**.

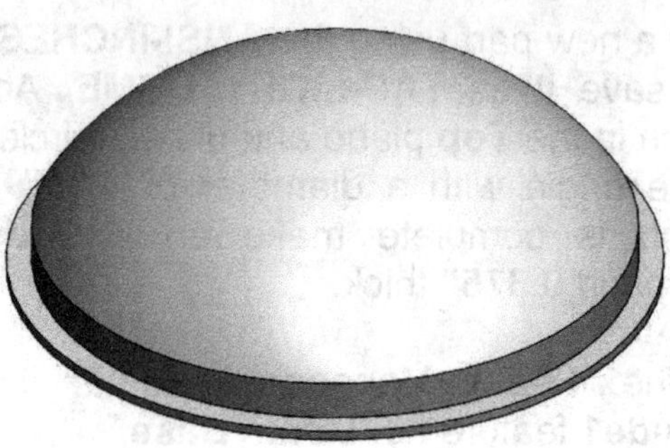

In the next step you will add four holes in the lip of the dome for fastening screws to hold the body and the lens together. The holes in the lip of the dome are placed at uneven angles. This is done to ensure that the dome can only be placed on the body of the light fixture in one position.

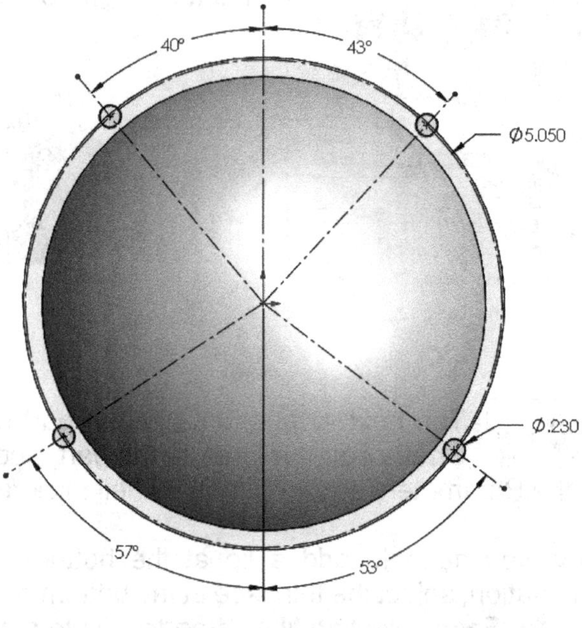

Change to a **Top** view, select the top face of the lip, add a new sketch, and draw a vertical centerline starting at the origin. Next add **four** construction lines starting at the origin radially out, making sure not to capture any automatic geometric relations; in other words, do not snap a line to any automatic reference and dimension their angle to the vertical as indicated.

Next draw a circle starting at the origin, and while it is selected, activate the **For construction** option in the properties, and finally dimension the diameter **5.05"**. This construction bolt circle will be used to locate the screw holes in the lip, since it is not the same diameter as the outer size of the lip.

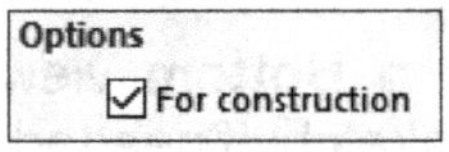

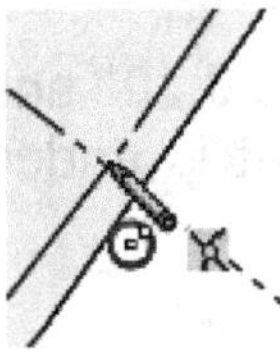

Draw **four** circles at the intersections of the construction circle and the radial lines; after you select the **Circle** command you will see an inference icon next to the cursor when you approach an intersection to start each circle. Dimension one of them **0.230"** diameter and add an **Equal** relation between all four circles to fully define the sketch.

Now use the **Extruded Cut** command and make a cut using the Through All end condition.

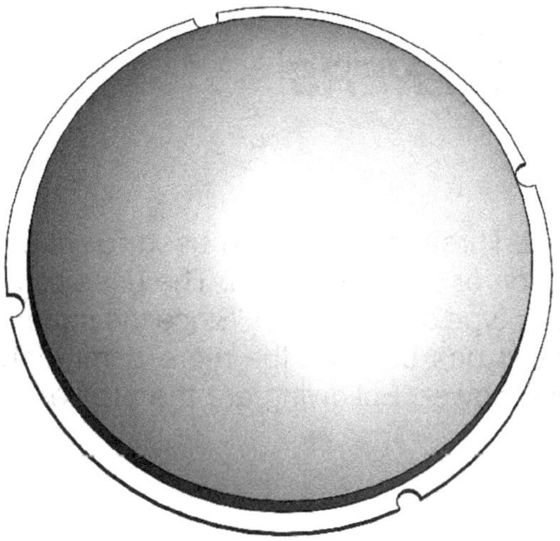

To assign a material to your part Right click on **Material**, select **Edit Material**, from the Plastics section select **ABS** and apply the material to your part.

Save your part as **Tap-Light Dome.sldprt**, make a new drawing with an **Isometric** view, add the missing annotations and save it as **Tap-Light Dome.slddrw**.

Exercise 4.3:
THREADS AND FASTENERS

THREAD TYPES

SQUARE THREADS & ACME THREADS These Threads have been used for many years to transmit power either from a turning motion or a linear motion. The uses have increased as more and more processes require this type of translation. Since the square thread is difficult to manufacture, the Acme thread has become the more commonly used thread profile for the purpose of transmitting power. Following are the profiles of an Acme Thread and a Square Thread profile.

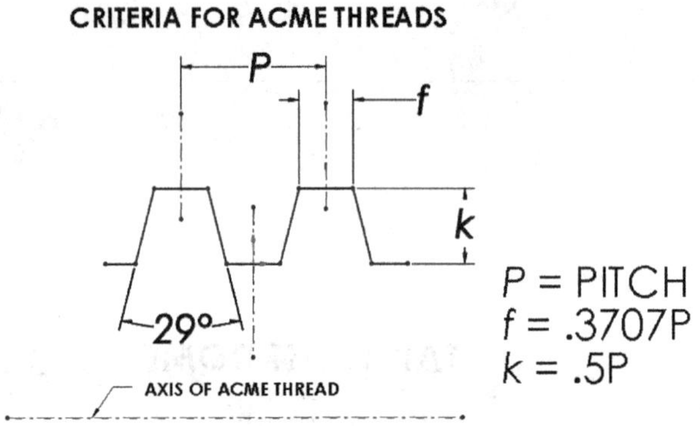

Acme Thread Profile

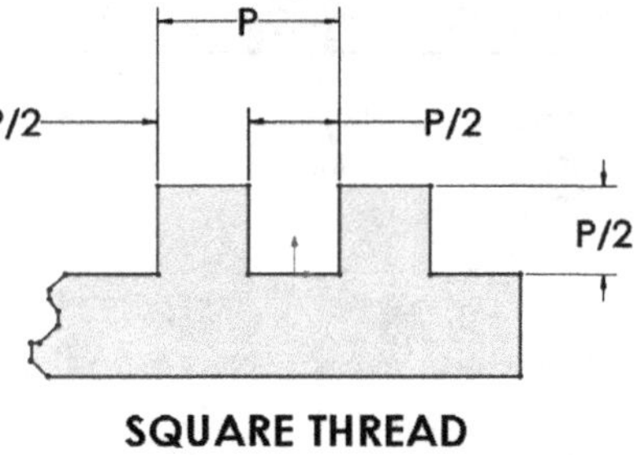

Square Thread Profile

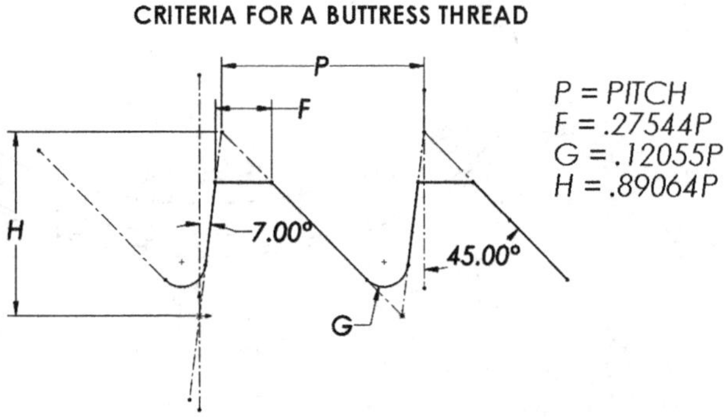

BUTTRESS THREAD This thread form is a nonsymmetrical thread that is used where exceptionally high stresses lie along the axis of the threaded shaft.

STANDARD V-THREADS This thread form is the basis for most nuts and bolts; however, within this thread form there are many deviations that are used for special applications. Several of the most common standards are the American Standards and the Metric Standards.

Within the American standards the most common series are "UNC" or Unified National Coarse (also known as "NC" National Coarse); "UNF" or Unified National Fine (also known as "NF" National Fine); and "UNEF" or Unified National Extra Fine (also known as "NEF" National Extra Fine). Within the Metric standards there is a coarse series and a fine series that are distinguished only by the different values of the pitch.

THREAD TERMINOLOGY

Following is a list of terms that are used extensively in association with threads:

- Major Diameter – the largest diameter of an internal or external thread.
- Pitch – The distance from one crest of a thread to the next crest.
- Pitch Diameter – the theoretical diameter at the point where the tooth width and gap width are equal.
- Minor Diameter - the smallest diameter of an internal or external thread.
- Threads per Inch – The number of threads in an inch that determines the pitch.
- Thread Depth – generally the difference between the major diameter and minor diameter divided by 2.

THREAD NOTES

Thread notes are the most critical portion of a thread representation. It gives all the information necessary for a machinist to produce the thread or the manufacturer to select the correct fasteners for assembly.

Following are shown an American thread note as well as a metric thread note.

AMERICAN BASIC THREAD NOTE

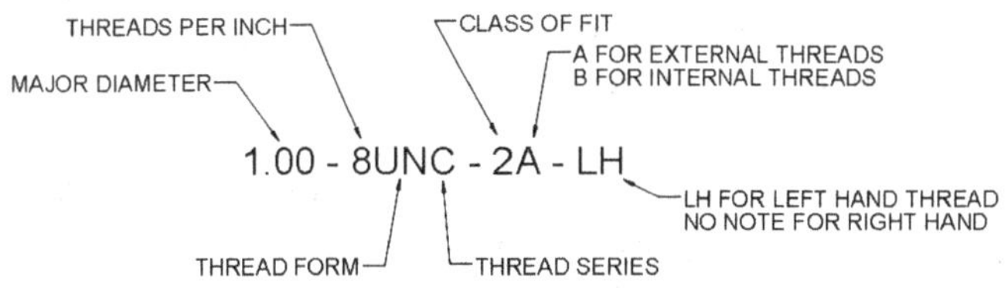

METRIC THREAD NOTE

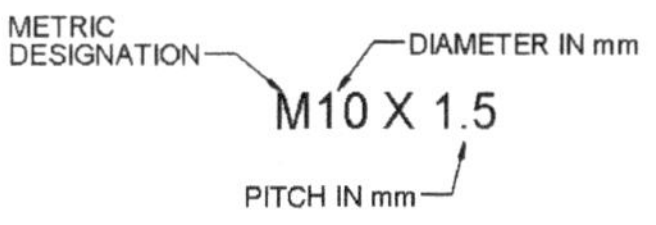

The dimensions of threads are based primarily on the number of threads per inch in the American system. The number of threads divided into one inch produces the pitch, which is also the basis in the metric system. In the exercise we will be using the criteria for internal and external threads. The following images show the criteria required for external and internal threads. The values for the different formulas are given in the various tables provided in Machinist or Engineering Handbooks.

CRITERIA FOR EXTERNAL THREADS

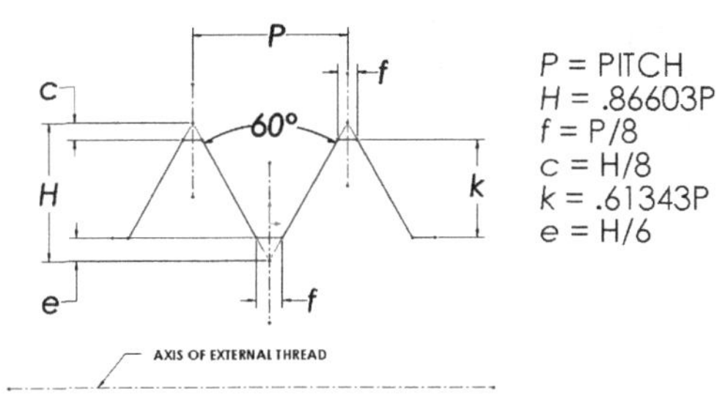

CRITERIA FOR INTERNAL THREADS

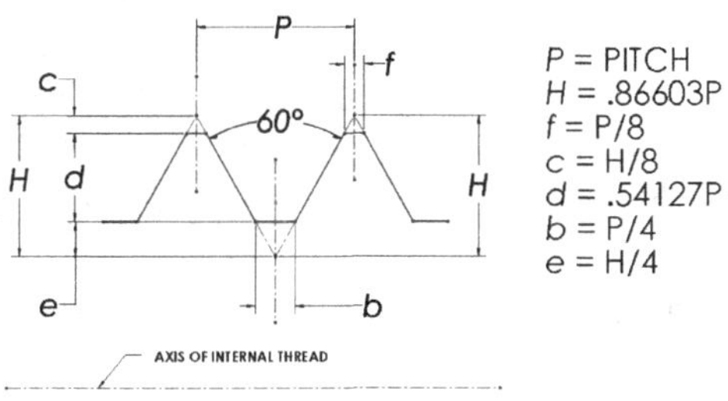

UNIFIED COARSE THREAD SERIES

PARTIAL TABLE

Fractional Sizes	Basic Major Diameter	Threads per Inch	Pitch	Minor Diameter Internal Threads	Thread Depth
1/4	0.250	20	0.0500000	0.1959	0.0325
5/16	0.313	18	0.0555555	0.2524	0.0361
3/8	0.375	16	0.0625000	0.3073	0.0406
7/16	0.438	14	0.0714286	0.3602	0.0464
1/2	0.500	13	0.0769231	0.4167	0.5000
9/16	0.563	12	0.0833333	0.4723	0.0541
5/8	0.625	11	0.0909091	0.5266	0.0590
3/4	0.750	10	0.1000000	0.6417	0.0650
7/8	0.875	9	0.1111111	0.7547	0.0722
1	1.000	8	0.1250000	0.8647	0.0767
1 - 1/8	1.125	7	0.1428571	0.9704	0.0928
1 - 1/4	1.250	7	0.1428571	1.0954	0.0928
1 - 1/2	1.500	6	0.1666667	1.3196	0.1083

Here is an example of a thread table and includes the criteria that will be used in **Exercise 4.3**.

Start a new part with the **ANSI-INCHES** template, and save it as **Threaded Bolt.sldprt.**

Designing a thread requires high precision. Select the menu **Tools, Options, Document Properties, Units** and change the units to **5 Decimal Places**.

Add a new sketch in the **Right** plane, draw a circle at the origin, dimension its diameter **1.00"** and finally extrude it **5.00"** to the right.

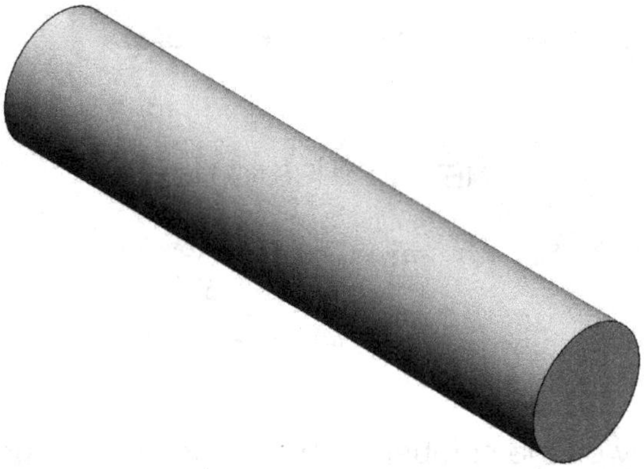

For proper clearances between interior and exterior threads, the exterior thread for a **1"** thread is undersized by a tolerance of between **0.9966"** and **0.9744"**. In our case we will use the mid-value of **0.9855"**.

Change to a **Left** view orientation, select the face on the left side of the shaft, add a new sketch and draw a circle with a **0.9855"** diameter.

From the Features Tab, select the **Extruded Cut** command, use the **Blind** end condition and make a cut **3.25"** deep. Activate the **Flip Side to Cut** option to make the cut outside of the profile.

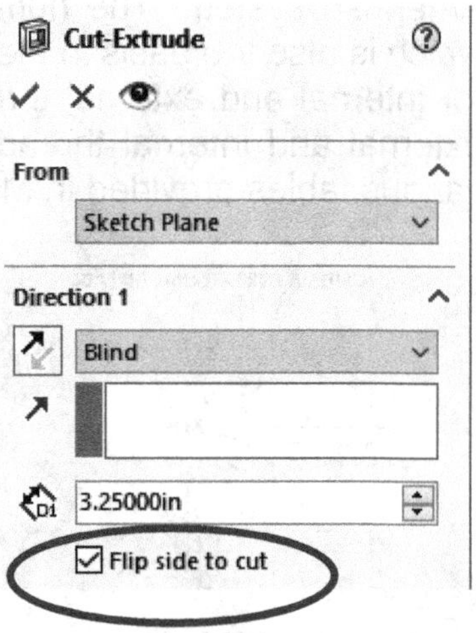

Now you need to add a **Chamfer** on the face of the threaded end. First you need to calculate the chamfer size using the formula for **Thread Depth** (k) for external threads found at the bottom of **Page 4-11**. Per the table, a **1.00"** diameter thread has a **0.125"** pitch.

Chamfer size = (0.61343 * 0.125) or **0.0767"**

From the Features tab select the **Chamfer** command, select the left side face of the threaded end, select the face and enter the **0.0767"** chamfer size. Click **OK** to finish.

The thread in this part will be **3.125"** long from the left side. To make the thread, it needs to start *away* from the left face the distance of the thread's **Pitch**. Per the previous table it should be **0.125"**. In this case you need to add a Reference Plane at this location.

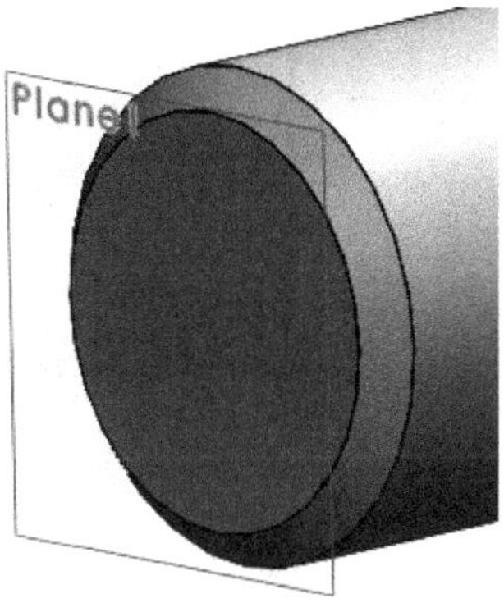

From the Features tab, click on **Reference Geometry** and select **Plane**. Select the left face of the threaded end, change the distance **0.125"**, make sure the plane is located on the left of the part and click **OK** to finish.

To make the thread you will need to make a **Sweep** feature, which requires a **Path** or curve, and a **profile**. The path will be a helix that will define the thread, and a profile with the thread's profile.

To make the helix, first you need to make a sketch with a circle. Select **Plane1** in the FeatureManager and add a new sketch. Select the larger diameter in the chamfer and click in the **Convert Entities** command.

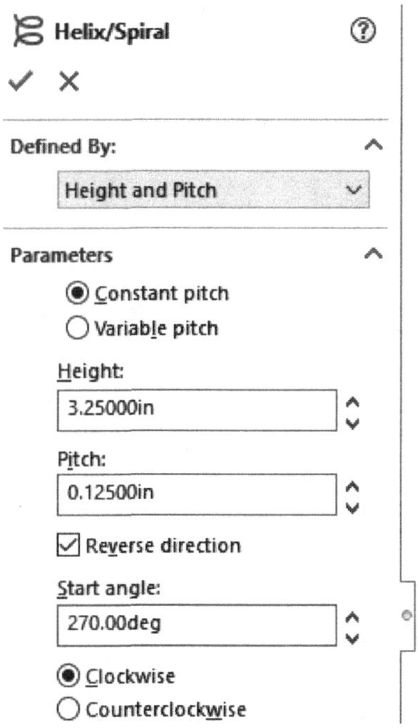

After the circle is projected, select the Features tab, click on the **Curves** command, and select **Helix/Spiral**, or go to the menu **Insert, Curves, Helix/Spiral**.

In the **Helix/Spiral** options use the **Height and Pitch** definition and select the **Constant pitch** option. Set the **Height** to **3.25"** and the Pitch to **0.125"**.

Activate the **Reverse Direction** checkbox to make the helix go into the part, set the **Start angle** to **270°** and make sure you are using the **Clockwise** option to create a right-hand thread. After entering all the parameters click **OK** to create the Helix.

The reason to start the helix at **270°** is to make the start of the curve coincide with the **Front** plane; this way you can add the thread's profile in the **Front** plane.

Change to a **Front** view, select the **Front** plane, and add a new sketch. **Zoom** in the left side of the helix, add a horizontal centerline coincident to the bottom of the threaded end, and a vertical centerline coincident to **Plane1**. Select the vertical centerline and go to the menu **Tools, Sketch Tools, Dynamic Mirror** to make the profile symmetrical. Draw and dimension the sketch as follows.

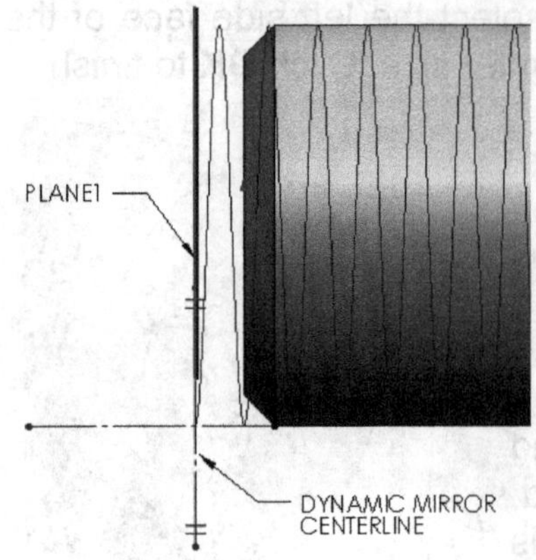

With the **Dynamic Mirror** active, draw one half of the profile, and the other half will be automatically created. To dimension the sketch, you must calculate the size of the thread profile using the formulas in **Page 4-11**, since this is an external thread. Remember the Pitch (P) is **0.125"**.

f = P/8 → 0.125"/8 = **0.0156"**
k = 0.61343 * P → 0.61343 * 0.125" = **0.07670"**

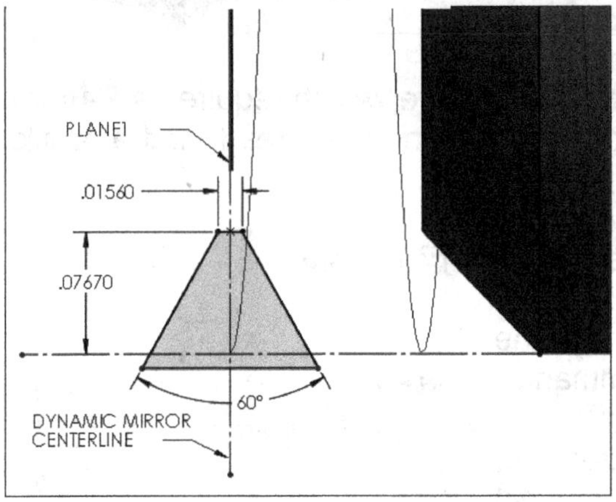

The finished sketch is shown here. Make sure the lower horizontal line is located slightly below the bottom of the threaded end.

The reason is to make sure the **Sweep Cut** feature extends past the solid body.

After the sketch is complete click on **Exit Sketch** or the **Rebuild** command.

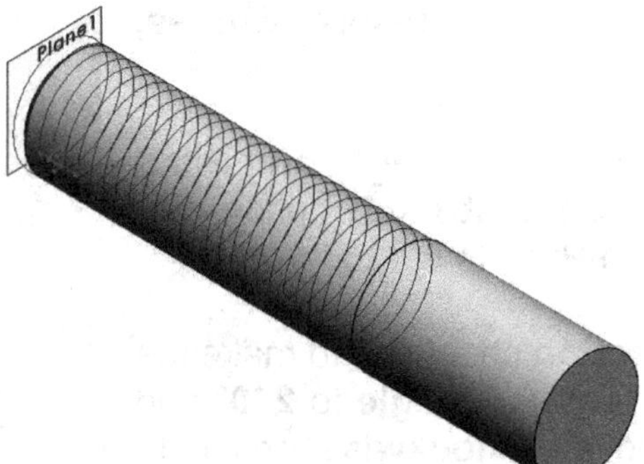

From the Features tab select the **Swept Cut** command and change to an **Isometric** view for visibility. In the **Profile** selection box add the thread profile sketch, and in the **Path** selection box add the Helix.

Review the preview and click **OK** to finish. To hide the helix curve, select it either in the FeatureManager or the graphics area, and click on Hide from the context toolbar. The finished thread will look as follows.

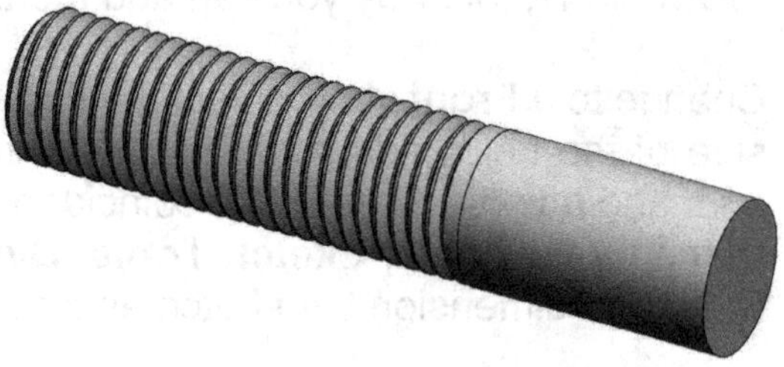

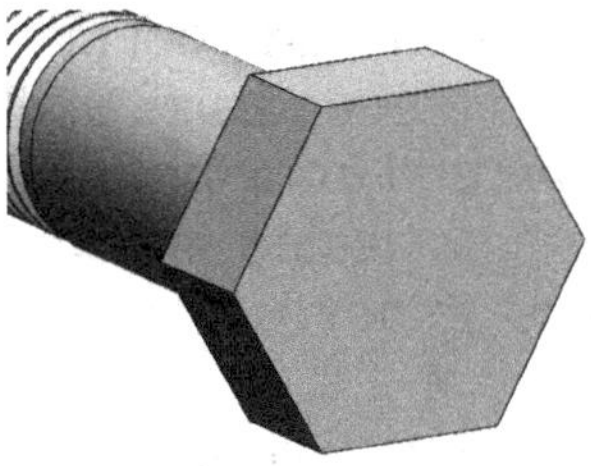

Now you need to make the head of the bolt. Change to a **Right** view orientation, add a sketch in the right face and draw a **Hexagon** at the origin.

Select one of the hexagon's lines, add a **Horizontal** geometric relation and dimension the distance between the top and bottom sides **1.50"**. The standard height for a **1"** screw head is **39/64"**. With the fully defined sketch make an **Extruded Boss**; in the extrusion's distance box enter **39/64"**. SOLIDWORKS will calculate the fraction and change it to **0.609375"**. Click **OK** to continue.

To complete the bolt's head, add a new sketch on the right face of the bolt, draw a circle and add a **Tangent** relation between the circle and one of the hexagon's edges. Click on **Exit Sketch** or **Rebuild** to continue.

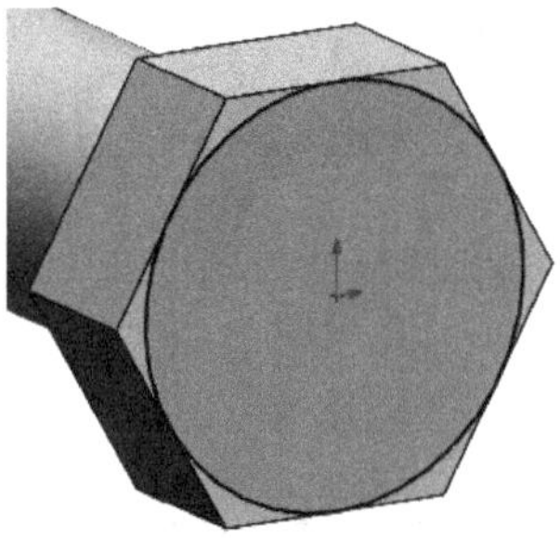

Now change to a **Top** view orientation, select the **Top Plane**, and add a new sketch. Draw the profile approximately as indicated, including a horizontal centerline going through the bolt's axis.

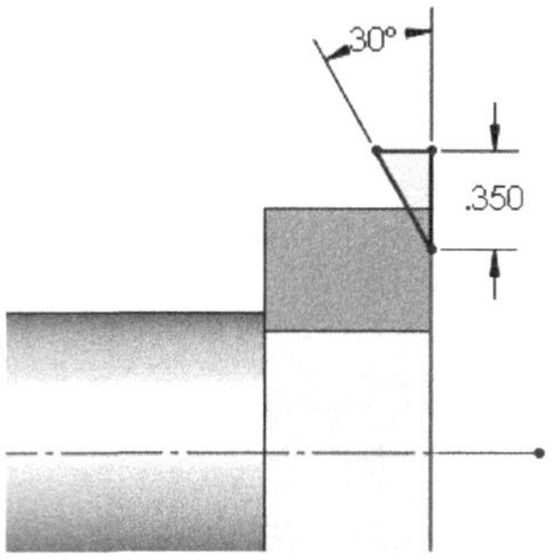

To finish the sketch, slightly rotate the part, select the previous sketched **circle** and the bottom **endpoint** of the triangle, and add a **Pierce** relation as shown.

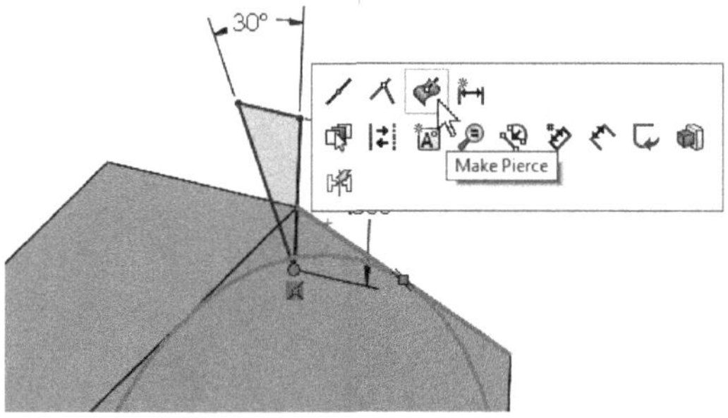

From the Features tab select the **Revolved Cut** command; the centerline is automatically selected and click **OK** to finish the bolt.
Change the material to 1045 Steel-Cold Drawn and save as **Threaded Bolt.sldprt**.

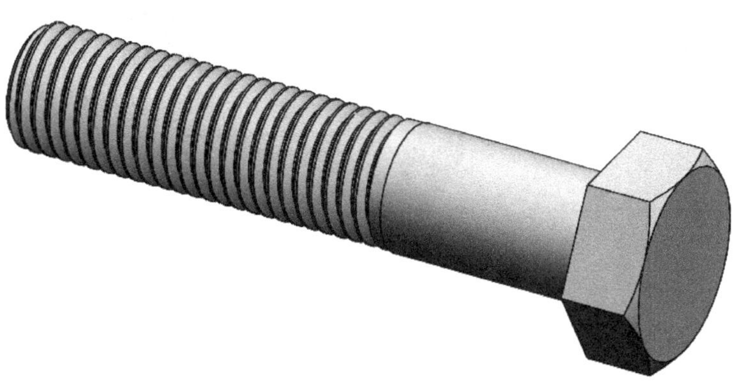

As an alternative to manually making a thread, SOLIDWORKS has a **Thread** Tool that allows you to add both Standard and Metric threads by specifying the thread size and location. To test it, select the **Cut-Extrude1** in the **FeatureManager** and from the pop-up menu select **Suppress**. The Suppress command will cause the selected feature, and all subsequent features that were created using it as a reference, to ***not*** be created.

By default, SOLIDWORKS shows the arrows to the features that will also be suppressed. The result will be a bolt without threads, and the suppressed features will be greyed out in the FeatureManager.

The Suppress command will be covered in more detail in subsequent chapters.

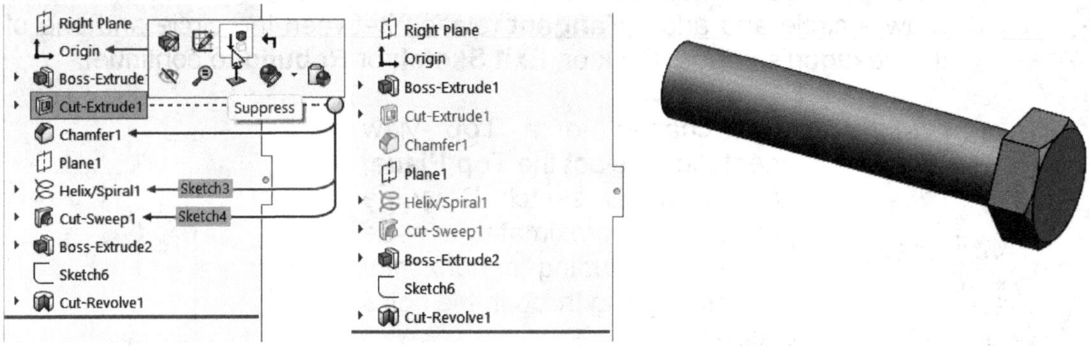

Now select the Chamfer tool and add the 0.0767" to the edge of the bolt, where the thread will begin.

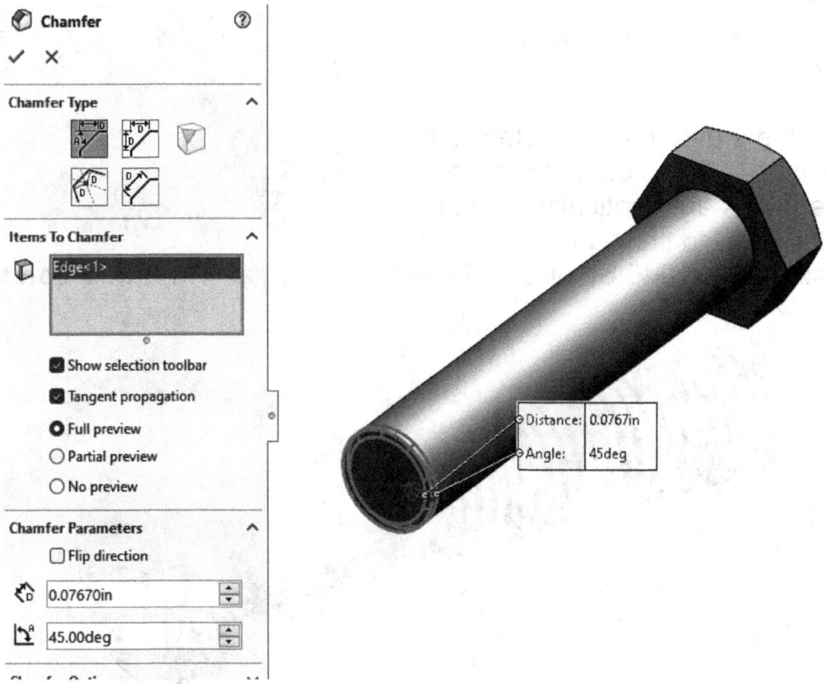

Select the **Thread** command from the menu **Insert, Features, Thread.** In the Thread Location select the larger edge of the chamfer you just made; activate the **Offset** option and reverse the direction to start the thread *away* from the selected edge as before. Set the offset distance to 1 thread pitch (0.125") + the Chamfer size, and round the value to **0.200"**.

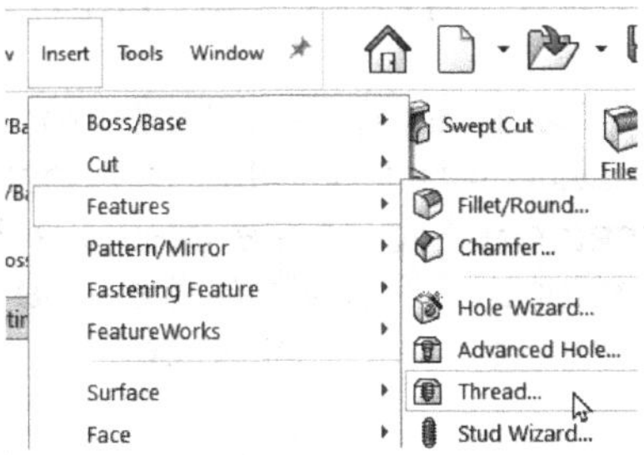

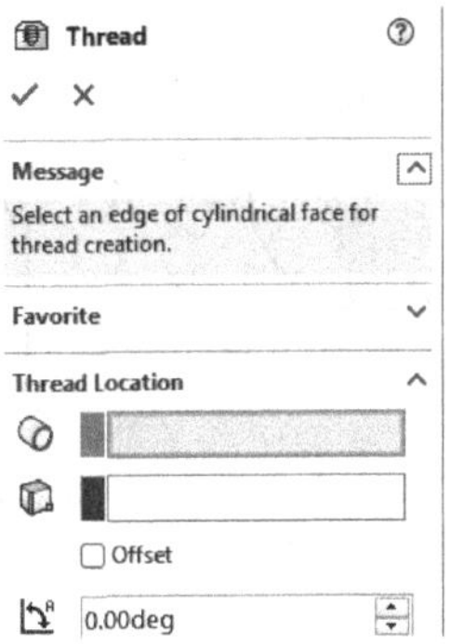

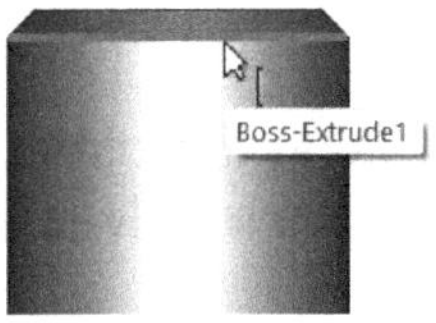

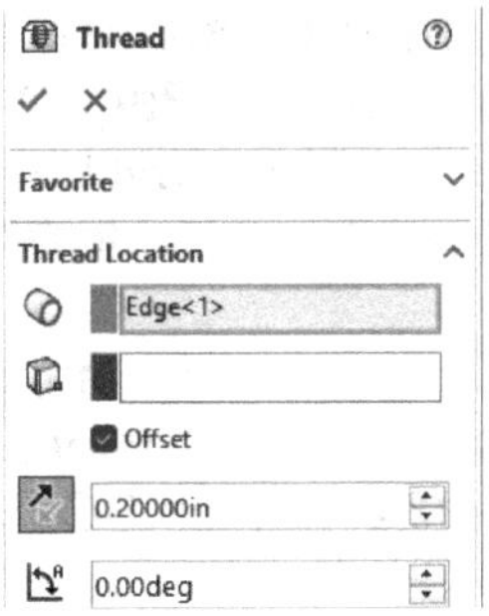

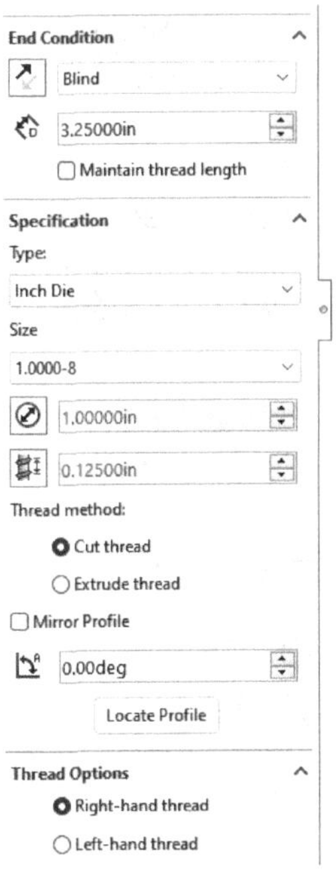

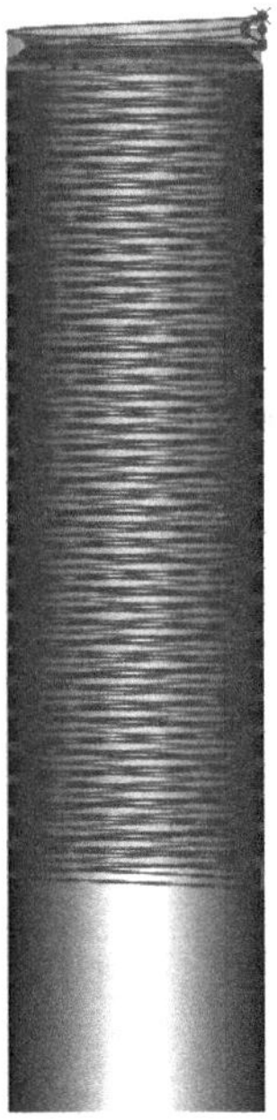

Set the End Condition to **Blind** and change the depth to **3.25"** long to compensate for the chamfer.

In the Specification select the **Inch Die** option to machine the thread from a solid cylinder, and set the size to **1.000-8**, meaning a **1.00"** thread with 8 threads per inch, in other words, a 0.125" pitch.

The Thread method should be "**Cut thread**". This option will cut the thread from the bolt, making the bolt's diameter as the thread's major diameter. Selecting the "**Extrude thread**" would add an external thread using the bolt's diameter as the root of the thread. Other options include making a Right-hand or Left-hand thread. Click Okay to finish.

CONSTRUCTION OF A CORRESPONDING NUT

There are many standards to produce threads and fasteners. When you design fasteners, it is advisable to consult official standards for accuracy. In the case of a 1" diameter nut to fit the previous bolt, the hexagon of the nut is 1.5 times the major diameter of the shaft. Make a new part using the **ANSI-INCHES** template and change the units to **5** decimal places.

Select the **Right Plane**, add a new sketch and draw a hexagon. Select one side, add a horizontal relation and dimension the distance between sides **1.500"**. Referring to the table on **Page 4-12**, draw a circle at the origin; it's diameter will be the Minor Diameter for Internal Threads, equal to **0.8647"**.

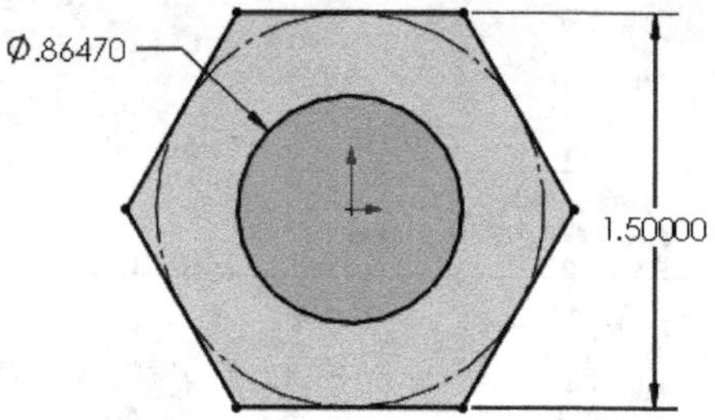

To complete the first feature, make an **Extruded Boss** using the **Midplane** end condition, and using the dimension from the standards table set the depth to **55/64"**, which SOLIDWORKS resolves to **0.859375"**.

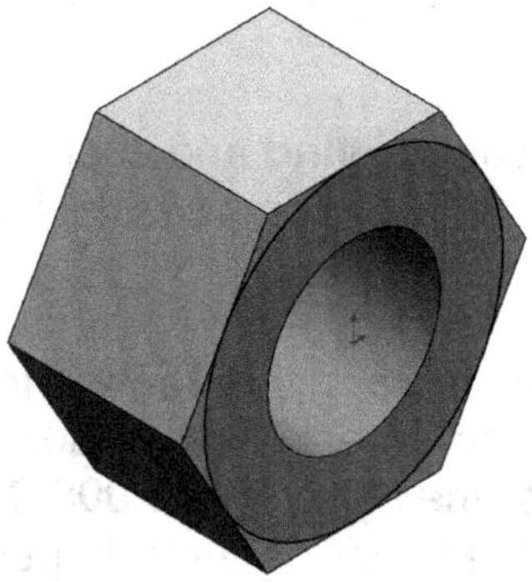

Select the right face of the base feature, add a sketch, and draw a circle at the origin. Just like with the bolt's head, make the circle **Tangent** to an edge, and **Rebuild** the model, or **Exit** the **Sketch** to continue.

As before, change to a **Top** view orientation, select the **Top Plane**, and add a new sketch. Draw and dimension the sketch profile approximately as shown here, including the centerline at the origin.

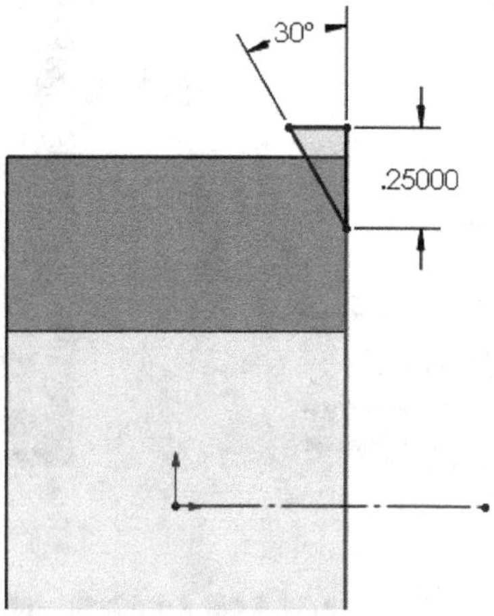

As you did before, slightly rotate the part, select the bottom endpoint of the triangle and the circle sketch, and add a **Pierce** relation to fully define the sketch.

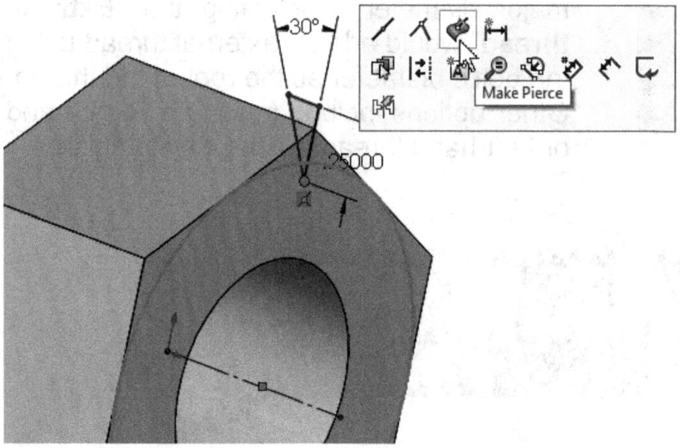

Select the **Revolved Cut** command and make a **360°** cut around the centerline. To duplicate the cut on the other side, select the **Mirror** command from the Features tab and mirror the **Cut-Revolve1** about the **Right Plane**.

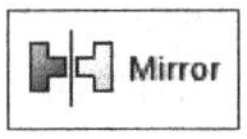

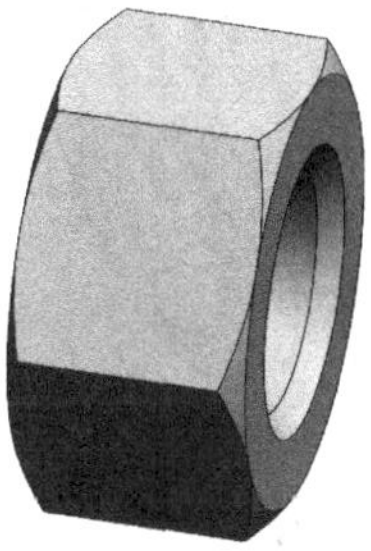

The next step is to add a **Chamfer** in both ends of the hole equal to the Thread Depth. According to the chart in **Page 4-12** for a **1"** diameter thread it must be **0.0767"**. Select the **Chamfer** command, set the size, and to add both edges at the same time, select the inside face of the hole. Click **OK** to finish.

Now, as you did before, add a reference plane 1 Pitch distance (**0.125"**) away from the nut. Select the **Reference Geometry** and click on **Plane**. Select the right face of the nut, set the distance to **0.125"** and click **OK**.

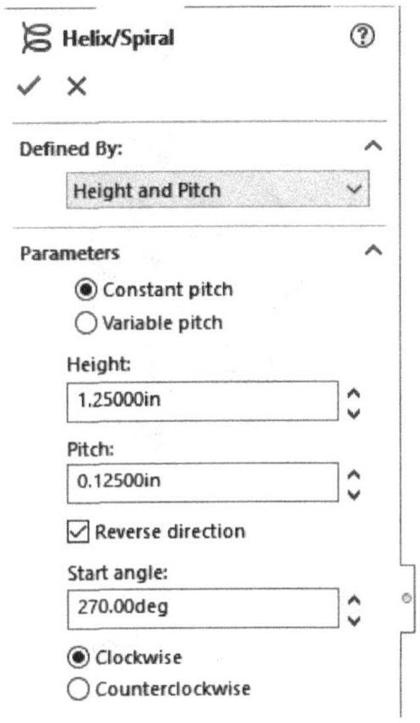

To make the thread's **Helix**, add a new sketch on **Plane1**, select the smaller edge inside the nut and click on **Convert Entities** to project it.

Activate the Features tab, select the **Curves** command and click on **Helix and Spiral**.

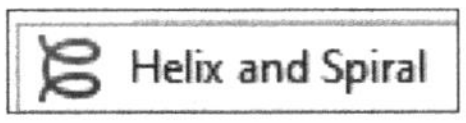

Just as you did previously with the bolt, in the **Helix/Spiral** options use the **Height and Pitch** definition and select the **Constant pitch** option. Set the **Height** to **1.25"** and the Pitch to **0.125"**.

Activate the **Reverse Direction** checkbox to make the helix go into the part, set the **Start angle** to **270°** to make the start of the helix at the **Front Plane** and make sure you are using the **Clockwise** option as shown. Click **OK** to create the Helix.

The next step is to draw the thread's profile sketch. Change to a **Front** view and set the Display Style to **Hidden Lines Visible**. Add a new sketch in the **Front Plane** and draw the following sketch.

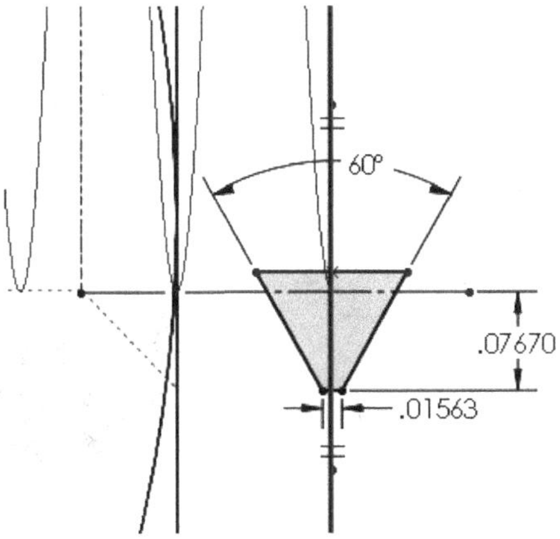

Remember to add a horizontal and vertical centerline. To make the sketch symmetrical, activate the **Dynamic Mirror** command about the vertical centerline, which is coincident to the **Plane1**. The sketch dimensions were calculated from the formulas provided in the table found in **Page 4-12**.

When the sketch is complete click on **Exit Sketch** or **Rebuild** the part to continue.

Change back to a **Shaded with Edges** display form for visibility, and select the **Swept Cut** command from the Features tab.

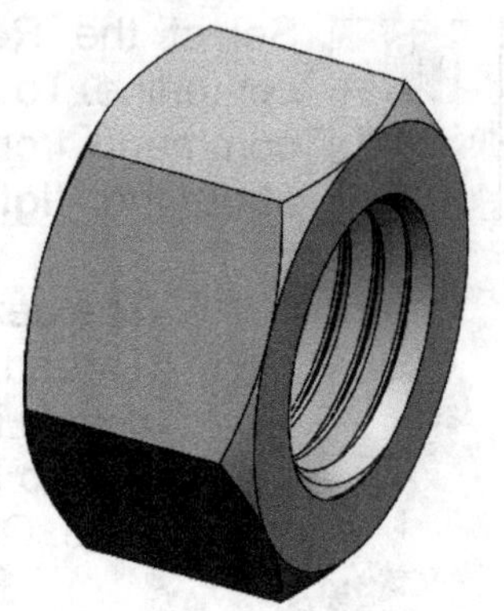

In the **Profile** select the thread profile sketch, and in the **Path** select the helix curve. Click **OK** to finish and hide the helix and the auxiliary plane. Change the material to 1045 Steel-Cold Drawn and save your part as **Threaded Nut.sldprt**.

Alternate method to create a Threaded Nut using the Thread command.

Just as we did before with the bolt, select the **Plane1** feature in the FeatureManager and Suppress it. The plane, helix and Cut-Sweep will be suppressed.

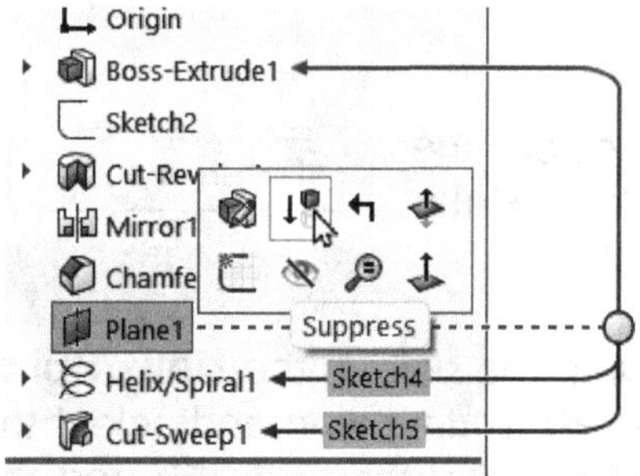

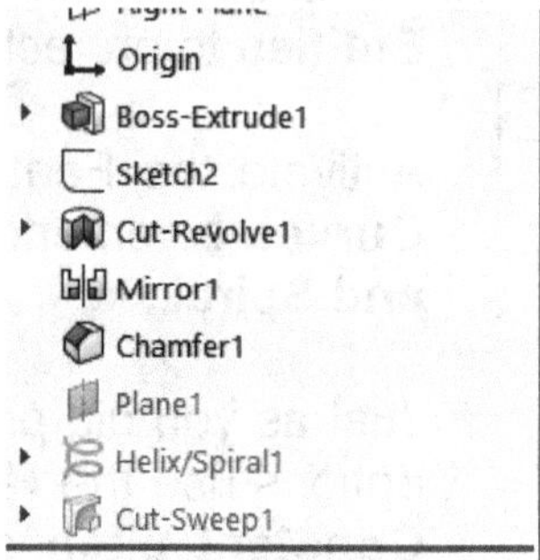

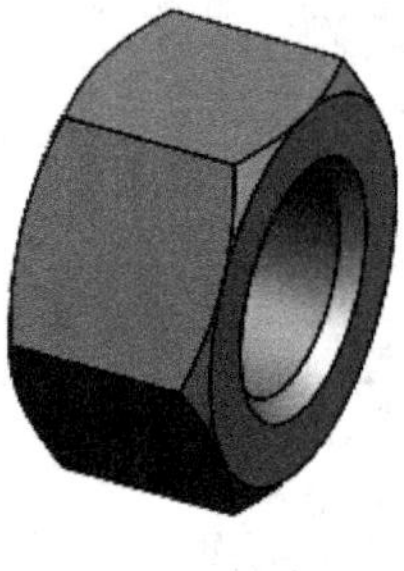

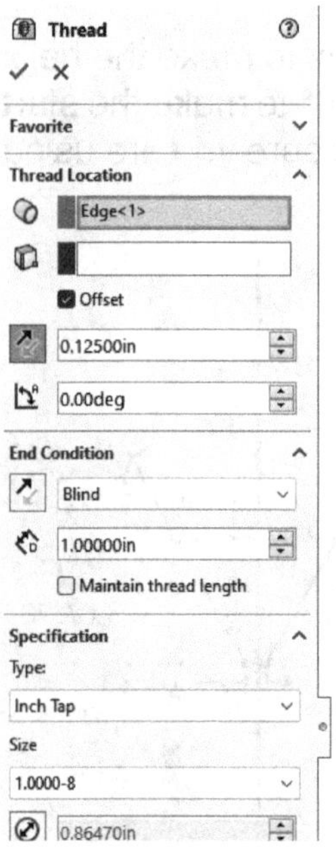

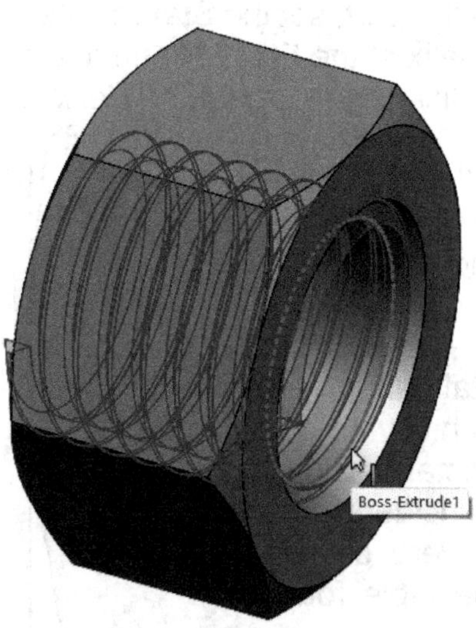

Use the **Thread** command as before and select the chamfer's minor diameter for the thread location. The selected edge must be on a cylindrical surface.

Activate the **Offset** option. Set the direction and distance to start the thread one thread-pitch (0.125") *away* from the nut.

Change the thread's depth to **1.00"** to make sure it goes through the nut, set the **Type** to **Inch Tap** and **Size** to **1.000-8** as before to match the bolt.

Click OK to finish the thread.

To finish this exercise, make a new drawing using the template in Inches. Add an **Isometric** view of the **Threaded Bolt** and the **Threaded Nut** parts as indicated.

Save your drawing as **SCREW THREADS.slddrw** and print a copy to submit to your lab instructor.

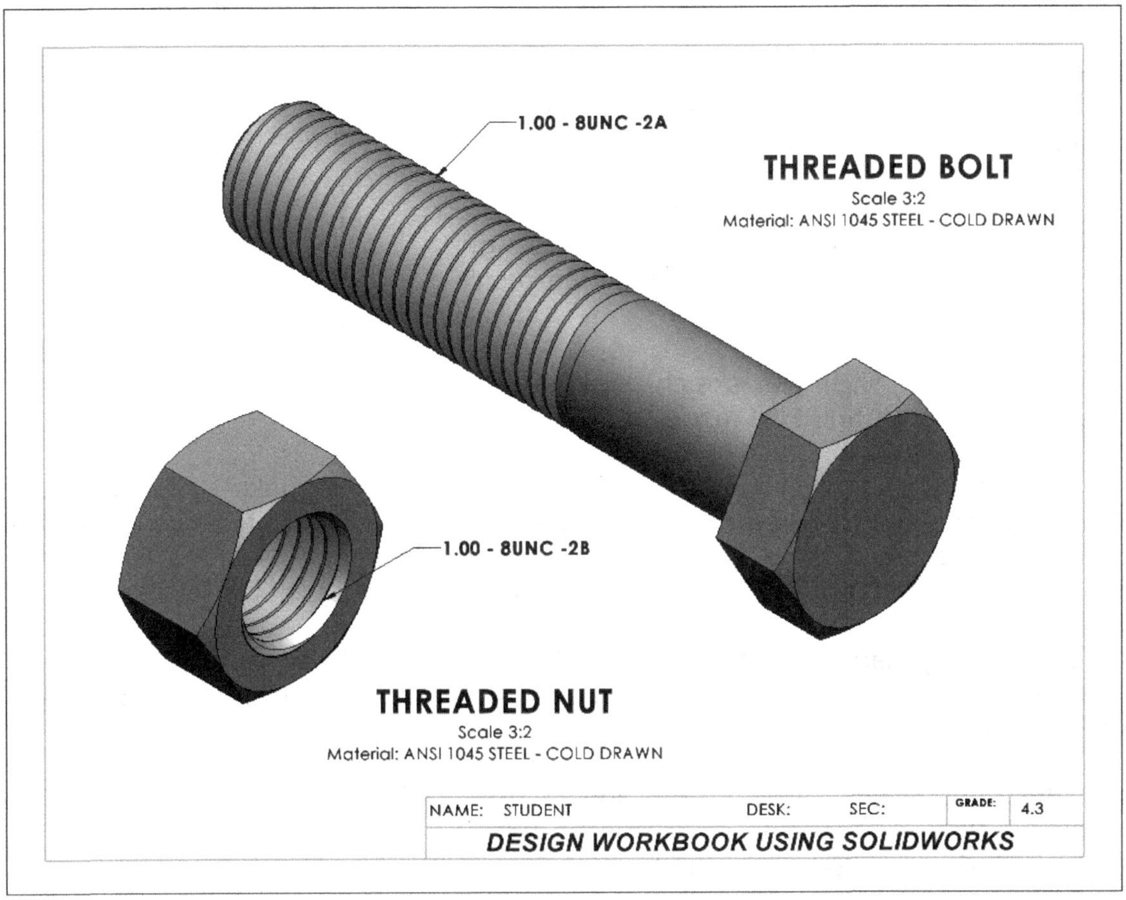

Exercise 4.4: Jack Stand

In this Exercise 4 you will learn the basic options to create a **Lofted Boss** feature. A loft is a feature that changes shape between multiple different cross sections. The sections can be model faces or sketch profiles and optionally can have guide curves to better control the resulting feature. To learn this feature you will create a Jack Stand.

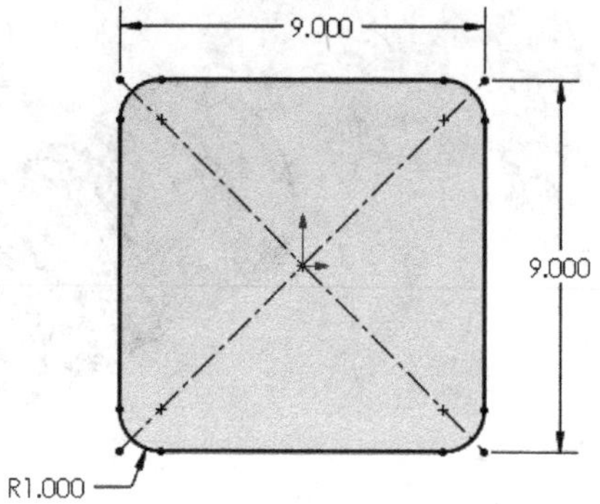

Make a new part using the **ANSI-INCHES** template and save it as **Jack Stand.sldprt.** Select the **Top Plane** and add a new sketch.

Using the **Center Rectangle** option from the **Rectangle** command and the **Sketch Fillet**, draw the profile indicated.

When you finish the sketch, click on **Exit Sketch** or **Rebuild** the model to continue.

To create the second section of the **Loft** feature you need to make a new **Auxiliary Plane**. Change to an **Isometric** view for better visibility, click on **Reference Geometry** command and select **Plane**. From the fly-out FeatureManager select the **Top Plane**, set the offset distance to **6.50"** above the existing sketch and click **OK** to create it.

Select the newly made **Plane1**, change to a **Top** view orientation and add a new sketch. Draw a circle centered on the origin, dimension it **4.0"** diameter and click on **Exit Sketch** or **Rebuild**.

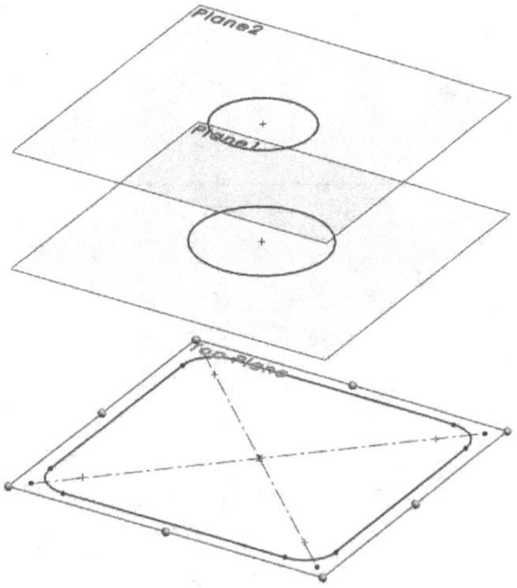

For the third section of the **Loft** add a new **Auxiliary Plane**. Change to an **Isometric** view, click on **Reference Geometry** and select **Plane** again.

Select the **Top Plane** for reference, set the offset distance to **10.0"** above the **Top Plane** and click **OK** to finish.

Add a sketch on **Plane2** and change to a **Top** view for visibility. Draw a circle centered in the origin, dimension it **3.0"** in diameter and **Exit** the sketch. Change to an **Isometric** view to see the resulting sketch profiles as shown here.

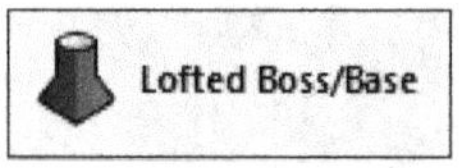

With the profiles completed, activate the Features tab, and click on the **Lofted Boss** command.

To better visualize the **Loft** operation, change to an **Isometric** view. In the **Profiles** section select the three profiles made starting from the bottom and going up as shown.

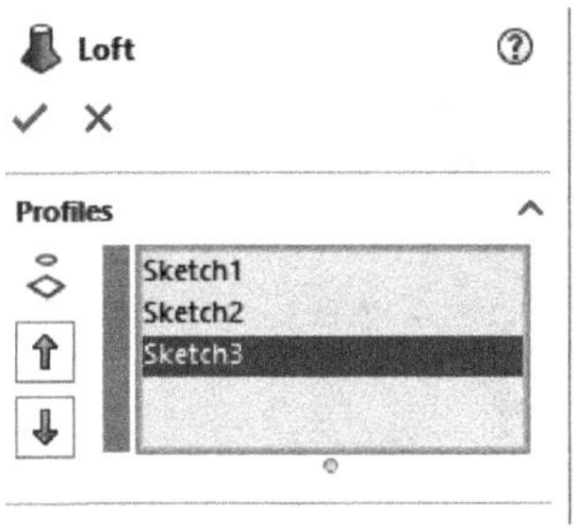

When you select the profiles, remember the location where you select a profile will be connected to the location where you select the next profile. For example, if you select the profiles on opposite sides, this will cause the **Loft** feature to twist, and possibly fail.

As you select the sketches, a preview of the loft will be shown on the screen. If it looks like this, click **OK** to complete the loft.

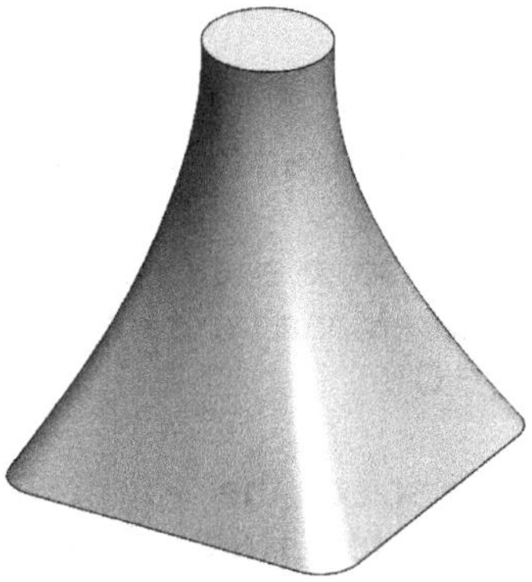

To make the jack stand lighter select the **Shell** command from the Features tab. Set the wall thickness to **0.1875"** and rotate the part to select the bottom face of the part to remove it. Click **OK** to finish.

The next operation is to reinforce the top face and add a hole for a threaded adjustment. Change to a **Top** view, select the top face, and add a new sketch.

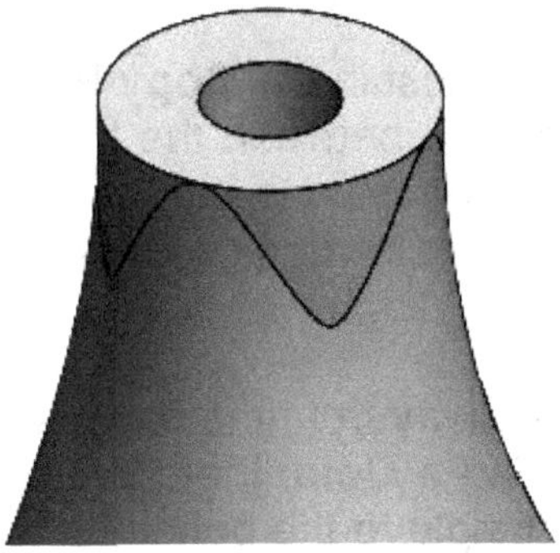

Use **Convert Entities** to project the edges of the top face and make a **Boss Extrude** going down **3.0"**. Select the top face again, add a new sketch and draw a circle of **1.25"** diameter.

Use the **Extruded Cut** command with the **Through All** end condition and click **OK**.

The last feature to add is a triangular cut on the sides of the jack stand to make it lighter. Change to a **Front** view and add a new sketch on the **Front Plane**.

To make the profile symmetric, first draw a triangle and then add a vertical centerline from the origin to the bottom line. Add a **Midpoint** relation between the top endpoint of the centerline and the horizontal line, making the bottom line symmetric.

Now add a **Coincident** relation between the top endpoint of the triangle and either the origin, or the vertical centerline.

Add the dimensions indicated, and use the **Extruded Cut** command, with the **Through All-Both Directions** end condition and click **OK** to finish.

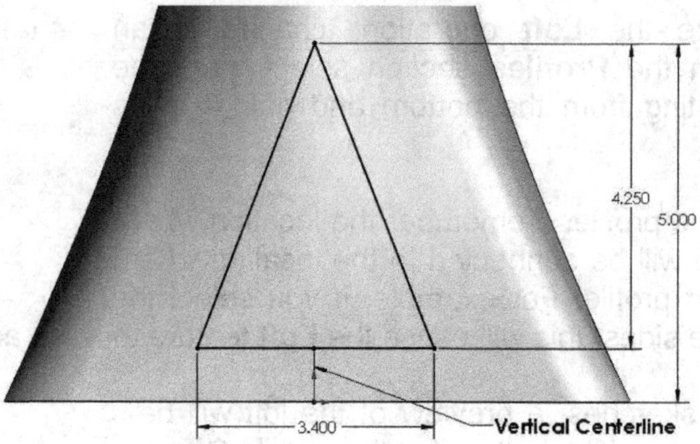

Now you need to add the same cut extrude feature perpendicular to the **Right Plane**.

To reuse the same sketch you just made, expand the last **Cut-Extrude** in the FeatureManager, select the Sketch and from the menu click **Edit, Copy** (default shortcut **Ctrl+C**).

Now select the **Right Plane** in the FeatureManager, change to a **Right** view orientation and click in the menu **Edit, Paste** (default shortcut **Ctrl+V**). Select the new sketch in the FeatureManager or the graphics area, and from the context toolbar select **Edit Sketch**.

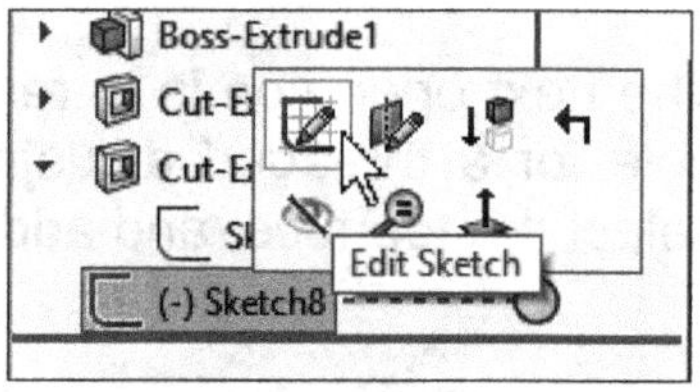

The reason to edit the sketch is because some information was lost in the **Copy-Paste** process. Even if the sketch *looks* correct, the coincident relation between the bottom endpoint of the centerline and the origin was lost, as well as the distance between the bottom of the part and the top of the sketch, making the sketch **Under defined** and colored *blue*, which is *not* the desired sketch state.

To fix it **click-and-drag** the bottom endpoint of the centerline ***away*** from the origin and move it back onto the origin. By doing this, you will be adding a **Coincident** relation again. Add the missing **5.0"** dimension from the top of the triangle to the bottom of the part. Now that the sketch is Fully Defined, make the **Extruded Cut** using the end condition **Through All-Both Directions** to finish.

To complete the design, change the material to **ANSI-1020 Steel-Cold Rolled** and save your model as **JACK STAND.sldprt**.

To finish this exercise, make a new drawing, add an **Isometric** view and the missing annotations. Save your drawing as **JACK STAND.slddrw** and print a copy for your lab instructor.

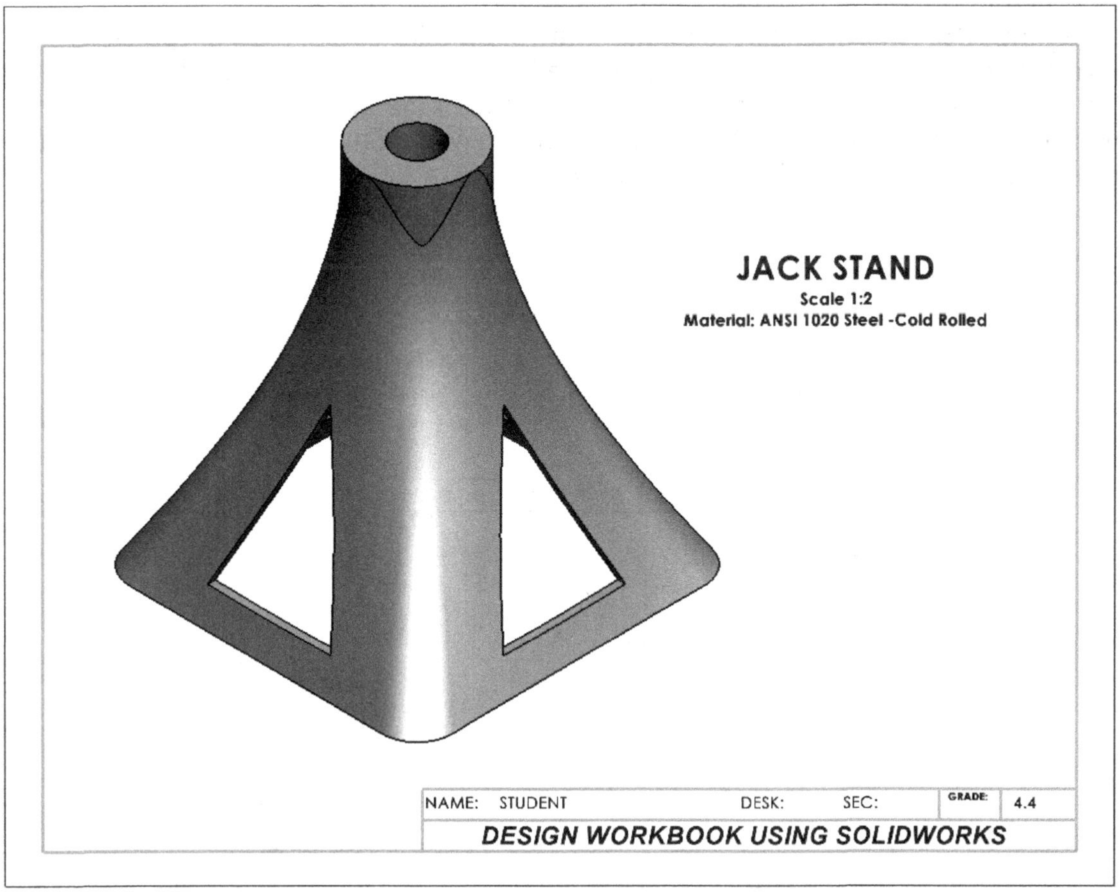
JACK STAND
Scale 1:2
Material: ANSI 1020 Steel -Cold Rolled
NAME: STUDENT
DESK:
SEC:
GRADE:
4.4
DESIGN WORKBOOK USING SOLIDWORKS

Supplementary Exercise 4-5: TAPE DISPENSER

Using the **ANSI-INCHES** part template draw the profile of the Tape Dispenser in the **Front Plane** and extrude it **0.75".** Select the front face and add a fillet of **0.05"**. Select the back face and the two interior faces indicated in the second image and Shell the part with a thickness of **0.10"**. Make a new drawing with a Shaded with Edges **Isometric** view and save it as **Tape Dispenser.slddrw**.

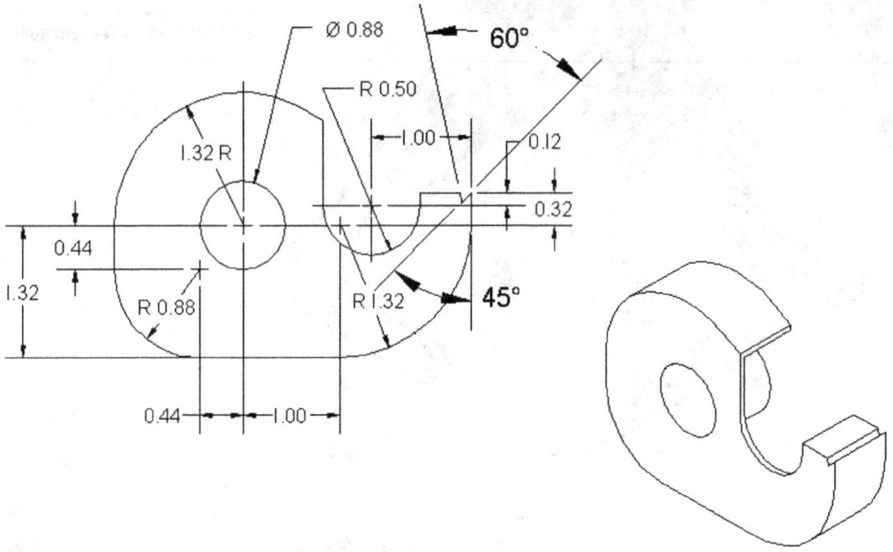

TAPE DISPENSER

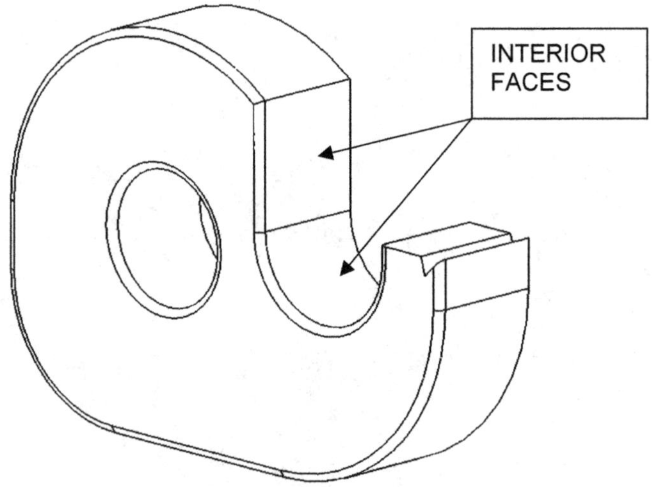

Supplementary Exercise 4-6: FUNNEL

Build a model of the Funnel below. Begin in the **Top Plane**. To get the desired model, each section must be lofted separately. Create auxiliary planes at the indicated distances and add the profiles needed for each Loft feature. When the loft features are complete, select the top and bottom faces and shell the model to a wall thickness of **0.10"**. Make a drawing with an **Isometric** view and save it as **FUNNEL.slddrw**.

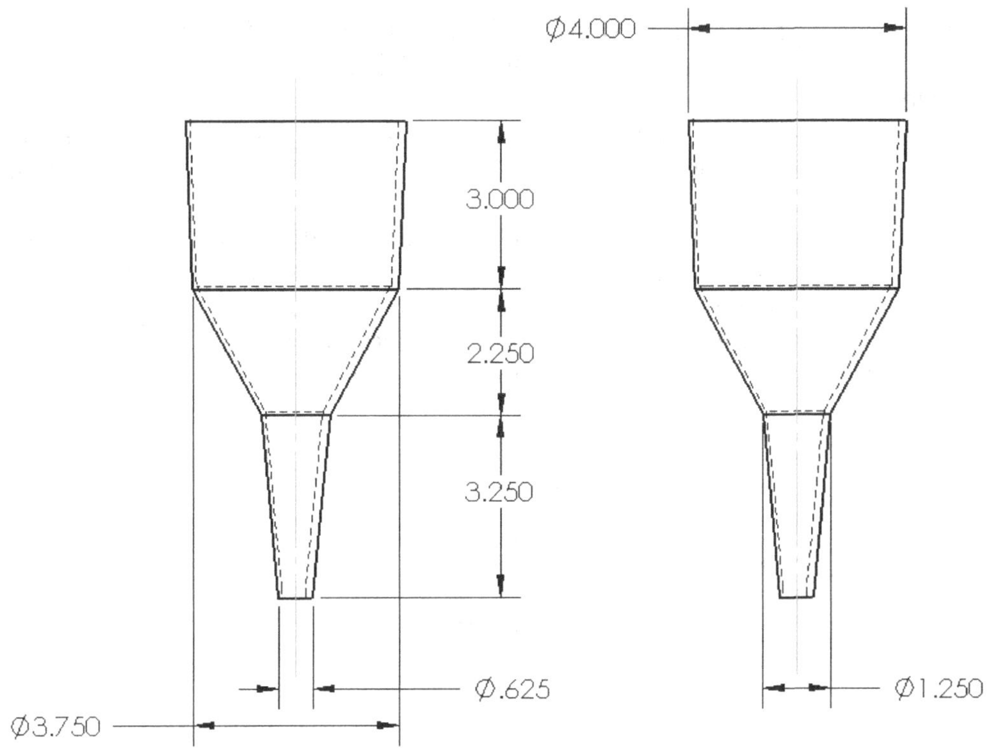

Supplementary Exercise 4-7: TWO INCH THREAD - UNC

1. Make a 2" diameter shaft 5 inches long.
2. The major screw thread diameter is from 1.9868 – 1.9424. Cut the shaft down to 1.9646 for three inches.
3. Screw thread note - 2-4 ACME-2A X 3.
4. Add a chamfer equal to the Thread Depth (the "k" value in the Definition of Terms) on the end of the shaft that will be threaded.
5. Add a reference plane one pitch "P" distance away from the end of the shaft.
6. Select the reference plane and project the 1.9646 diameter circle using Convert Entities.
7. Make a Helix using the following parameters – height and pitch. The height will be 3.25" and the pitch will be 0.25", with the starting angle being at 270 degrees – clockwise.
8. Select the Front Plane and draw a centerline along the lower edge of the shaft and a vertical centerline coincident with the auxiliary plane. Select the vertical Centerline and use the Dynamic Mirror function.
9. Sketch the thread's profile according to the given parameters.
10. Use the Features – Swept Cut function to cut the thread.

ACME THREAD SPECIFICATIONS					
Nominal Size	Basic Major Diameter	Threads Per Inch	Pitch	Thread Depth d = .5P + 0.01	Width of Space at Bottom of Thread W = .3707P - .0052
2	1.9646	4	0.25	0.135	0.087475

CRITERIA FOR ACME THREADS

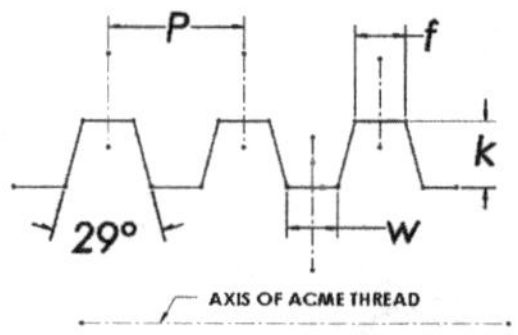

P = PITCH
f = .3707P
k = .5P +0.01
w = .3707P - .0052

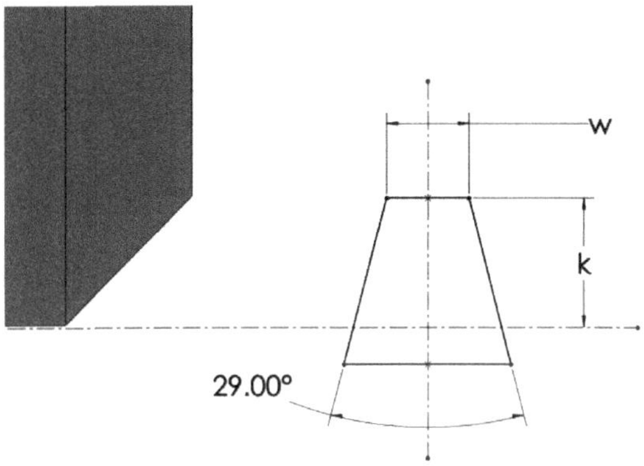
w
k
29.00°

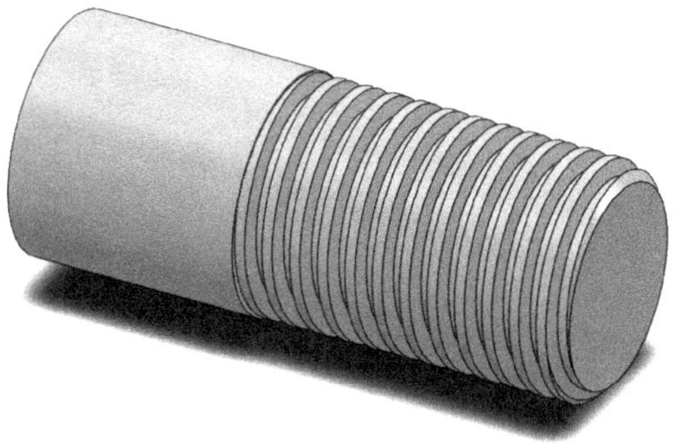

NOTES:

Design Workbook Lab 5
Assembly Modeling

In this exercise you will learn how to make an assembly from multiple components, including how to use the basic tools available, mate components, add Exploded Views, and make assembly drawings with Bill of Materials and identification balloons.

ASSEMBLIES

So far you have been making individual parts; however, your designs will likely be made of multiple assembled components. To make a new assembly, just like a part or drawing, you must select an assembly template, add components to it and join them together using mates to reference how one component's geometry is connected to another. Just like parts and drawings, you can make multiple assembly templates with different options like units, precision, dimensioning standard, etc., following the same procedure used to create **Part Templates** in **Page 1-6**, except you will save them with the Assembly Template type (*.asmdot) to your templates folder.

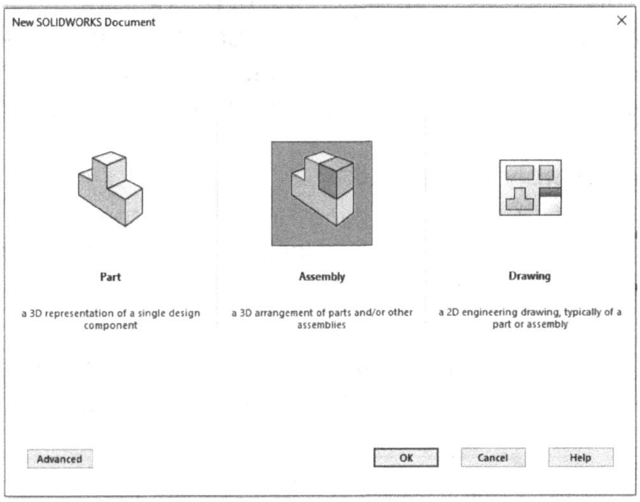

To start a new assembly, select the **New** command, or use the menu **File, New** and select the "Templates" tab if you are using the **Advanced** option, or the Assembly document type if you are using the **Novice** option.

In an assembly, every component has six **Degrees of Freedom** (**DOF**): three **translations** about the **X, Y and Z axis**, and three **rotations** about the **X, Y and Z axis**, and each of them can be either free or constrained based on how a component is related (**mated**) to other components. Partially constraining a part's DOF is the basis for assembly motion in SOLIDWORKS. The first component added to an assembly has all six degrees of freedom automatically constrained or **Fixed**.

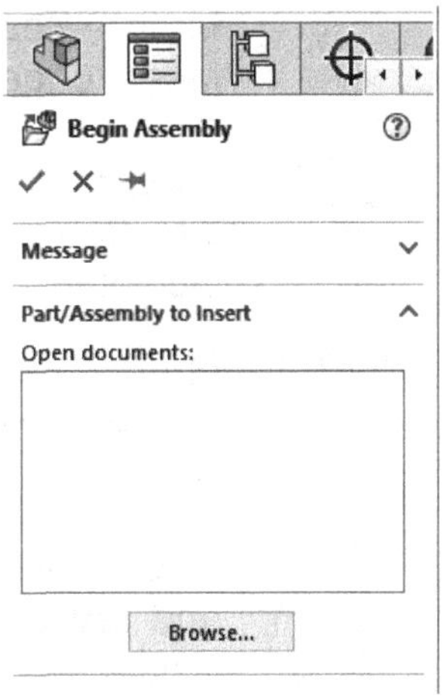

Making a new assembly using the default **Assembly** template will automatically start the **Begin Assembly** (or **Insert Component**) command shown here and ask you to browse and select the first assembly component.

After you have selected the first component, click **OK** to add the component to the assembly. By doing this, the *part's origin* will be made coincident with the *Assembly's origin*, the part's **Front** plane will be coincident with the Assembly's **Front** plane, and likewise for the **Top** and **Right** planes.

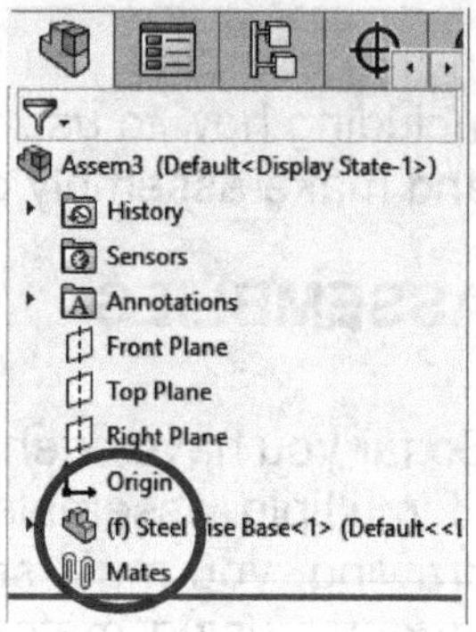

It is usually good practice and convenient to align the first part's default Planes to the assembly's Planes for reference, especially when mating other components in the assembly.

After the first component is added, its name will have the prefix "**(f)**" in the FeatureManager, as seen here, indicating it is **Fixed** in place, meaning all six **Degrees of Freedom** (**DOF**) are constrained in space, and the component cannot be moved or rotated.

This is the reason why it is a good idea for the first component to be the main part of the assembly, or the one other parts and sub-assemblies will be attached to. For example, if you are assembling a bicycle, the frame would be the best option as a first component because every other part will be attached to it and will move relative to it.

Assembly Commands

When you are working in the assembly environment, the CommandManager automatically changes to include assembly specific tools. Some of the most commonly used commands include:

Edit Component	**Edit Component** – Used to modify a component or sub-assembly while working in the assembly. Useful to make changes to match other assembly components.
Insert Components	**Insert Components** – Add an existing part or sub-assembly to the assembly.
Mate	**Mate** – Position two components relative to each other using component faces, edges, planes, axes, vertices, sketch geometry, etc.
Linear Component Pattern	**Linear Component Pattern** – Patterns components in one or two linear directions. Includes a drop-down menu with additional pattern types including Circular, Curve Driven, Mirror components, etc.
Move Component	**Move Component** – Moves a component within the degrees of freedom defined by its mates. Includes a drop-down menu to also rotate the component.
Show Hidden Components	**Show Hidden Components** – Temporarily hide all visible components and reveal hidden components. Select a hidden component to make it visible.

Assembly Features	**Assembly Features** – Creates different assembly features, including cuts, simple holes, fillets, chamfers, hole series, weld beads, etc.
Reference Geometry	**Reference Geometry** – Allows you, just like in components, to add reference geometry including Plane, Axis, Coordinate System, Point, and Mate Reference.
Exploded View	**Exploded View** – Creates a view of the assembly components in an exploded state.

MATE TYPES

To make an assembly, you must add multiple components (parts and/or sub-assemblies) and MATE them together. To add Mates, you can select component's faces, edges, vertices, sketch elements, planes, auxiliary geometry (planes, axes, etc.), origins, etc. Based on the different component entities selected, the valid mate references will be presented.

The most common Mates include:

Mate Type	Selections
Coincident	Places the selected Faces, Edges, Planes, Vertices, etc. coincident and parallel. Two Coincident vertices will share the same point.
Parallel	Places the selected flat Faces, linear Edges or Planes Parallel to each other, but not necessarily coincident.
Perpendicular	Places the selected flat Faces, Linear Edges, or Planes at 90° to each other.
Tangent	Places the selected entities Tangent to each other. At least one of the selected items must be a circular edge, cylindrical, conical, or spherical face.
Concentric	Places the two selected entities so that they share the same axis.
1.000in	**Distance**: Places the selected entities Parallel at a defined distance from each other.
135.00deg	Angle: Places the selected entities at the specified angle from each other.
Symmetric	Places the selected items at an equal distance from a plane of symmetry.
Width	Places two selections of one component centered between two selections of a second component.

VIEWING ASSEMBLIES

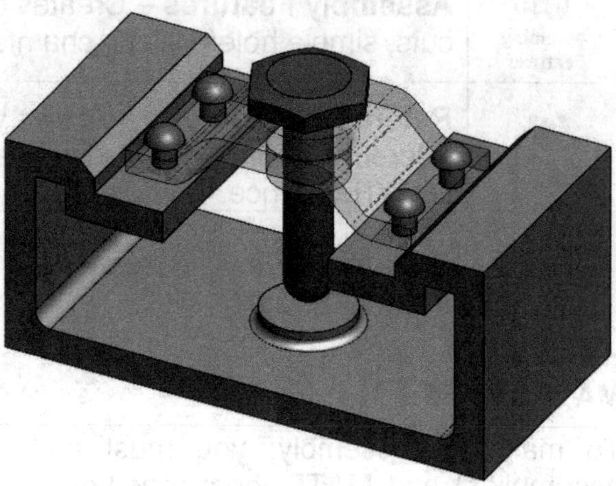

When you work with assemblies including multiple components, viewing specific components, assembly details or making selections may be difficult. One way to improve component visibility is by hiding components or making them transparent.

To hide a component select it in the FeatureManager or the screen, and from the context toolbar select the **Hide Component** command.

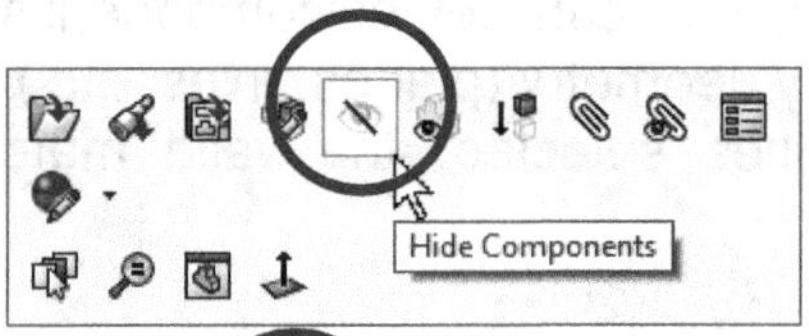

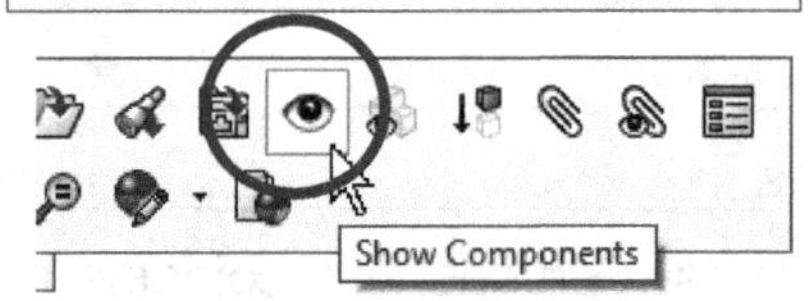

To Show a hidden component, select it in the FeatureManager and click in the **Show Component** command.

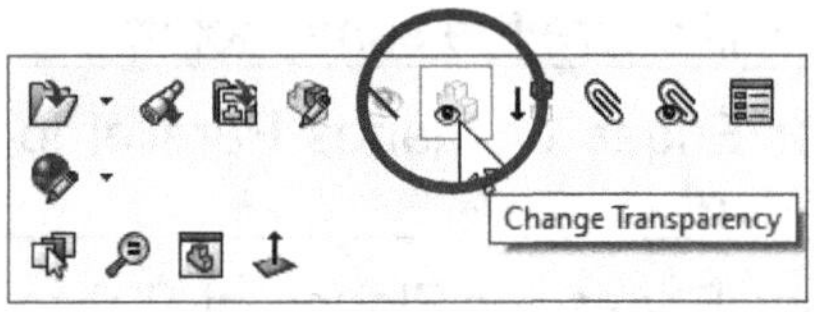

To make a component transparent, or change back to opaque, select the **Change Transparency** command. The result is seen above.

ASSEMBLY EXPLODED VIEW

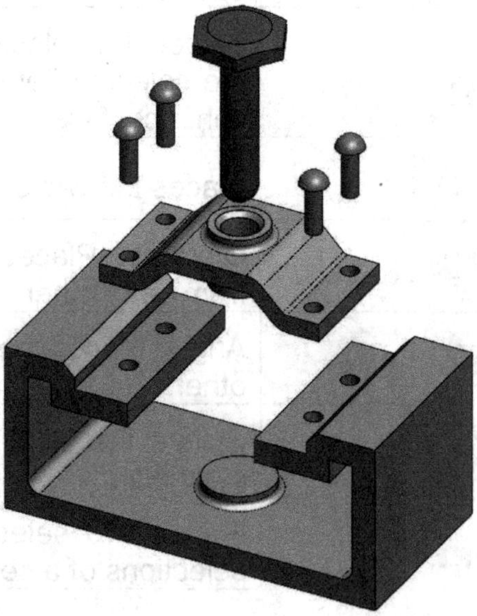

Assembly components can be exploded using the **Exploded View** command from the assembly toolbar for visualization and documentation purposes. Using the **Exploded View** command, you can define the parameters for each step of the exploded view to document your designs and/or create assembly documentation.

You will learn more about creating an Exploded View of an assembly in **Unit 8**, including animating an exploded view.

Exercise 5.1: Terminal Support Assembly

The Terminal Support Assembly has four components: The **Frame**, the **Wing Base**, a **Pin** and four **Rivets**. You will start by designing the parts, starting with the **Frame**. When all the parts are completed, you will make a new assembly, add the components, and assemble them using the **Mate** command.

FRAME

Make a new part using the **ANSI-INCHES** template and save it as **Frame.sldprt**. Add a new sketch in the **Front Plane**, draw a vertical centerline starting at the origin, activate the **Dynamic Mirror** command and draw the profile indicated next. When the profile is complete, make an **Extruded Boss**, and set the depth to **4.00"** using the **Mid Plane** end condition.

Change to a **Top** view, select the top-most face of the model and add a new sketch. Draw a horizontal and vertical centerline at the origin, draw and dimension a circle, and use the **Mirror Entities** command to create the other three holes indicated. Use the **Extruded Cut** command using the **Up to Next** end condition; using this option will only cut the part until the next face.

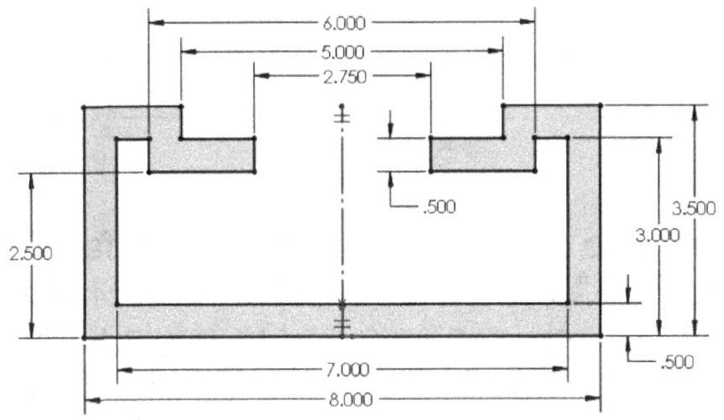

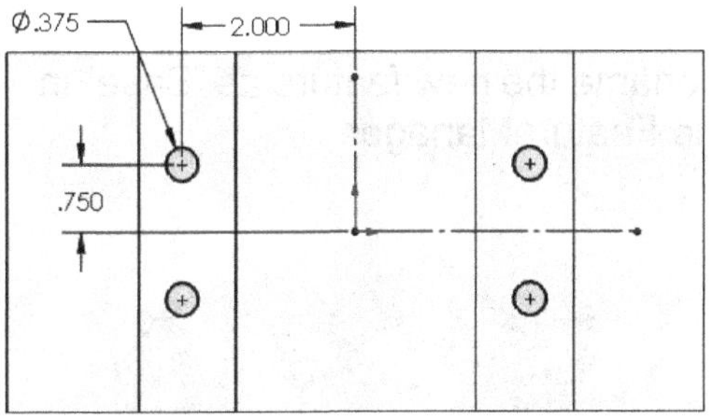

If needed, change back to a **Top** view, select the face at the bottom of the Frame part, add a new sketch and draw a circle at the origin. Dimension its diameter **1.50"** and make an **Extruded Boss 0.25"** going up.

Finish the part by adding a **0.125"** fillet at the base of this extrusion, a **0.25" Chamfer** to the two top inner edges and a **0.25"** fillet to the bottom inner edges of the Frame. The complete part is shown here.

Change the part's material to **AISI 4340 Steel, annealed** and save the finished part as **Frame.sldprt**.

WING BASE

Make a new part using the **ANSI-INCHES** template and save it as **WING BASE.sldprt**. Add a new sketch on the **Front** plane and draw the profile indicated. Be sure to add a **Parallel** geometric relation to the angled lines. Note there is no line under the origin.

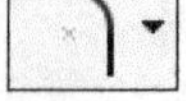

Use the **Sketch Fillet** command to round the small radius corners with a **0.125"** radius, and the large radius corners with a **0.500"** radius, as shown.

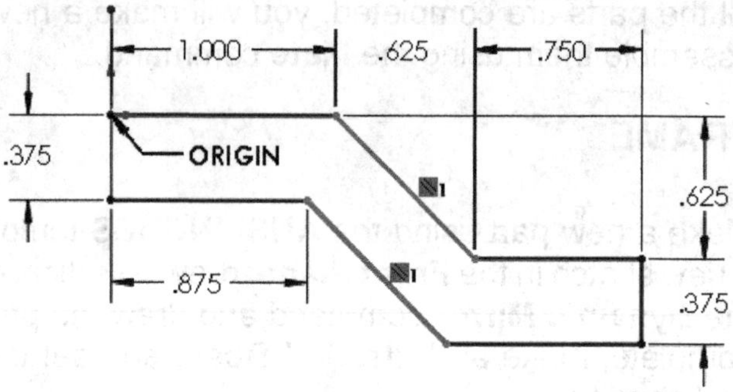

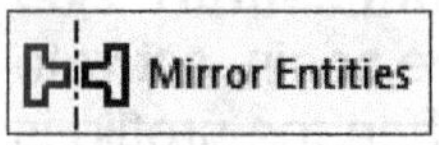

To complete the sketch, window-select the entire sketch geometry including the centerline going through the origin and select **Mirror Entities** to finish the sketch.

Select the **Extruded Boss** command and make a **2.500"** extrusion using the **Mid Plane** end condition and click **OK** to continue. Your part will look as follows.

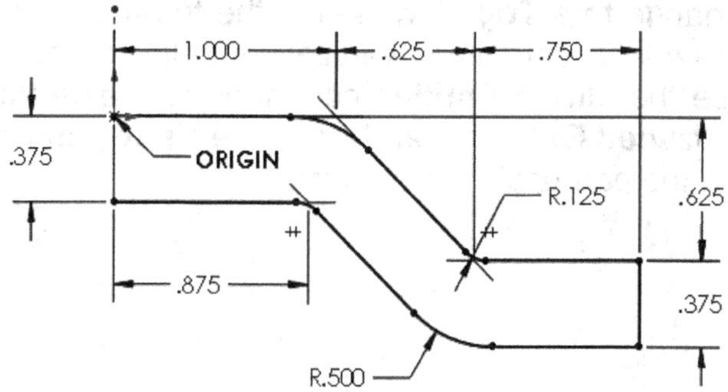

Rename the new feature as "Base" in the FeatureManager.

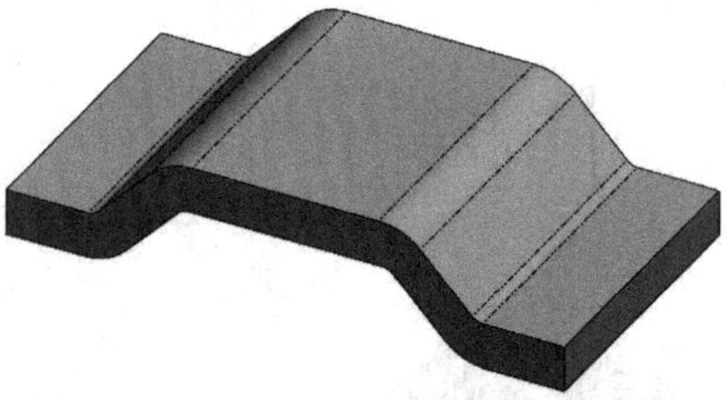

Change to a Top view, select the top face of the "Base" feature and add a new Sketch. Draw a circle centered at the origin with a diameter of **1.250"**, select the **Extrude Boss** command using the Blind End Condition, and set **Direction 1** (up) to **0.25** inch. Set **Direction 2** (down) to **1.00"** and click **OK** to finish. Rename this feature "Round Boss" to continue.

Select the top face of this "Round Boss", add a new sketch and draw a circle concentric with the round face, centered at the origin, with a diameter of **0.75"**.

Select the **Extruded Cut** command, use the **Through All** end condition and click **OK** to finish. Rename this new hole feature as "Center Hole". Your model should now look like this.

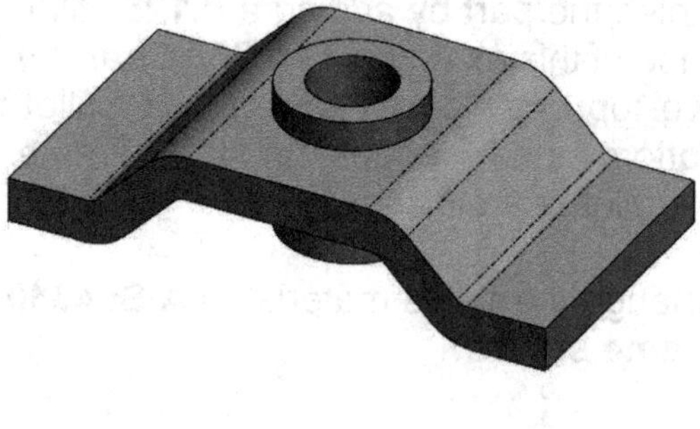

Now you need to add a chamfer and fillets to your model. Select the **Fillet** command from the Features tab, set the radius to **0.125"** and select the two edges indicated next.

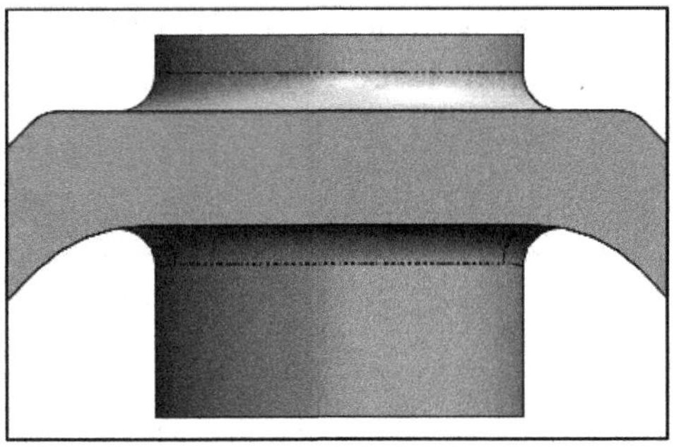

If needed, change the Display Style to Hidden Lines Visible to make selection easier.

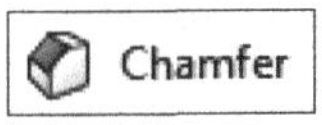

To add the **Chamfer**, select the top inside edge of the "Center Hole" and from the context toolbar select the **Chamfer** command.

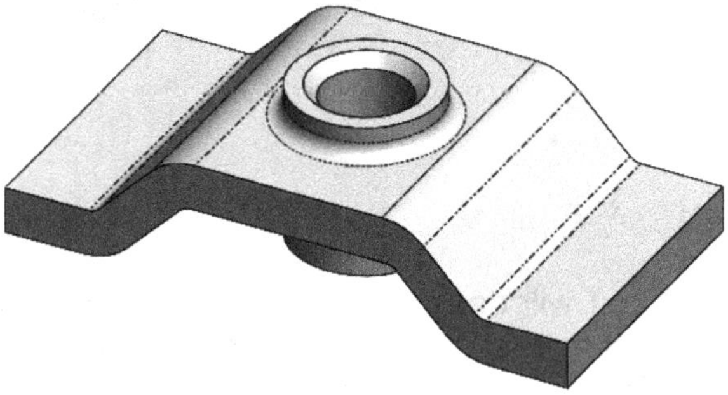

Set the **Chamfer Type** to Angle-Distance, change the distance to **0.100"** and the angle to **45°**, and click **OK** to finish. Your resulting model is shown next.

The last feature to add to the model are the four holes for the rivets. Change to a Top view orientation, select the top face of either wing side and add a new sketch.

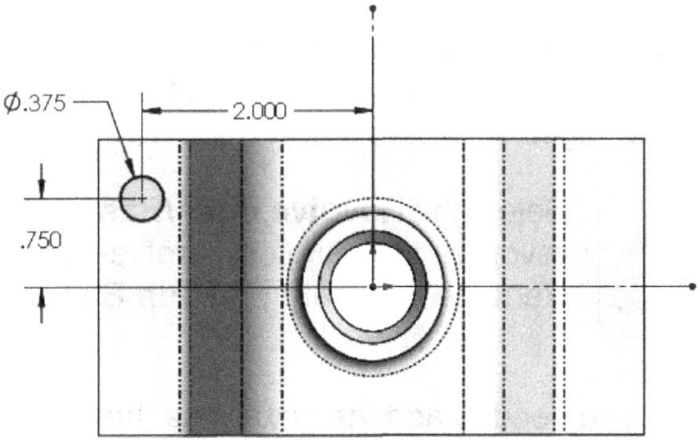

Draw and dimension the profile shown, including a horizontal and a vertical centerline starting at the origin.

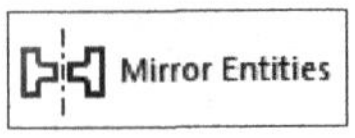

To better maintain the design intent to make sure the part remains symmetrical, even if we introduce a dimensional design change, use the **Mirror Entities** command to mirror the circle about the vertical centerline first, and then select both circles and mirror them about the horizontal centerline. The finished fully defined sketch is shown here.

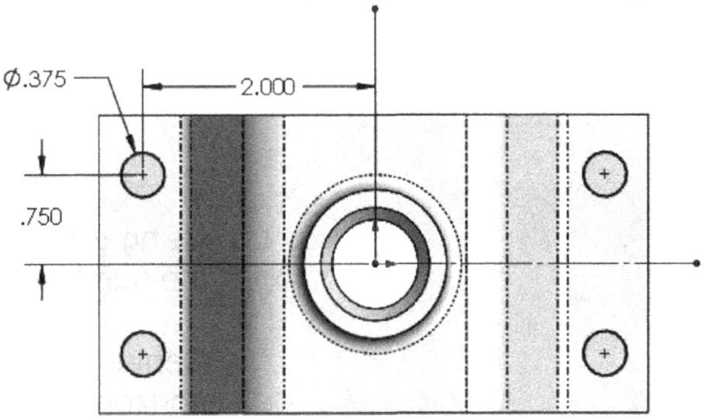

To complete this feature, select the **Extruded Cut** command, use the **Through All** end condition and click **OK**. Rename the feature as "Rivet Holes."

Set the part's material to **Chrome Stainless Steel** and save it as **Wing Base.sldprt**. Close the part to continue.

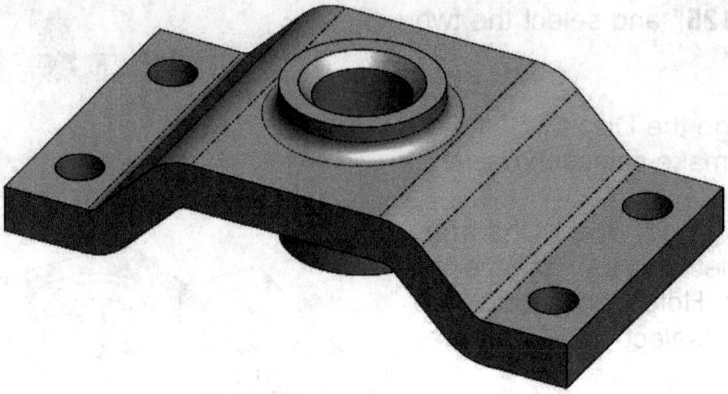

PIN

Next you will design the center pin. Make a new part using the **ANSI-INCHES** template and save it as **Pin.sldprt**.

Add a new sketch in the **Front** Plane and draw the profile indicated. Remember to add a vertical centerline at the origin.

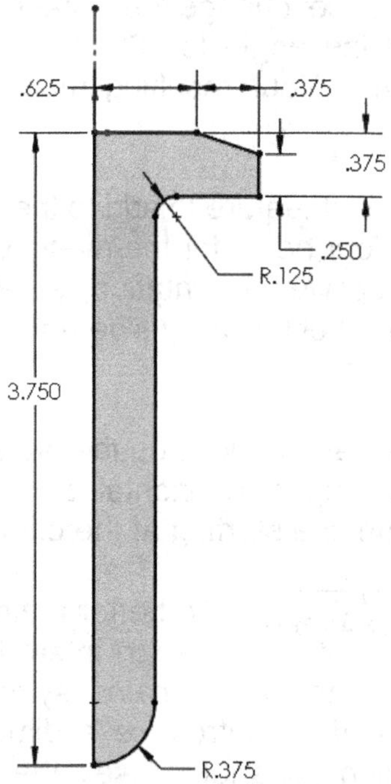

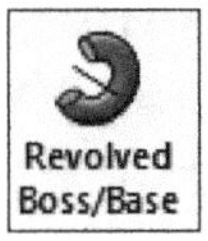

Select the **Revolve Boss/Base** command and revolve the sketch profile **360** degrees. Rename this feature as **Pin Base.**

Now you need to add the hexagonal head feature to the top of the Pin. Change to a **Top** view, select the top flat face of the Pin and add a new sketch.

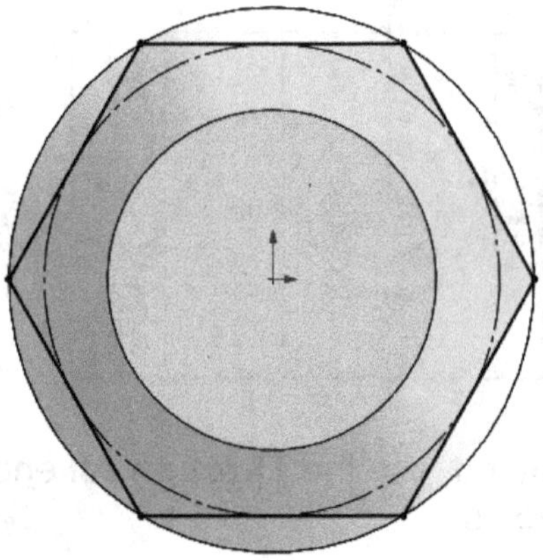

Using the **Polygon** tool draw a hexagon by changing the number of sides to 6.

Center the hexagon at the origin and drag a vertex horizontally out to the right to make it coincident with the edge at the perimeter of the Pin as shown. Finally select the top (or bottom) horizontal line and add a **Horizontal** relation to fully define the sketch.

Change to an **Isometric** view and select the **Extruded Cut** command. Set the end condition to **Through All** and activate the **Flip side to cut** option to cut outside of the sketch profile. Click OK to complete the Cut Extrude and rename this feature **Hex Cut**.

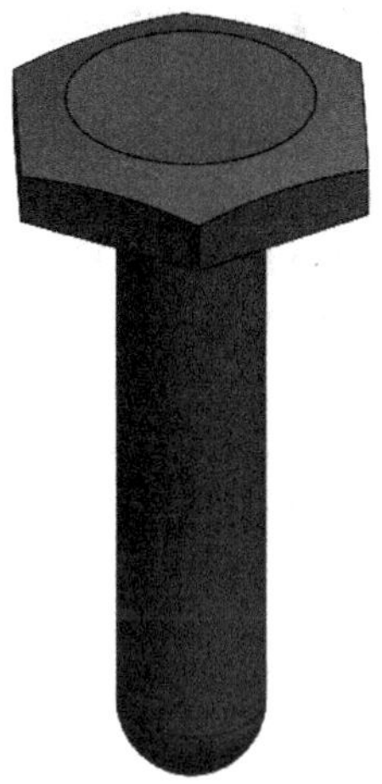

Finally, change the part's material to **2.1020 (CuSn6)** from the **DIN Copper Alloys** material section. Save your part as **PIN.sldprt** and close the file.

RIVET

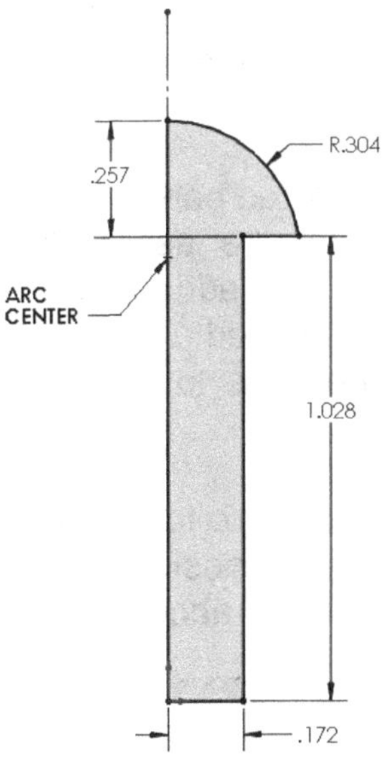

The **Rivet** is a simple part, and you can create it in a number of ways. The method chosen here is to simply sketch a half-profile and then revolve it around a centerline. There are different types of rivets (flat head, pan head, etc.). In this exercise, you will design a "Button Head" rivet.

Start a new part using the **ANSI-INCHES** template and save it as **Rivet.sldprt**. Add a sketch in the **Front Plane** and draw the profile shown; remember to add the centerline.

Use the **Centerpoint Arc** tool to draw the arc, making sure the arc's center is located below the head of the rivet.

Select the **Revolve Boss/Base** command and revolve the sketch **360** degrees.

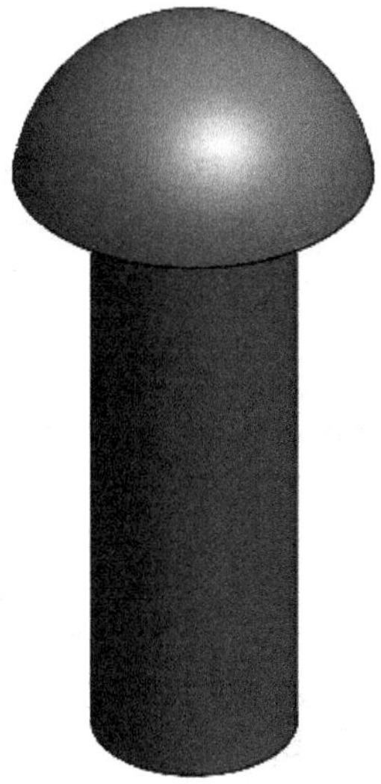

Rename this feature as **Rivet Base** in the FeatureManager and change the part's material to **2.0090 (Cu-DHP)** from the **DIN Copper Alloys** Materials list.

Save the part as **RIVET.sldprt** and close the part. Now that you have made all the parts needed you are ready to start the assembly.

TERMINAL SUPPORT ASSEMBLY

There are multiple ways to make a new assembly. In this lesson we'll use the new assembly's automatic command to add components to the assembly. Before you start, open the **Frame**, **Wing Base**, **Pin**, and **Rivet** parts. The reason to open the parts before starting the assembly is to facilitate their selection.

Select the **New** command from the main toolbar, or the menu **File, New.** In the **New SOLIDWORKS Document** window select the **Assembly** template and click **OK** to continue.

When the new assembly is created, the **Begin Assembly** command is automatically loaded.

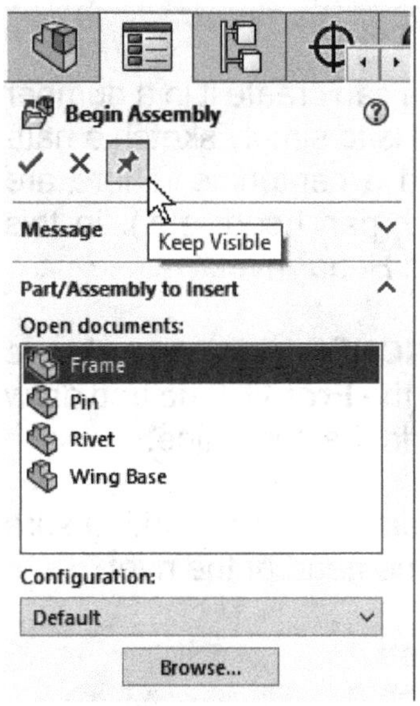

At the top of the command options, select the Keep Visible option to keep the dialog open. In the **Open Documents** list select the **Frame** part (or **Browse** to locate it if it is not currently open) and click **OK** to add it to the assembly.

By adding the **Frame** this way, the selected part's default planes are automatically aligned with the assembly's default planes. Remember *the first part* added to the assembly is **fixed** and cannot be moved unless its constraints are deleted. It will be the base to which all other parts are mated.

The reason why this is desirable is to be able to leverage the assembly's planes to maintain your design intent, especially if the assembly has symmetry about any of those planes.

With the **Begin Assembly** command still open, select the **Wing Base** part, rotate the assembly, and click in the screen to locate it above the **Frame**.

Repeat the same process with the **Pin** and finally import the **Rivet** four times, locating them approximately as shown.

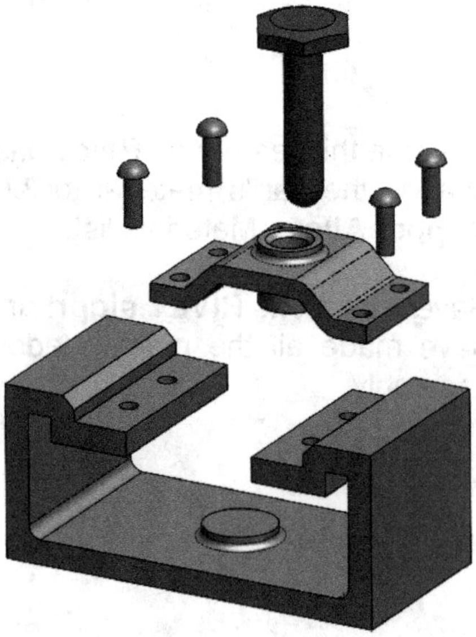

After you have added all components to the assembly click **OK** to continue. The next step is to accurately locate all the components using the **Mate** command.

The first component to be assembled to the **Frame** will be the **Wing Base**. In the Assembly tab select the **Mate** command. First select the inside face of a hole of the **Wing Base** (not the edge) and the inside face of the corresponding hole in the **Frame**. Both faces are added to the Mate Selections list.

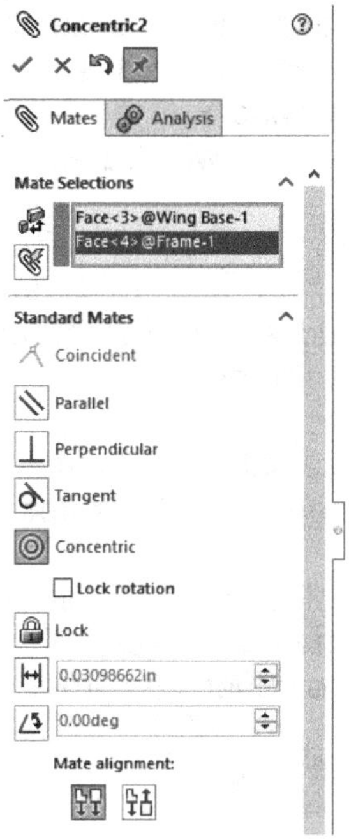

By selecting two cylindrical faces, SOLIDWORKS automatically pre-selects the **Concentric** mate as the most likely option and

moves the **Wing Base** (the part that can move) towards the **Frame** (the Fixed part). Click **OK** to add this mate and continue.

Click and drag the **Wing Base**; notice how it rotates about the hole's axis and moves up and down. The next mate you will add is to restrict the vertical motion of the **Wing Base**. With the **Mate** command still visible, select the flat face of the Frame with the holes, and the bottom face of the **Wing Base**. Since both faces

are planar, the **Coincident** mate type is pre-selected. Click **OK** to add the mate to continue.

When you click and drag the **Wing Base**, notice it only rotates about the hole's axis. To restrict this motion, select the right side faces of the **Wing Base** and the **Frame**, and select the **Parallel Mate**.

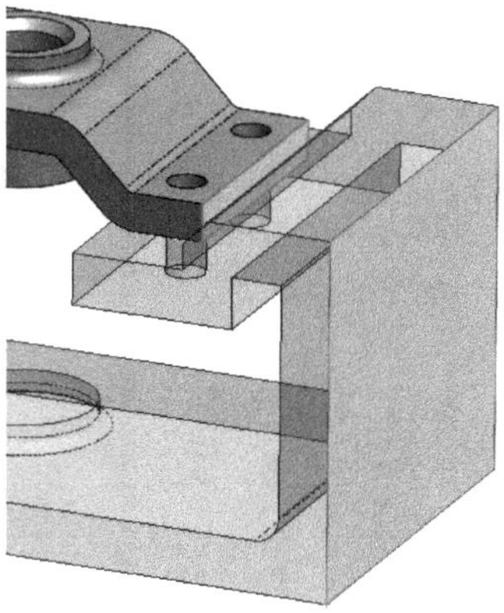

Optionally, you could select the cylindrical faces of a second pair of holes to add a **Concentric** mate. The potential problem with this approach is that if the hole's centers are not *exactly* at the same distance, SOLIDWORKS would not be able to add the mate, over defining the assembly and generating an error message. Instead, the **Parallel** mate is more forgiving and can absorb small dimensional differences.

Now you need to mate the **Pin** to the **Wing Base**. Select the inner cylindrical face of the Center Hole and the outer cylindrical face of the **Pin**. Just as before with the holes, the **Concentric** mate is pre-selected. Click **OK** to add the mate. If you click and drag the **Pin**, it will move along the axis. Now select the face at the top of the **Wing Base** and the matching face at the bottom of the **Pin**. Zoom and rotate the assembly as needed to select the faces to be mated and add a **Coincident** mate.

While still using the **Mate** command, add four more **Concentric** mates between the **Rivets** and their corresponding holes, and finally add a **Coincident** mate between the bottom face of each **Rivet** head to the **Wing Base**. Close the **Mate** command; the finished assembly is shown next.

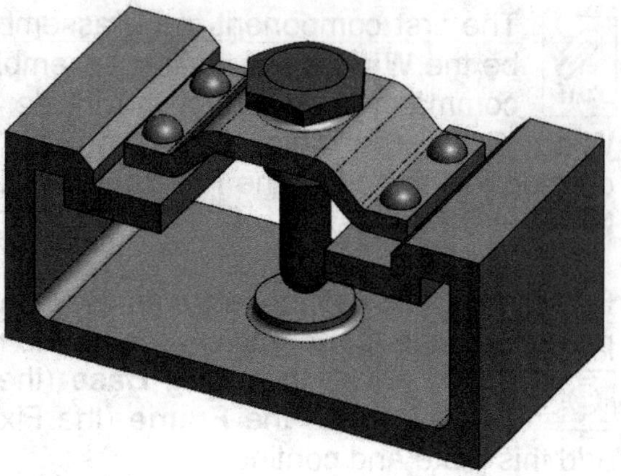

Save the assembly as **TERMINAL SUPPORT.sldasm** and make a new drawing with an Isometric view, using the Shaded with Edges display style.

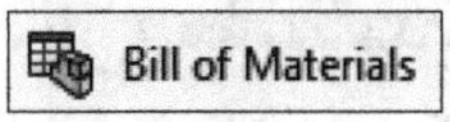

To complete the drawing, you need to add a **Bill of Materials** and **Identification balloons**.

Pre-select the Isometric view and click in the menu **Insert, Tables, Bill of Materials**.

Select the options **Parts Only** and **Display all configurations of the same part as one item**. Click **OK** to accept the parameters and click to locate the **Bill of Materials** in the upper left corner of your drawing.

To add identification balloons, select the Annotation tab and click in the **Balloon** command, or from the menu **Insert, Annotations, Balloon**. In the Settings section, select the **Circular** option, and **Item Number**, as shown.

After setting these options, click on each component of the assembly and locate the corresponding balloon in the drawing. You can select a part's face, edge or vertex as needed.

Optionally change the balloon's font size. Select the four balloons and click in "**More Properties…**", turn off the "**Use document font**" option and use the **Font** command to change the balloons to use a **16-point** size font.

After completing the balloons, select the **Note** command from the Annotations tab, and add a note next to each balloon with the component's name. Set the font size to also use a 16-point.

As an alternative to changing the font style individually, you can select the menu **Tools, Options, Document Properties**, and select the corresponding section to set the default font for Annotations, Balloons, Notes, etc.

Remember that every option changed in the Document Properties section can be saved to an assembly template.

Save your drawing as **TERMINAL SUPPORT.slddrw** and **Print** a copy of the finished drawing for your lab instructor.

Do not discard your assembly files; they will be used in a later lesson to learn how to create an **Exploded View** and animation.

ITEM NO.	PART NUMBER	DESCRIPTION	QTY.
1	Frame		1
2	Wing Base		1
3	Pin		1
4	Rivet		4

TERMINAL SUPPORT
ASSEMBLY
Scale 1:2

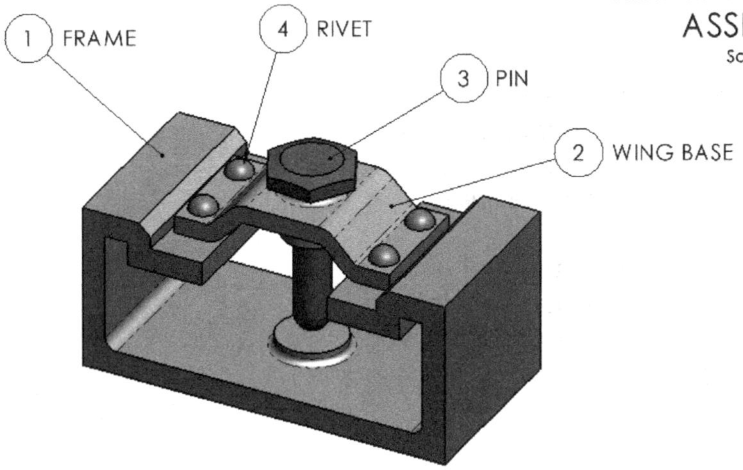

NAME:STUDENT	DESK:	SEC:	GRADE:	5.1
DESIGN WORKBOOK USING SOLIDWORKS				

Exercise 5.2: PULLEY Assembly

The Pulley Assembly is a simple design used to hoist objects with a mechanical leverage advantage. It has four major components: An **Eye Hook**, a **Pulley Sheave**, a **Spacer**, and a **Base Plate**. It is assembled using small rivets that are peened on one end to secure them and hold the components together. You will start by designing the **Base Plate**, which is the part that drives the overall dimensions of the other components.

BASE PLATE

Start a new part using the **ANSI-Inches** template. Add a new sketch in the **Front Plane** and draw the profile shown.

Since it is symmetrical, you can use the **Dynamic Mirror** command. Remember to add Tangent relations between the lower arc and the inclined lines.

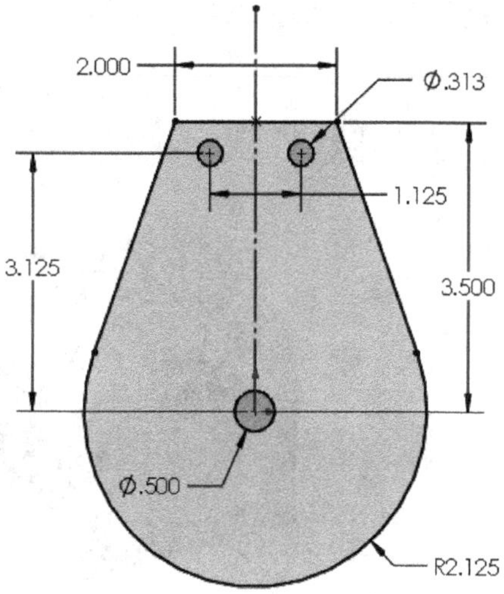

Select the **Extruded Boss** command and extrude the profile to **0.125"**. Change the part's material to **AISI 4340 STEEL, NORMALIZED** and save your part as **Base Plate.sldprt**.

PULLEY

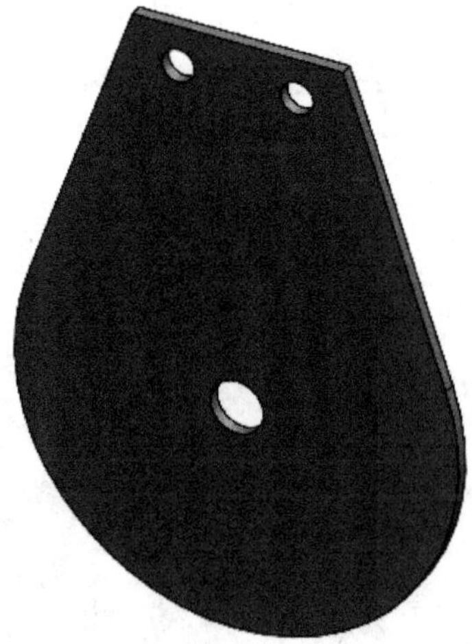

Start a new part using the **ANSI-INCHES** template, add a sketch in the **Right Plane** and draw the sketch indicated.

Draw a vertical **Centerline** for the **Dynamic Mirror** command, and a horizontal centerline to use it as the axis of rotation. Use the **Dimension** tool to fully define the sketch. Use the menu **Tools, Options, Document Properties, Units** to change the number of significant digits to four decimal places.

Now select the **Revolve Boss** command and make the revolved feature **360** degrees about the horizontal centerline. You now have the pulley part of the assembly.

Set the part's material to **2.0375 (CuZn36Pb3)** from the **DIN Copper Alloys** section. Finally, save your part as **Pulley.sldprt** to finish.

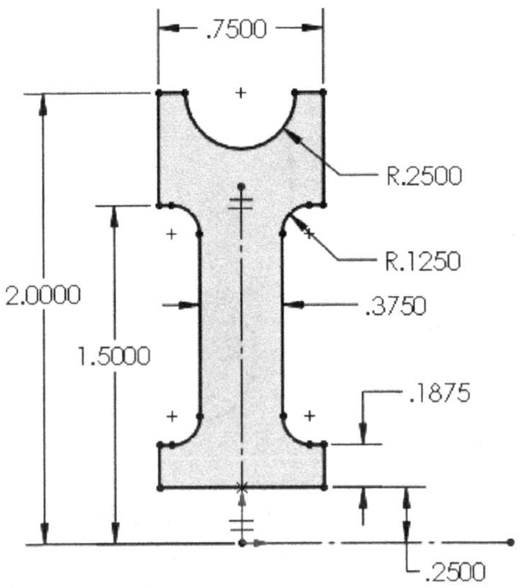

SPACER

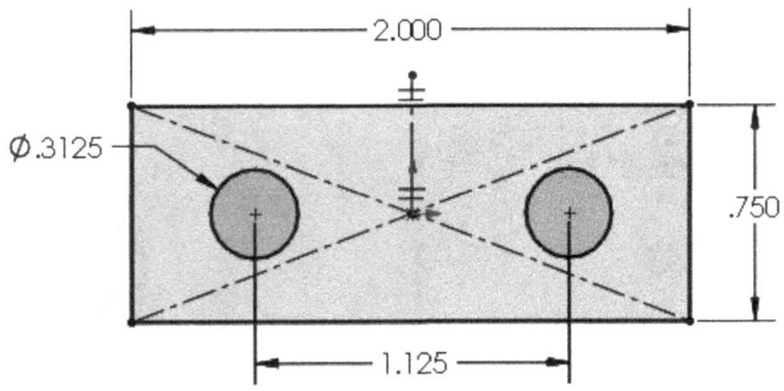

Start a new part using the **ANSI-INCHES.prtdot**, add a new sketch in the **Front Plane** and draw the profile shown here. Use the **Center Rectangle** and **Circle** commands and dimension as indicated.

Use the **Dynamic Mirror** to make the two circles symmetric and add a **Horizontal** relation between the center of one circle and the origin to fully define your sketch.

When the sketch is complete, extrude it using the **Mid Plane** end condition a distance of **1.00"**.

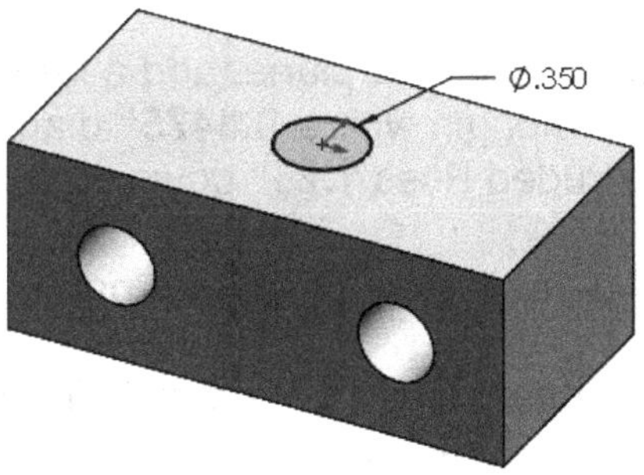

Remember the **Mid Plane** option will extrude the sketch half of the specified distance in each direction, making the extrusion symmetrical about the sketch plane.

Next, change to a **Top View** and add a new sketch on the top face. Draw a circle in the origin with a diameter of **0.350"** and make an **Extruded Cut** using the **Through All** end condition.

Set the material to **AISI Type A2 Tool Steel** and save your part as **Spacer.sldprt** to finish.

EYE HOOK

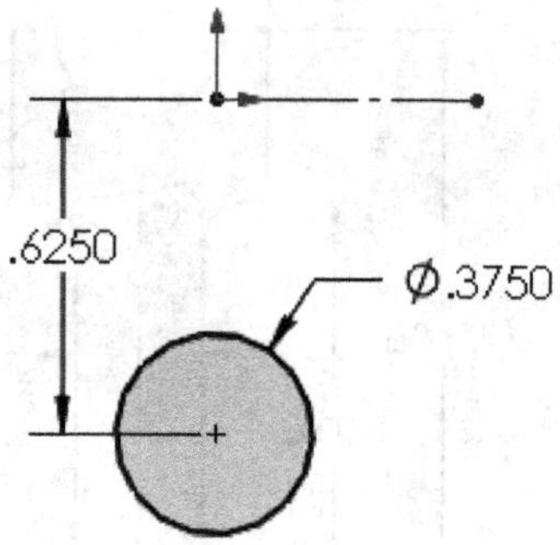

The next part for the assembly is the **Eye Hook**. It can be made by making a revolved boss and an extrusion.

Start a new part in inches and set the dimensions to four decimal places. Add a new sketch in the **Right Plane** and draw the profile shown. Add a **Vertical** relation between the origin and the circle's center to fully define your sketch.

Select the **Revolved Boss** command and revolve the circle **360** degrees to create the top part of the Eye Hook.

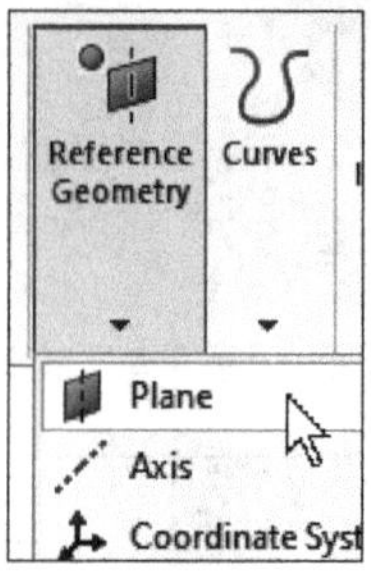

To make the bottom half you need to add an auxiliary plane. Select **Reference Geometry** and click on **Plane**. From the fly-out FeatureManager select the **Top Plane** (or pre-select it before starting the **Plane** command), set the distance to **0.625"** and activate the "Flip Offset option to create the new plane below. Click **OK** to continue.

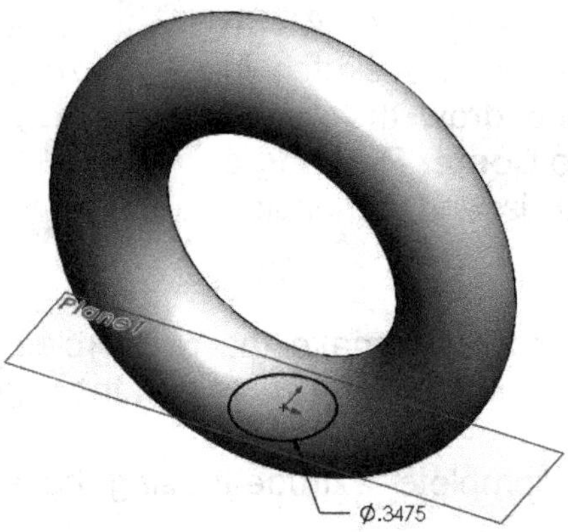

Select the new plane, add a new sketch and draw a circle in the origin with a **0.3475"** diameter, as shown. Make an Extruded Boss **1.25"** downward.

The completed **Eye Hook** part is shown here.

Change the part's material to **Tin Bearing Bronze** from the **Copper Alloys** library and save it as **Eye Hook.sldprt**.

RIVETS

The last parts you have to make are the **Small** and **Big Rivets**. These are standard components whose dimensions can be found in engineering handbooks.

For both components start a new part using the template in Inches, add a new sketch in the **Front Plane** and draw the corresponding profile shown next for the **Big Rivet** and **Small Rivet**, respectively. Remember to add a centerline and make a 360° **Revolved Boss** to complete the parts.

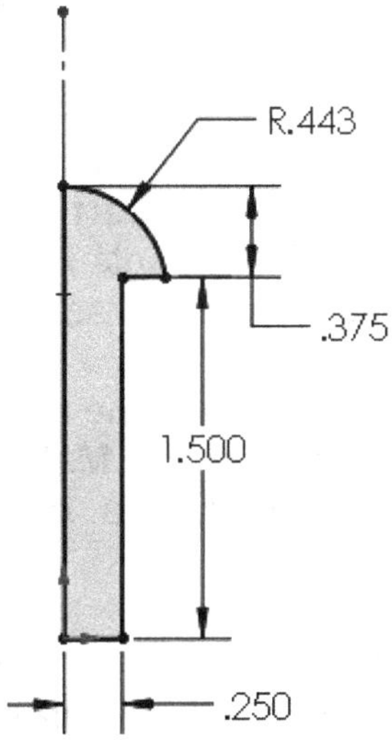

Sketch for Big Rivet

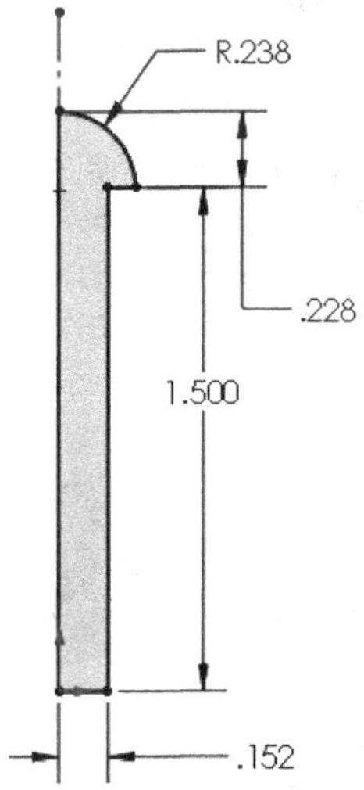

Sketch for Small Rivet

Set the material on both parts to **2.0060 (Cu-ETP)** from the DIN Copper Alloys list found in SOLIDWORKS DIN Materials. To finish, save each part as **Big Rivet.sldprt** and **Small Rivet.sldprt**, respectively.

SWIVEL EYE BLOCK ASSEMBLY

Now you are ready to start the new assembly. In the previous exercise you opened all the parts needed, and then added them from the **Begin Assembly** command. In this exercise you will use a different approach.

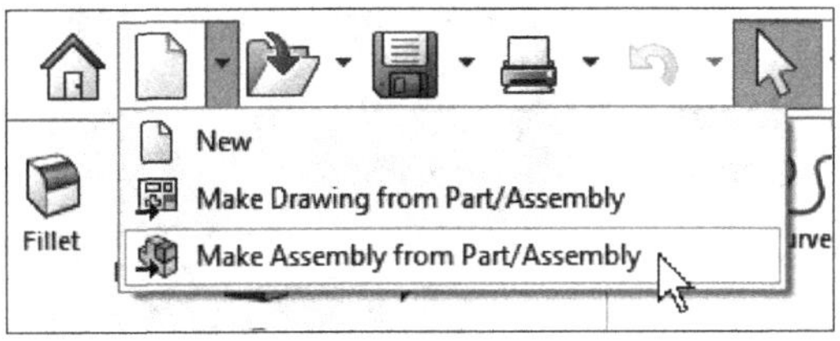

The first component added to this assembly will be the **Spacer**. Open the **Spacer** part, and select the menu **File**, **Make Assembly from Part**, or from the **New** command's drop-down list select **Make Assembly from Part/Assembly**.

Select the default **Assembly** template and click **OK** to continue (if you made an assembly template in Inches you can use it, too).

You are immediately presented with the **Begin Assembly** command. Make sure the **Spacer** is selected in the **Open Documents** list and click **OK**. Remember the **Spacer** will be added to the assembly and its default planes will be aligned with the assembly's default planes.

If needed, go to the menu **Tools, Options, Document Properties** and set the dimensioning standard to **ANSI**. Go to the **Units** section and change the assembly's units of measure to **Inches** with three decimal places and save your assembly as **Swivel Eye Block.sldasm**.

In this exercise we'll add the components to the assembly and mate them one at a time. From the Assembly tab select the **Insert Component** command, select the **Base Plate** from the Open Documents list, or browse to locate it. Then click in the graphics area to add it to the assembly.

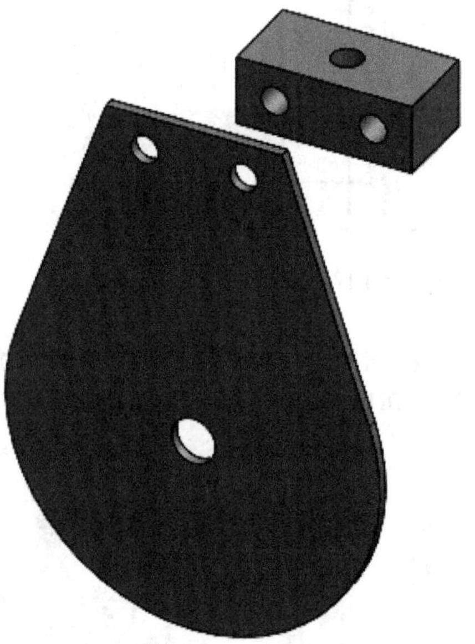

Now you need to mate the **Base Plate** to the **Spacer**. Select the **Mate** command from the Assembly tab or the menu **Insert, Mate**.

Select the upper left hole's cylindrical face of the **Base Plate** and the corresponding **Spacer's** hole. Since both faces are cylindrical, the **Concentric** mate is automatically pre-selected. Click **OK** to add this mate. Repeat the same mate using the faces of the other pair of holes.

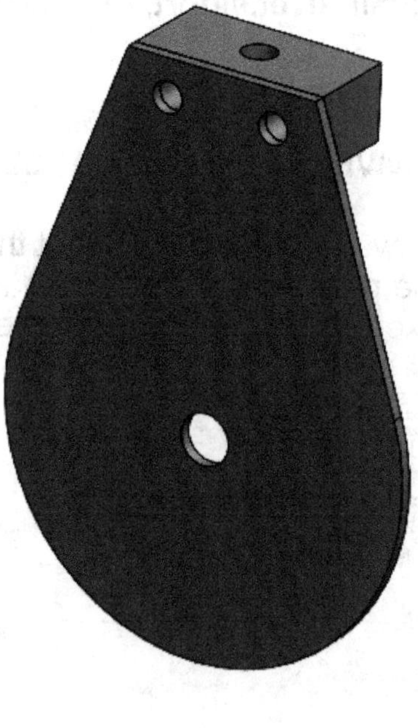

If needed, click and drag the **Base Plate** away from the **Spacer**, and select the inside face of the **Base Plate** and the outside face of the **Spacer**.

Add a **Coincident** mate, click **OK** and close the **Mate** command. The **Base Plate** is now fully defined about the **Spacer**.

Use the **Insert Components** command to add a second copy of the **Base Plate**, click to add it to the assembly and mate it to the other side of the **Spacer** using the same mates as before. **Zoom** and **Rotate** the assembly as needed to select the faces to be mated.

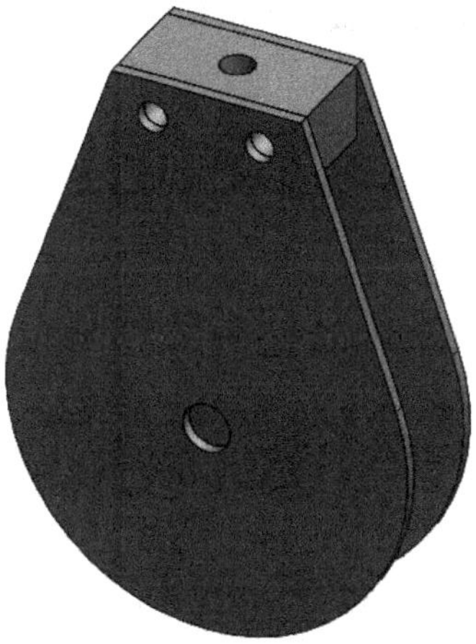

Close the **Mate** command to continue. Your assembly should look as follows.

Now you need to add the Pulley to the assembly. Use the **Insert Components** command and locate it in the assembly.

Select the **Mate** command and add a **Concentric** mate between the face at the center hole of the **Base Plate** and the cylindrical face at the center of the **Pulley**.

After this mate is added, the pulley can move along the hole's axis and is not centered about the **Base Plates**.

There are different ways to center the **Pulley**:

You can add a **Distance Mate** between model faces, but this approach would probably not maintain your design intent to keep the **Pulley** centered if the **Spacer** or the **Pulley** change in size.

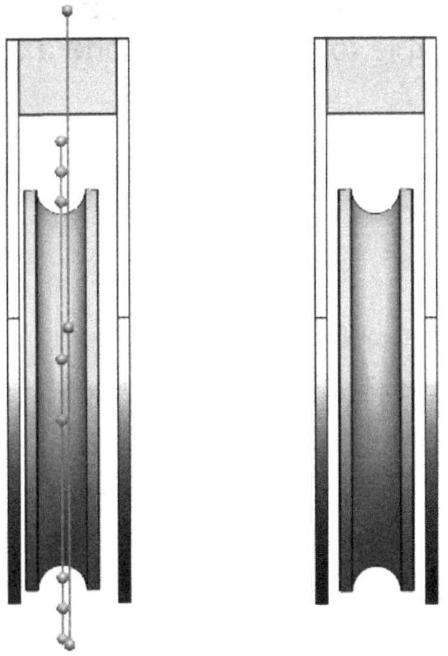

A different approach is to take advantage of the symmetry built into your parts. In this case, you can add a **Coincident Mate** between the **Pulley's Front Plane** and the assembly's **Front Plane**, which is coincident to the **Spacer's Front Plane**.

Change to a **Right** view orientation, expand the FeatureManager, and pre-select the Assembly's **Front Plane**, and the **Pulley's Front Plane**.

Click in the **Mate** command and add a **Coincident** Mate. Now your Pulley is centered and will remain centered even if the size of the **Spacer** or the **Pulley** changes.

Next you need to add the **Eye Hook** to the assembly. Use the **Insert Component** command and locate it above the **Spacer**. Use the Mate command to add a Concentric Mate between the top hole in the **Spacer** and the circular shaft of the **Eye Hook**.

Change to a **Right** view and drag the Eye Hook enough for it to pass through the Spacer, as shown.

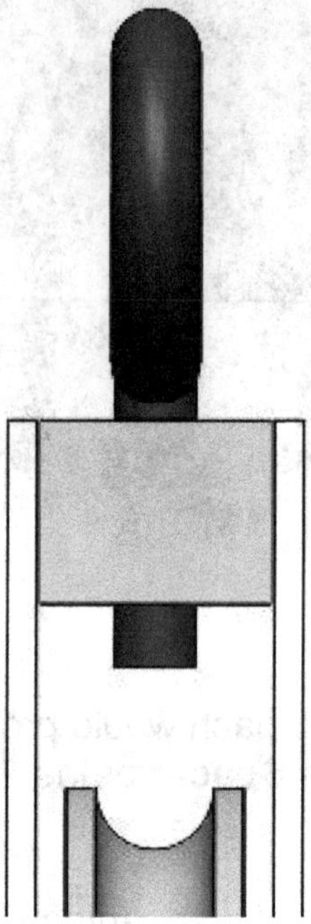

Rotate the view to see the bottom of the **Eye Hook** and the **Spacer**, select the bottom faces of both parts and add a **Distance** Mate. Set the distance to **0.125"**. Preview the mate and make sure the **Eye Hook** passes through the **Spacer**; if it does not, click in the **Flip Dimension** option to fix it.

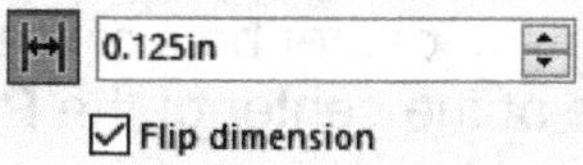

When finished, close the **Mate** command and expand the **Mates** folder in the FeatureManager. You will see all the mates added so far to your assembly, as shown.

Now that the major components have been assembled, you can see the **Spacer** is Fixed in place, indicated by the letter (f), the **Base Plates** are fully defined, and the **Pulley** and **Eye Hook** have one un-constrained degree of freedom indicated by the (-) sign next to their name in the FeatureManager, and are able to rotate about one axis.

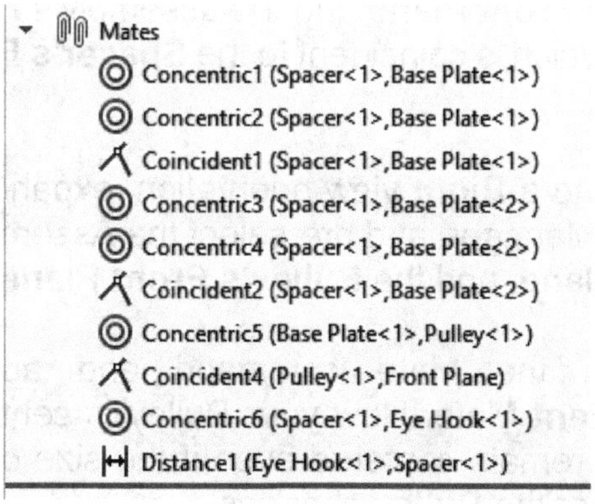

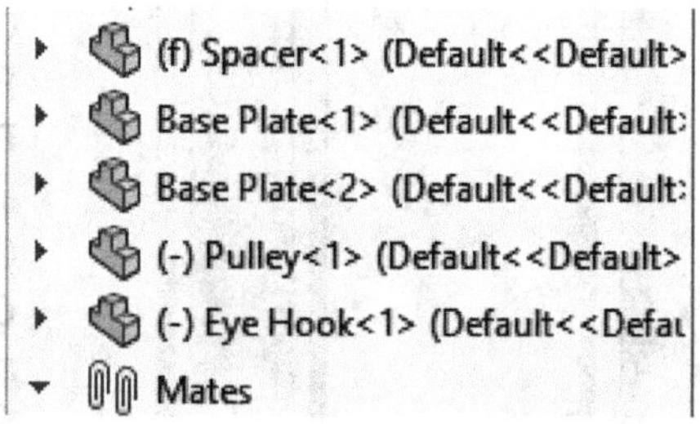

To verify their motion, click and drag them to see their rotation. The motion will be easier to visualize in the **Eye Hook**.

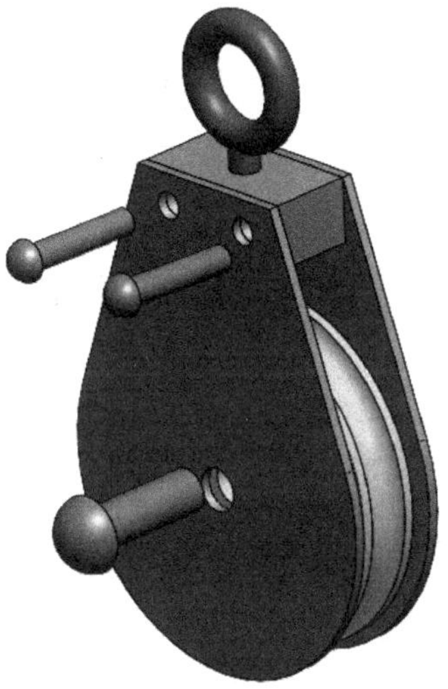

To finish the assembly, now you need to add the rivets. Use the **Insert Components** command to add two **Small Rivets** and one **Big Rivet**.

After adding the rivets, you can right-click to rotate them on the screen and then click-and-drag to locate them approximately as shown.

Now use the **Mate** command to assemble the three rivets. Each rivet will have a **Concentric** mate between the rivet's shaft and their corresponding hole, and a **Coincident** mate between the bottom face of the rivet's head and the **Base Plate**.

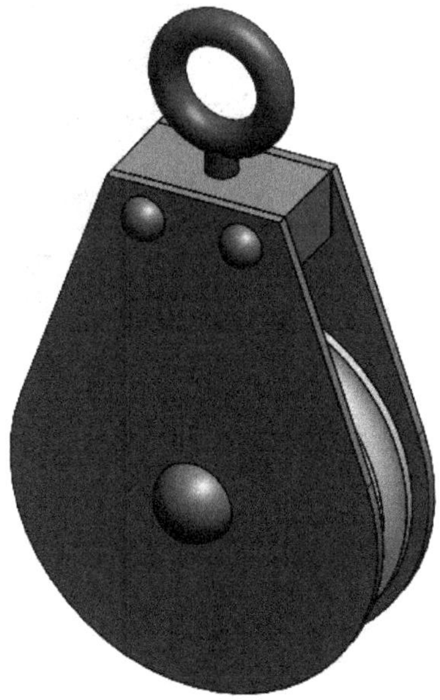

After the Rivets have been mated, your assembly should look like this.

Save your assembly as **Swivel Eye Block.sldasm** to finish.

<u>Manufacturing Note:</u> *If you rotate the assembly, you will see the Rivets extending about 0.25" on the other side. The way these rivets are assembled is with a peen operation. Peening the end of the rivet with a hammer or automatic press causes the flat end of the rivet to deform in a spherical shape, preventing the rivet from coming out of the hole.*

Next make a new drawing and add an Isometric view using the Shaded with Edges display style.

In the Drawing sheet, select the menu **Tools, Options, Document Properties**, and under the **Annotations** section select **Balloons** and change the **Font** to Arial Regular, and a Height of **18 pt**. Click **OK** to accept the changes.

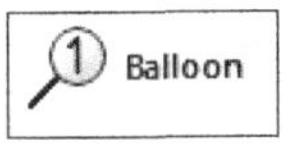

From the Annotations tab select the **Balloon** command and select each component to add their respective **Item Number**. After adding the balloons, select the **Note** command, use the No Leader option and add the part name using an **18 pt** font next to the corresponding balloon.

Now select the assembly view and click in the menu **Insert, Tables, Bill of Materials**. Use the option Select Parts only, Display all Configurations of the same part as one item and click **OK**. Locate the Bill of Materials in the upper right corner of your drawing.

Save your drawing as **PULLEY ASSEMBLY.slddrw** and **Print** out a copy for your lab instructor.

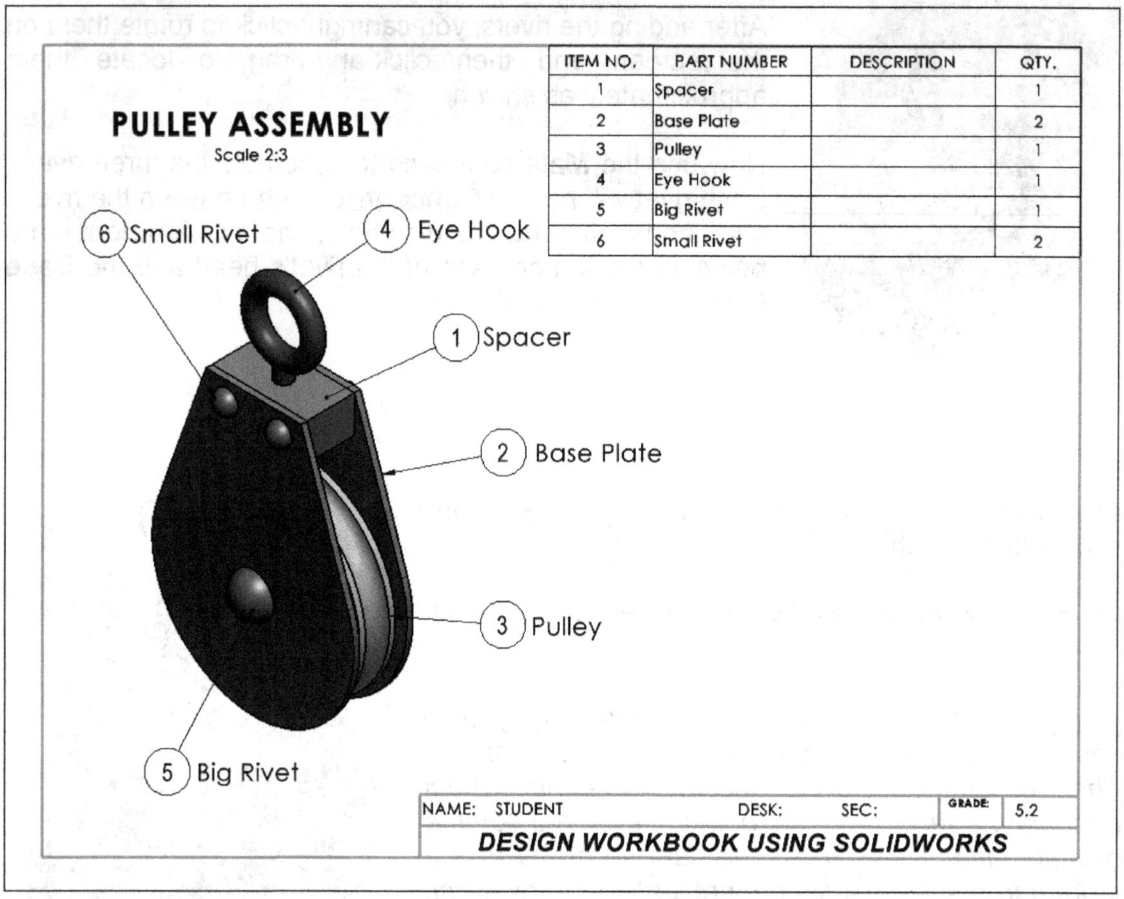

ITEM NO.	PART NUMBER	DESCRIPTION	QTY.
1	Spacer		1
2	Base Plate		2
3	Pulley		1
4	Eye Hook		1
5	Big Rivet		1
6	Small Rivet		2

Supplementary Exercise 5-1: CASTER ASSEMBLY

Create the parts of the Caster Assembly of the Option specified by your instructor in the Table provided.

1. Create the Assembly indicated by your instructor following the table and make the corresponding Assembly Drawing with a Bill of Materials.

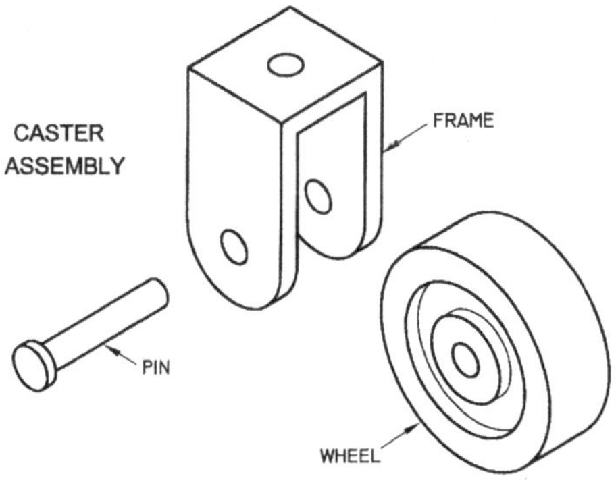

	CASTER DIMENSION OPTIONS	
OPTION	WHEEL DIA. (D)	SADDLE HEIGHT (H)
A	2.00	1.50
B	2.20	1.60
C	2.40	1.70
D	2.60	1.80
E	2.80	1.90
F	3.00	2.00

ALL DIMENSIONS IN INCHES

DESIGN PROJECT I

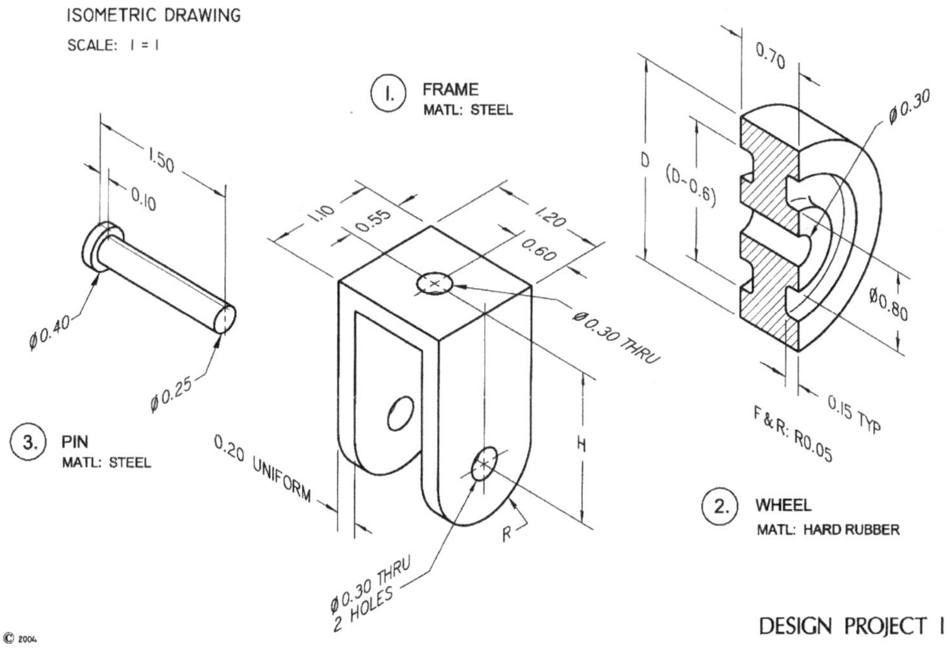

Supplementary Exercise 5-2: PULLEY ASSEMBLY

Create the parts of the Pulley Assembly of the Option specified by your instructor in the Table provided.

1. Create the Assembly indicated by your instructor following the table and make the corresponding Assembly Drawing with a Bill of Materials.

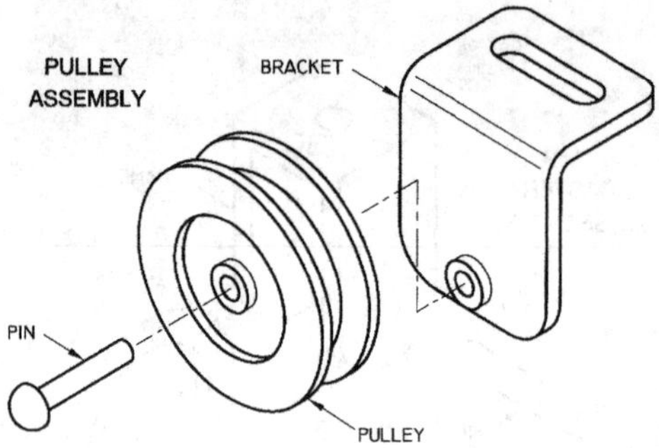

	PULLEY DIMENSION OPTIONS	
OPTION	PULLEY DIA. (D)	BRACKET HEIGHT (H)
A	2.00	1.50
B	2.20	1.60
C	2.40	1.70
D	2.60	1.80
E	2.80	1.90
F	3.00	2.00

ALL DIMENSIONS IN INCHES

DESIGN PROJECT II

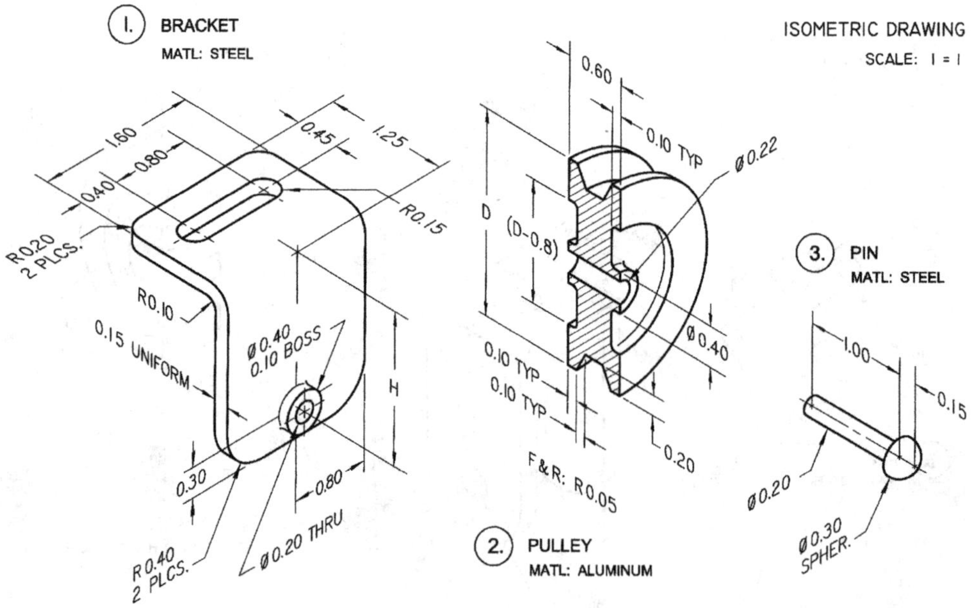

DESIGN PROJECT II

Design Workbook Lab 6
Part Evaluation and Configurations

In this lesson you will learn how to use the tools available to calculate a component's **Mass Properties**, and how to create a family of parts using design tables. One of the biggest advantages of a parametric solid model is that you can accurately calculate mass properties without the need to fabricate a physical model, as well as digitally **Measure** every characteristic of your design. Additionally, you will learn how to generate multiple similar but different components using a **Design Table**, where you can control different design dimensions using an Excel spreadsheet.

THE MEASURE TOOL

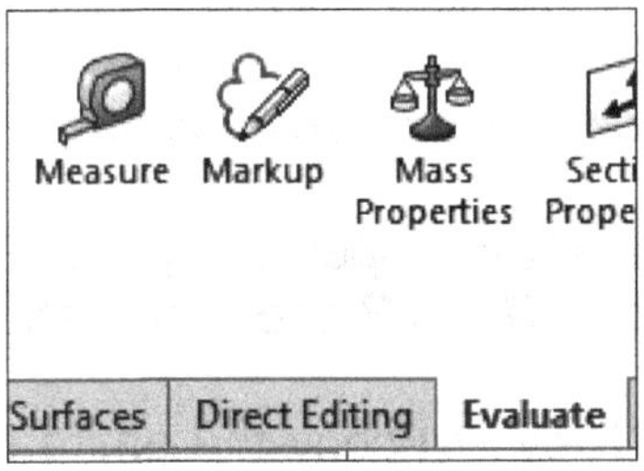

The **Measure** tool can be found in the menu **Tools, Evaluate, Measure**, or from the Evaluate tab in the FeatureManager. When you use the Measure tool, the cursor will change to a ruler; at this time, you can select the entity (or entities) to measure in the screen.

You can measure 2D sketch lines or 3D solid model geometry, including:

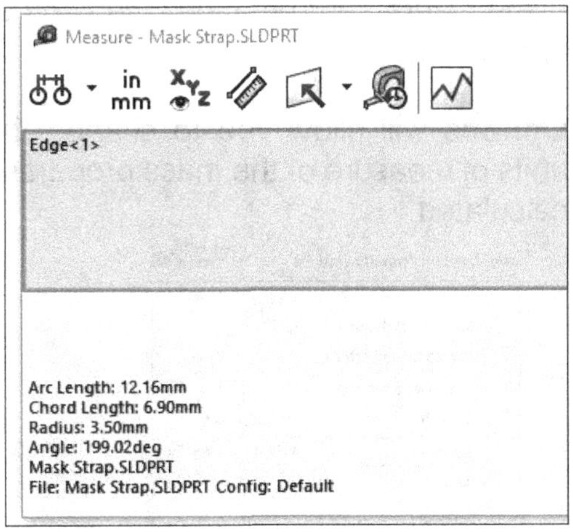

Sketch Line/3D model edge returns the length of the entity.

Sketch Arc/3D model circular edge returns the length of the arc and the diameter. If the arc is a full circle, the length is the circumference.

3D Model Face returns the area and the total length of all model edges connected to the face.

Pair of Edges/Faces/Vertices returns the distance/angle/total length/total surface area of the selected items.

THE MASS PROPERTIES TOOL

Also located in the Evaluate tab or in the menu **Tools, Evaluate, Mass Properties**, this tool allows you to calculate the mass properties of a part or assembly.

Before you can calculate the mass properties accurately, you must set the part's material, including the density. **Table 6-1** includes the density of some common materials. Once the part's material is defined, you can then select the **Mass Properties** command.

The result will be reported in the "Mass Properties" Report shown. The mass properties and their units of measure reported include the following (see **Page 6-11** for their definitions):

- **Part Name**
- **Density (lbs/in^{3})**
- **Mass (lbs)**
- **Volume (in^{3})**
- **Surface Area (in^{2})**
- **Center of Mass (in)**
- **Moments of Inertia (lbs*in^{2})**

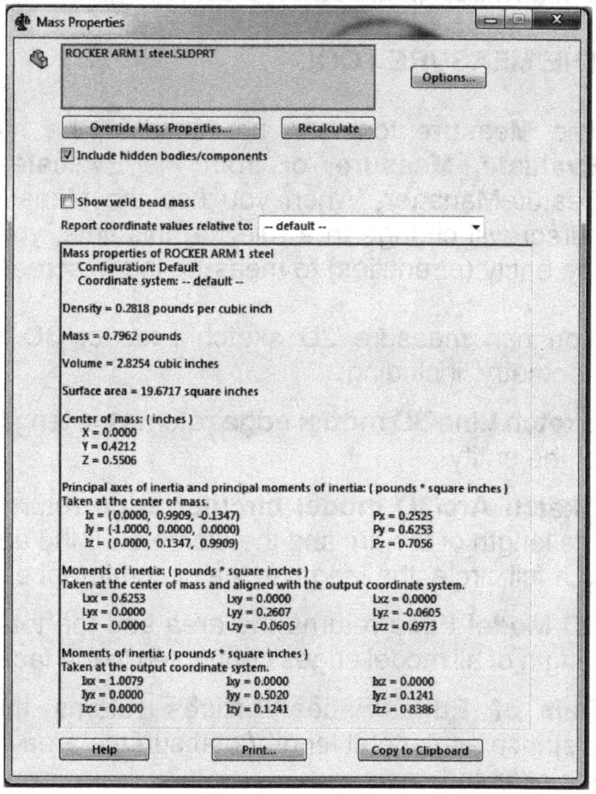

After the component's "Mass Properties" are calculated, you can review the results in this window. The options available include:

Print will open the print dialog to print the "Mass Properties" Report.

Copy will copy the entire report to the computer clipboard. Then you can paste this information in other applications.

Options will allow you to change the units of measure of the mass properties calculated.

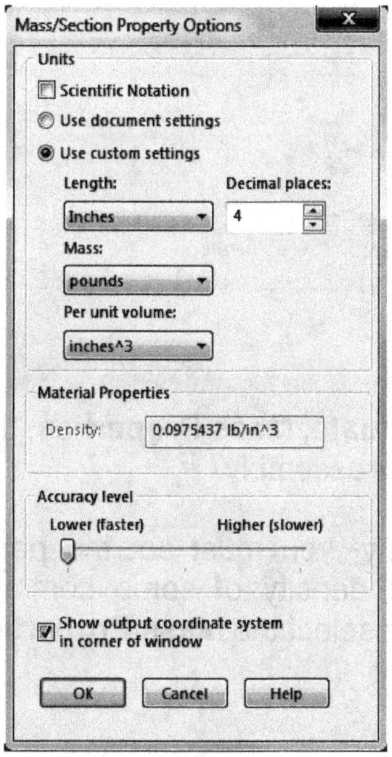

Recalculate will recalculate the mass properties after other options are changed.

ABOUT THE MASS PROPERTIES UNITS

Some mass properties calculated are based on the model's geometry and are independent of the material's density (assumed to be uniform).

Examples of these properties are volume and center of mass (centroid). Other properties, such as mass and moments of inertia, are dependent on the material's density. **Table 6-1** lists the density of some materials commonly used in engineering.

You can use these density values to set the material properties of your part. Weight and Mass are often confused. Below is an example of the correct way to calculate them for a unit one-inch cube.

<u>Note</u>: The "Density" value used by SOLIDWORKS is actually the value of the "Unit Weight" listed below in **Table 6-1**. This lack of uniformity in terminology contributes to the general confusion on this matter.

Example Calculation: Material is Mild Steel (density is 0.728×10^{-3} lbs-sec^2/in^4)

Weight = (unit weight)x(volume) = (0.281 lbs/in^3) x (1.00 in^3) = 0.281 lbs.

Mass = (density)x(volume) = (0.728×10^{-3} lbs-sec^2/in^4)x(1.00 in^3) = 0.728×10^{-3} lbs-sec^2/in.

= (weight) / (gravity) = (0.281 lbs) / (386 in/sec^2) = 0.728×10^{-3} lbs-sec^2/in.

Table 6-1. Unit Weights and Densities of Common Materials.

Material	**Unit Weight (lbs/in^3)**	**Density (lbs-sec^2/in^4)**
Aluminum	0.097	0.251×10^{-3}
Brass	0.307	0.794×10^{-3}
Chromium	0.259	0.671×10^{-3}
Copper	0.323	0.837x10-3
Magnesium	0.063	0.163x10-3
Plastic	0.036	0.093×10^{-3}
Rubber	0.041	0.106×10^{-3}
Steel	0.281	0.728×10^{-3}
Titanium	0.163	0.422×10^{-3}

Exercise 6.1: Rocker Arm Mass Properties

For Exercise 6.1, you will build the rocker arm, which is designed to rotate about a principal axis. You will then copy the first rocker arm data into a second rocker arm file and make a change in its geometry. Mass Properties calculations will be performed on both models and the results will be compared to each other.

ROCKER ARM 1

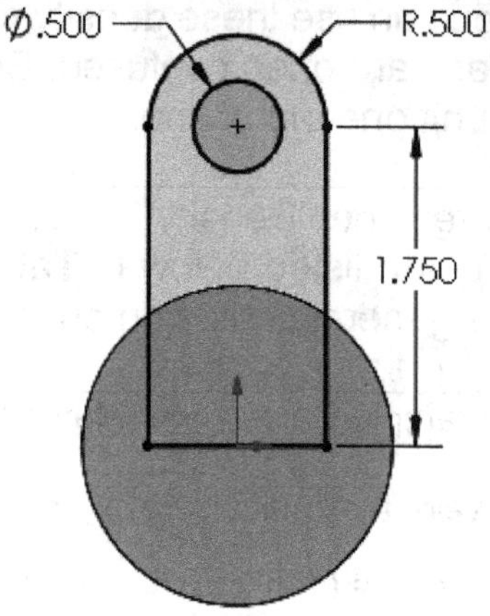

Start a new part using the **ANSI-INCHES** template, add a sketch in the **Front Plane** and draw a circle with a diameter of **1.75"** centered in the origin. When complete **Extrude** it **1.25"** out from the **Front Plane**.

Next add a new sketch in the **Front Plane** and draw the profile shown. This will be extruded to create the Rocker Arm. Make sure you add the following relations, and dimension as indicated:

Both **Vertical lines** are **Tangent** to the **Arc**.
The **Circle** is **Concentric** to **Arc**.
The **Horizontal** line has a **Midpoint** relation to the origin.

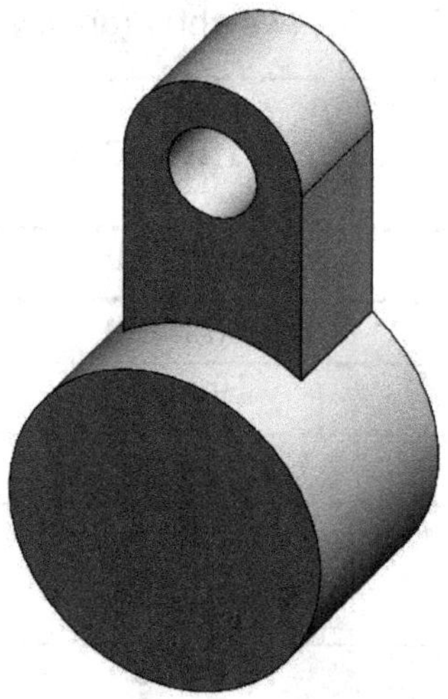

Now **Extrude** the sketch **0.75"** in the same direction as the first extrusion.

The last step to finish this model is to add a cutout for a keyway through the part. Change to a Front view, select the front face of the model, add a new sketch and draw the profile indicated next. Draw a circle and a rectangle, and then use the **Trim** tool.

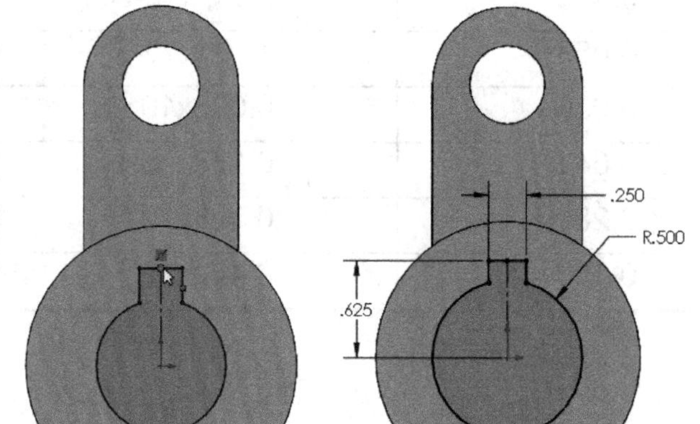

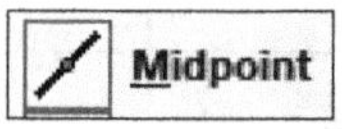

Before adding the dimensions, add a vertical centerline from the origin up to the top horizontal line, and add a **Midpoint** relation between the horizontal line and the top endpoint of the centerline; this way the keyway will remain centered about the origin.

Add the necessary dimensions to fully define the sketch, and make a **Cut Extrude** using the **Through All** end condition. Save your model as **Rocker Arm 1.sldprt** to continue.

ROCKER ARM 2

The **Rocker Arm 2** will be very similar to **Rocker Arm 1**. Use the **Save As** command and save the model as **Rocker Arm 2.sldprt**.

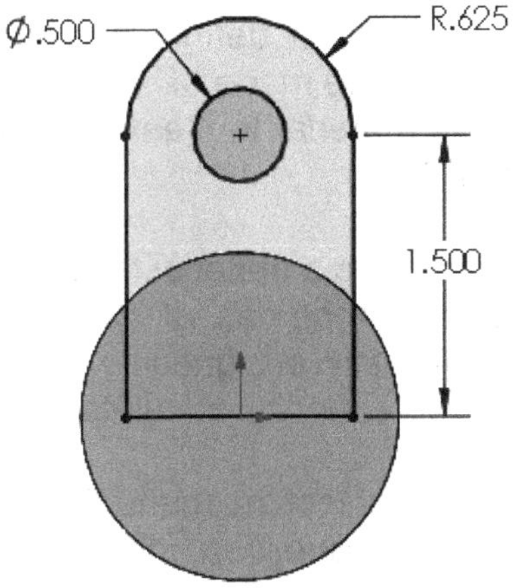

After saving it select the **Boss-Extrude2** feature and click on the **Edit Sketch** command from the context toolbar.

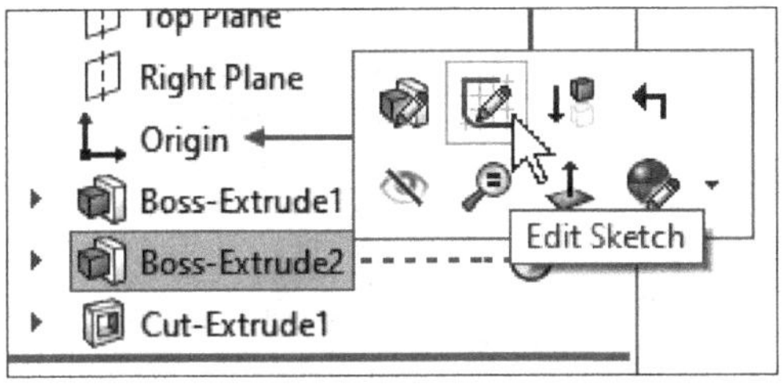

While editing the sketch, double click to change the arc's radius to **0.625"** and the height to **1.50"** as shown.

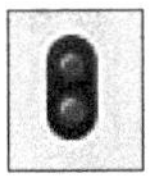

After making these dimensional changes, click on the **Rebuild** command to exit the sketch, and update the 3D model with the new dimensions.

Now you need to change the extrusion depth of the upright feature. Select the **Boss-Extrude2** feature in the FeatureManager tree.

Select the **Edit Feature** option from the context toolbar, change the extrusion's depth to **0.500** inches and click **OK** to rebuild the model. The solid model will now be rebuilt with the new extrusion depth.

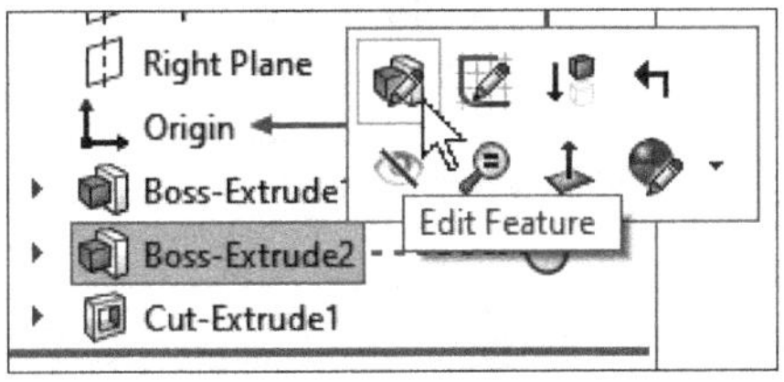

Save your completed **Rocker Arm 2** model to continue.

MASS PROPERTIES ANALYSIS OF ROCKER ARM 1

To start your mass properties calculation, open the **Rocker Arm 1** part. After opening the file, you need to change the units to four decimal places. Select the menu **Tools, Options, Document Properties, Units**, change the **Mass/Section Properties** Length units to **4 decimal places** and click **OK** to continue.

Now you must define the material the **Rocker Arm 1** will be made of. As you have done before, set the material to **Cast Carbon Steel** from the **Steel** library.

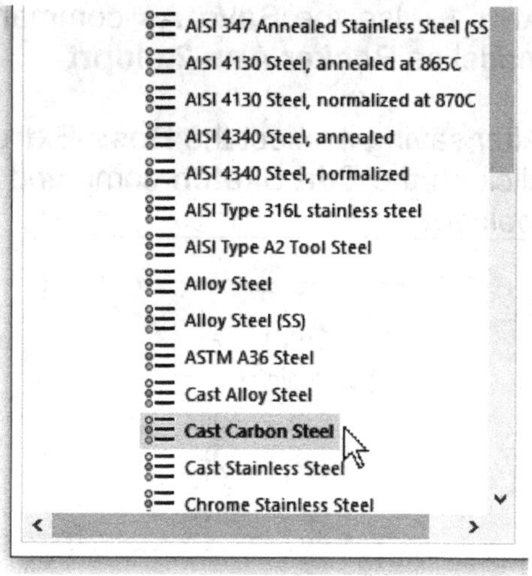

By assigning a material to your model, the correct physical properties of **Cast Carbon Steel** are automatically assigned to the model, which will help you correctly calculate its mass properties.

Select the menu **Tools, Evaluate**, **Mass Properties**, or from the Evaluate tab click on **Mass Properties**. The "Mass Properties" window is now shown with all properties calculated for **Rocker Arm 1**, made of **Cast Carbon Steel**.

The Mass Properties report includes:

- Density (assigned by the material selection)
- Mass
- Volume
- Surface Area
- Center of Mass
- Principal axes of inertia
- Principal Moments of Inertia
- Moments of Inertia (taken at center of mass)
- Moments of Inertia (taken at output coordinate system)

Because of the design and functionality of a Rocker Arm, a critical design parameter is the part's resistance to rotation about the axis of rotation (Z-axis), in this case concentric with the large cylindrical hole.

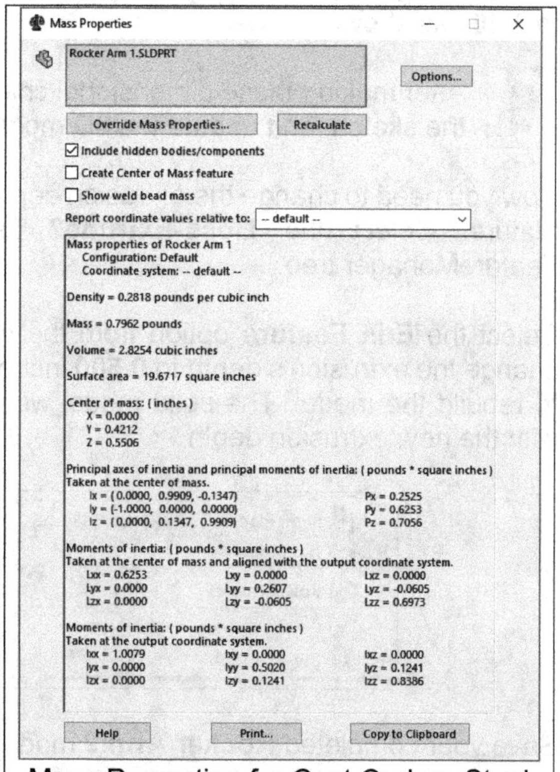

Mass Properties for Cast Carbon Steel
Rocker Arm 1

Based on this criterion, a part with a lower moment of inertia about the Z-axis would be a more desirable design because it would offer less resistance to rotation. To find the best option, you will evaluate both **Rocker Arm** designs using two different materials in each, to help you determine the optimum design.

From the Mass Properties report copy the full report using the **Copy to Clipboard** button and paste the information to a word processor. Personalize the report and add a title including the part's Name and Material, your Name, Seat number or assigned class ID. As an example:

"**Mass Properties of Rocker Arm 1.**
Material: Cast Carbon Steel.
Student Name, ID."

After you have personalized the report, save it as **Rocker Arm 1-Cast Carbon Steel.**

Return to SOLIDWORKS, close the Mass Properties report, and change the part's material to **1060 Alloy** from the **Aluminum** material section.

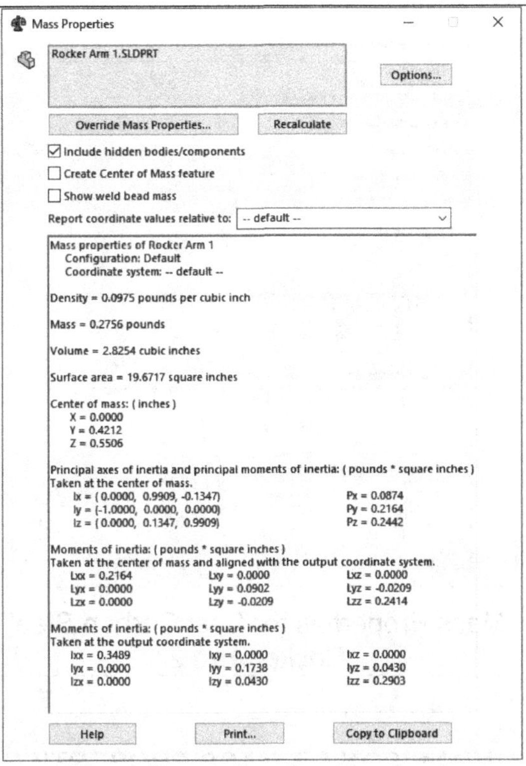

Mass Properties for Aluminum Rocker Arm 1

Once you have assigned the Aluminum material properties, select the Evaluate tab again, and re-calculate the part's **Mass Properties**. The report for the **Aluminum 1060 Alloy** is displayed.

As you previously did for the **Cast Carbon Steel**, copy the **Aluminum** mass properties to a word processor, personalize it and save it as **Rocker Arm 1-Aluminum**.

Now you need to repeat the same process with the second design. Open the part **Rocker Arm 2.** Set the material to **Cast Carbon Steel,** calculate the mass properties, copy the properties to your word processor, personalize your report and save it as **Rocker Arm 2-Cast Carbon Steel**.

Finally, change the **Rocker Arm 2** material to Aluminum, **1060 Alloy**, calculate the mass properties and prepare the corresponding report. Save the last report as **Rocker Arm 2 - Aluminum** to finish. The mass properties for the **Rocker Arm 2** part with the **Cast Carbon Steel** and **Aluminum** are shown next.

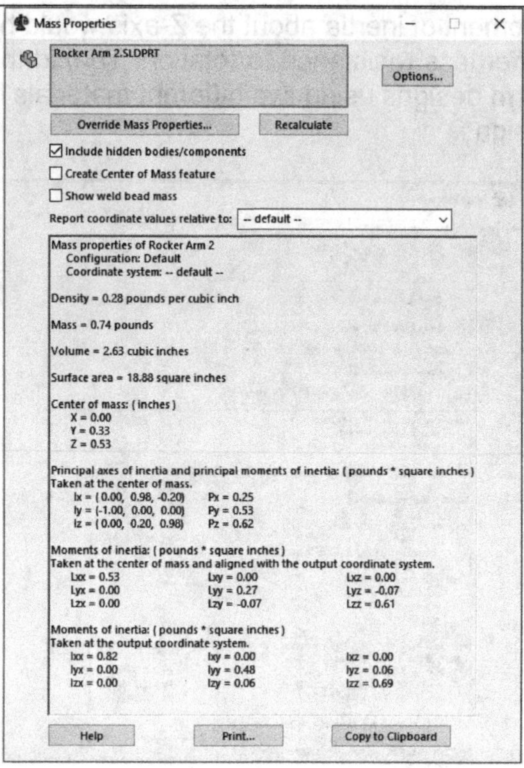

Mass Properties for Cast Carbon Steel Rocker Arm 2

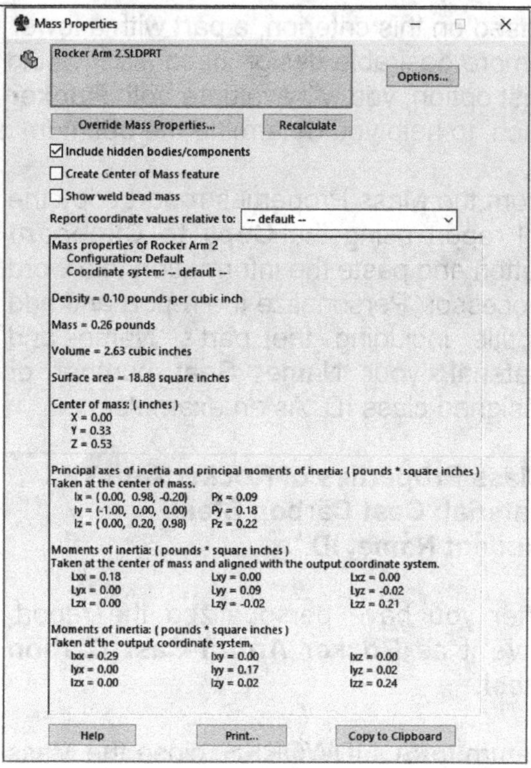

Mass Properties for Cast Carbon Steel Rocker Arm 2

COMPARISON OF MASS PROPERTIES FOR ROCKER ARMS 1 and 2

After you have completed the mass properties analysis for both Rocker Arms 1 and 2 with different materials, you can evaluate your results and determine which is the best design and material.

TASK: Identify the report with the best combination of Geometry and Material that is easiest to rotate (requires the least torque) about the central Z-axis and turn it in to your lab instructor.

CREATE A DRAWING WITH BOTH ROCKER ARMS

Now you will make a new drawing including both **Rocker Arm** parts. One way is to make an assembly with both parts, and then make an assembly drawing as in the previous lesson. To learn additional functionality, the option you will use in this lab will cover adding an individual drawing view of each **Rocker Arm** to the drawing sheet.

Start by creating a new drawing using the **ANSI-Inches** template and click in the **Model View** command from the Drawing tab, or the menu **Insert, Drawing View, Model**.

- If the **Rocker Arm** parts were open in SOLIDWORKS at the time of making the new drawing, they will be listed under the **Open Documents:** section. Select the **Rocker Arm 1** part and click on the blue **NEXT** arrow at the top right of the command's options.
- If the parts were not open, click on the **Browse** button, locate the **Rocker Arm 1** part, click **Open** and then **NEXT** as in the previous option.

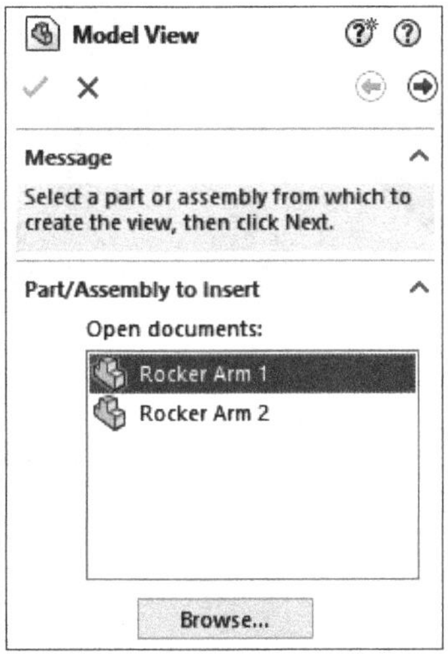

On the next page you can select the desired view orientation and options. For this exercise select the **Dimetric** view orientation, change the **Display Style** to **Shaded with Edges,** change the Scale to **3:2**, and if desired, turn On the **Preview** option. Click on the left side of the drawing sheet to locate the **Rocker Arm 1** view.

Repeat the previous steps to add the **Rocker Arm 2** drawing view on the right side of the drawing sheet, shown in the next page.

From the Annotations tab select the **Note** command and add a note above each view with the part's name and scale. For the part's name use a **Bold Font**, **20pt** high. For the scale's note use a **Regular Font**, **15pt** high.

Arrange the views and notes in your drawing sheet, save the finished drawing as **Rocker Arms** and print a copy to submit to your lab instructor, along with the four mass properties reports.

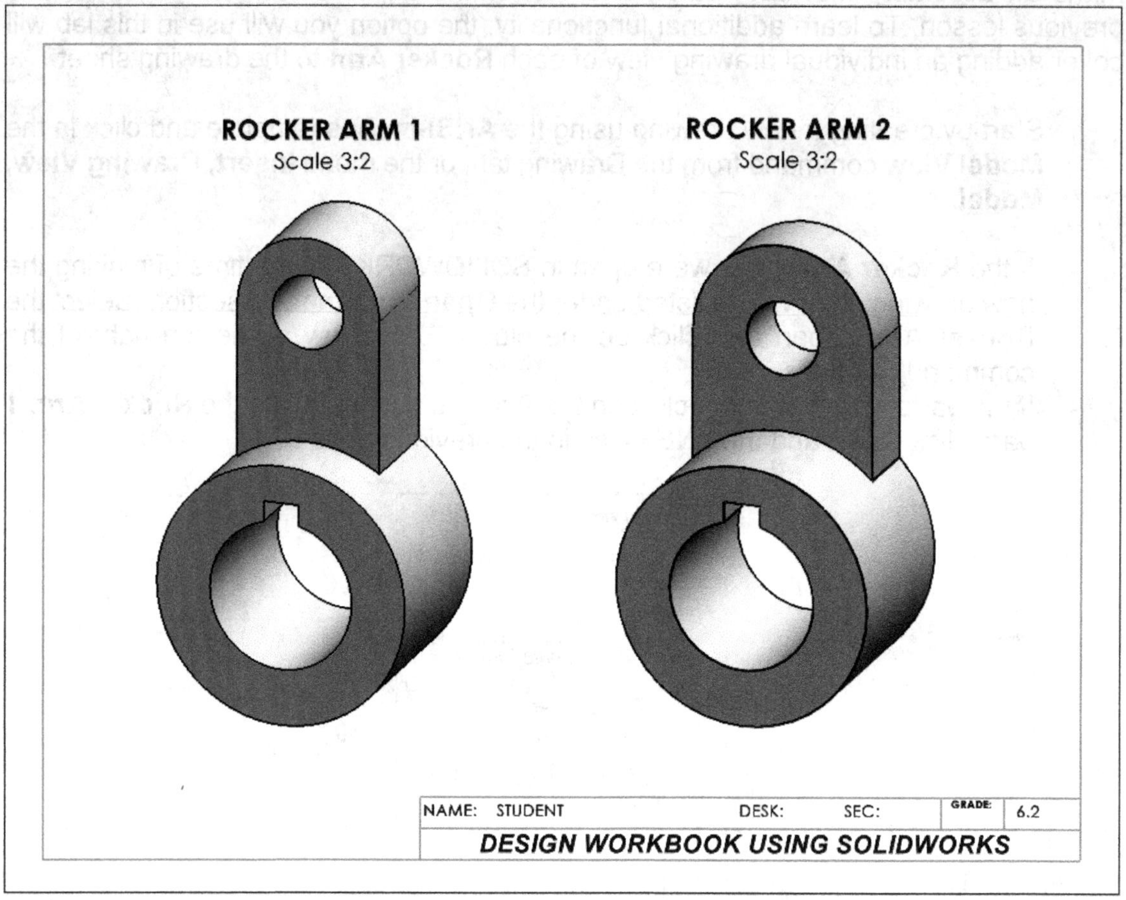

INFORMATION PROVIDED IN

SOLIDWORKS MASS PROPERTIES REPORTS

1. **DENSITY** is the mass or the weight per unit volume of the material the part is made from.

2. **MASS**: The mass of a body is the measure of its property to resist change in its steady motion. The mass depends on the volume of the body and the density of the material of which the body is made. In this case with SOLIDWORKS, mass is equivalent to weight.

3. **VOLUME**: The volume of a body is the total volume of space enclosed by its boundary surfaces.

4. **SURFACE AREA**: The surface area is the total area of the boundary surfaces defining the solid model.

5. **CENTER OF MASS**: Center of Mass (or Centroid) of a volume is the origin of coordinate axes for which first moments of the volume are zero. It is considered the center of a volume. For a homogeneous body in a parallel gravity field, mass center and center of gravity coincide with the centroid.

6. **PRINCIPAL AXES OF INERTIA AND PRINCIPAL MOMENTS OF INERTIA**: Principal moments of inertia are extreme (maximal, minimal) moments of inertia for a body. They are associated with principal axes of inertia which have their origin at the centroid and the direction of each usually given by the three unit-vector components. For these axes, the products of inertia are zero.

7. **MOMENTS OF INERTIA**: A moment of inertia is the second moment of mass of a body relative to an axis, usually X, Y, or Z. It is a measure of the body's property to resist change in its steady rotation about that axis. It depends on the body's mass and its distribution around the axis of interest.

Exercise 6.2: Socket Design Table

In Exercise 6.2, you will make a family of parts for a Socket, using a Design Table. A Design Table allows you to create multiple configurations (or variations) of similar parts or assemblies by specifying parameters in an embedded Microsoft Excel worksheet. These parameters include model dimensions and features that will vary from one configuration to the next. Once a design table is added, you can select the **ConfigurationManager** tab to review the different configurations, or variations. These configurations can be used in assemblies and drawings.

Make a new part using the **ANSI-INCHES** template and save it as **Socket.sldprt**. Add a new sketch in the **Top Plane** and draw the profile indicated. When you add dimensions, SOLIDWORKS automatically assigns them an internal name with the format:

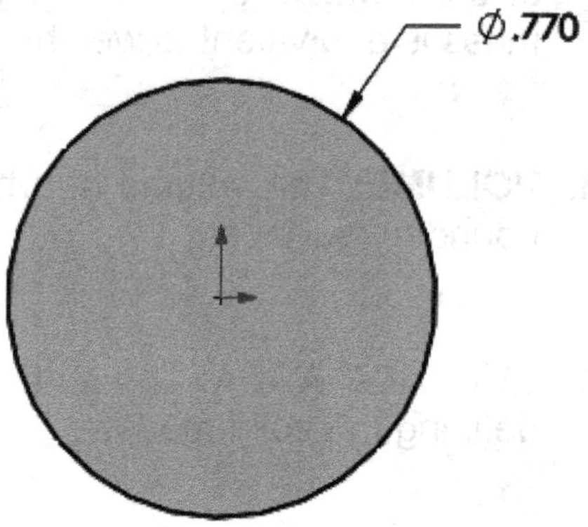

DimensionName @ FeatureName

The ***DimensionName*** is the name of a dimension; the automatically assigned names have the format D1, D2, etc. for each sketch or feature. ***FeatureName*** is the name of the Feature or Sketch the dimension belongs to. For example, the **0.770"** dimension was the first dimension added to *Sketch1*; therefore, it will be named "*D1@Sketch1*."

A Design Table uses SOLIDWORKS' dimension names to assign a different dimension value for each configuration.

When you select a dimension, its name and other properties are displayed in the PropertyManager. Select the **0.770"** dimension to display its properties. In the "Primary Value" section the dimension's name is listed as **D1@Sketch1** and shows the value of that dimension (**0.770 in**).

To make it easier to identify dimensions used in a Design Table, you must rename them. With the **0.770 in** dimension selected, in the name text box type:

"**Socket_Diameter@Sketch1**"

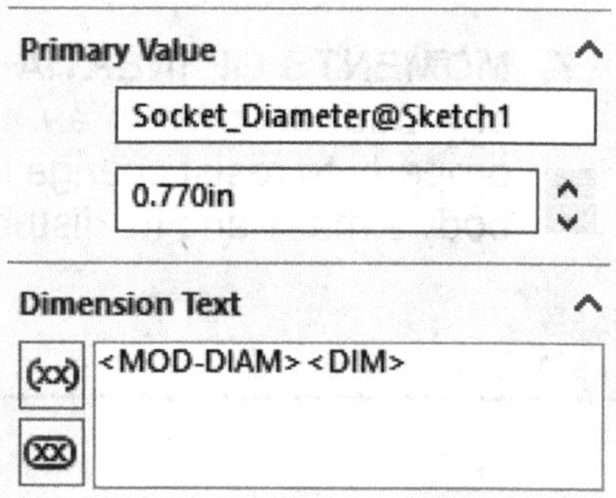

NOTE: You can enter only the part of the name before the "**@**" symbol; SOLIDWORKS will automatically fill in the rest of the name.

After renaming the dimension, extrude the sketch up **0.960”** to create the base feature.

When the first extrusion is created, its depth is also assigned a dimension. To make the model dimensions visible right-click in the **Annotations** folder of the FeatureManager and turn on the **Show Feature Dimensions** option, making all model dimensions visible until the option is turned off.

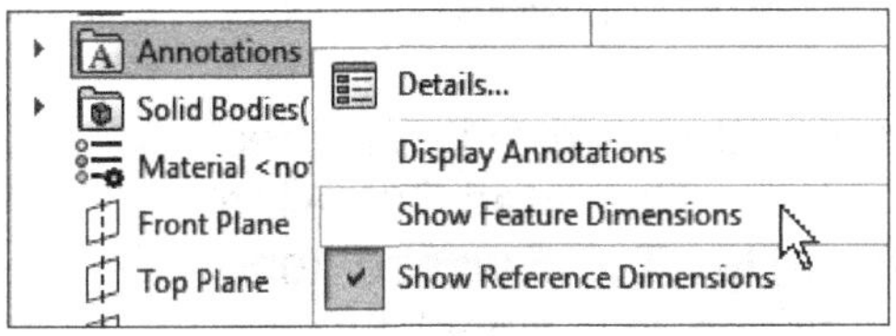

Select the **0.960”** dimension and change its name to

“Socket_Height@Boss-Extrude1”

in the dimension’s properties. Remember that *“Boss-Extrude1”* is the name of the first feature.

Right click in the **Annotations** folder and turn off the **Show Feature Dimensions** option.

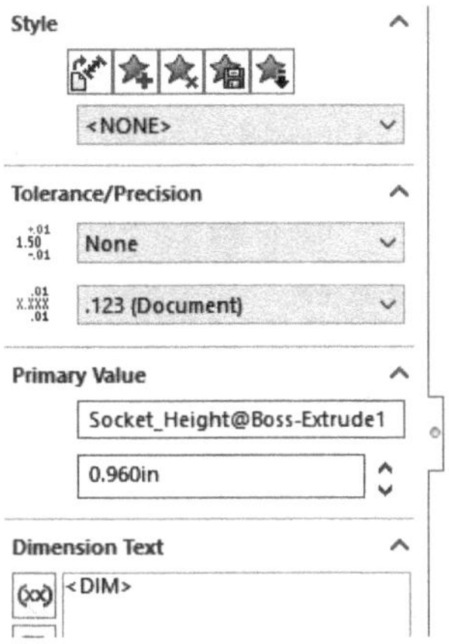

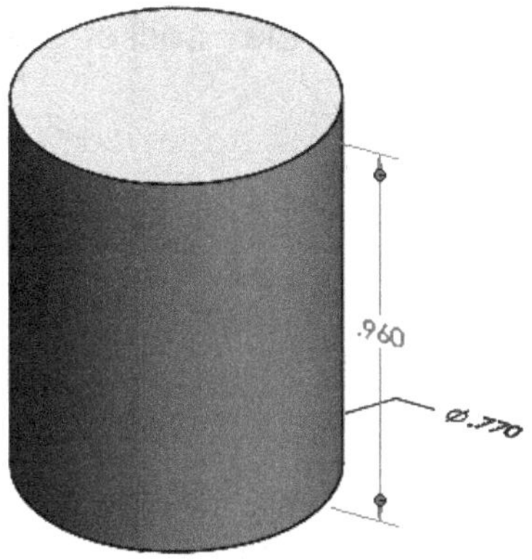

Select the **Fillet** command and add a **0.03”** fillet to the top edge of the first feature.

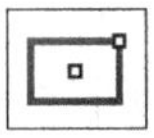

Now you will add the square hole used to attach the Socket to the ratchet. Change to a **Bottom** view orientation, add a new sketch in the bottom face and draw the profile shown using the **Center Rectangle** option.

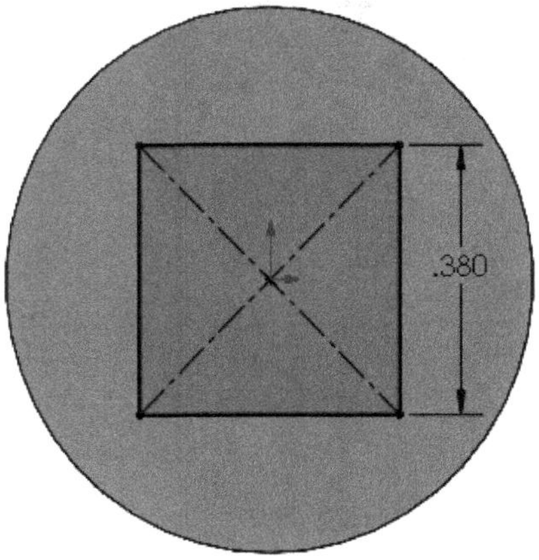

Add an **Equal** relation between a vertical and a horizontal line to fully define the sketch with a single dimension. When finished make an **Extruded Cut** of **0.450”** deep.

Change to a **Top** view, select the top face, and add a new sketch. Draw a **Hexagon** centered in the origin using the **Inscribed Circle** option and dimension the diameter **0.566”**. Select a hexagon line and add a **Horizontal** relation to fully define your sketch. Just as you did with the other dimensions, rename the diameter dimension as

Socket_Size@Sketch3.

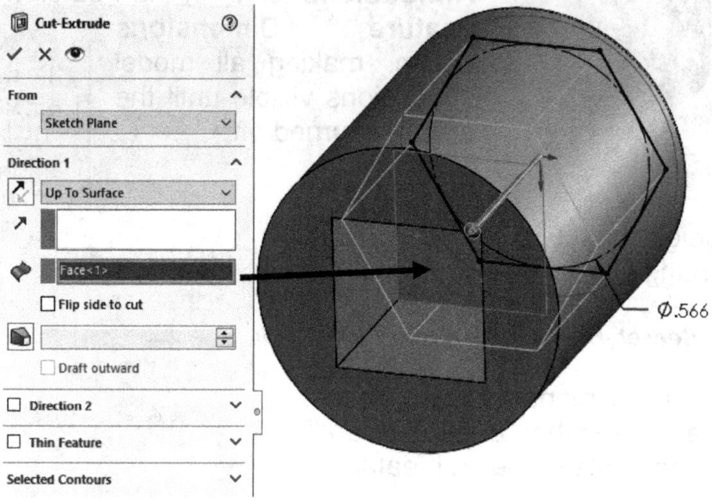

With the sketch finished make an **Extruded Cut** using the **Up To Surface** end condition and select the top face of the square hole on the bottom of the socket as shown.

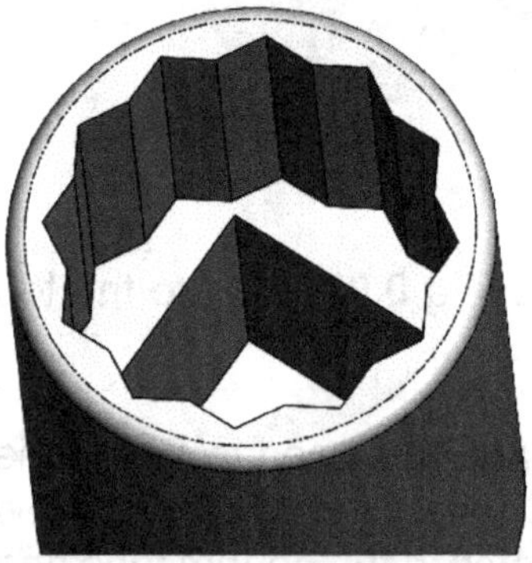

To complete the socket opening, select the **Circular Pattern** command from the Features tab, select a circular edge of the first extrusion as the pattern direction, and add the hexagonal cut in the **Features to Pattern** selection box.

Use the **Instance Spacing** option, enter **30°** and **2** copies. You will have to activate the **Geometry Pattern** option at the bottom. This option will make an exact geometric copy of the hexagonal copy, without calculating the **Up to Surface** end condition of the original hexagonal cut.

The last step in the construction of the socket is to add a **Chamfer** to all the short edges of the polygon on the top face of the Socket. You can window-select the edges to make selection easier.

Set the value of the **Chamfer** to **0.023”**. The finished socket is shown here. Save your model as **SOCKET.sldprt**.

With your finished part you are now ready to add a **Design Table** to create additional variations of the socket, called **Configurations**. Select the menu **Insert, Tables, Excel Design Table,** use the **Auto-Create** option, and click **OK** to continue. In the **Dimensions** window list, you will see all the dimensions available in the model.
This is why renaming the dimensions is a good idea.

Hold down the **Ctrl** key, select the following dimensions to add them to the **Design Table** and click **OK** to continue:

- **Socket_Diameter@Sketch1**
- **Socket_Height@Boss-Extrude1**
- **Socket_Size@Sketch3**

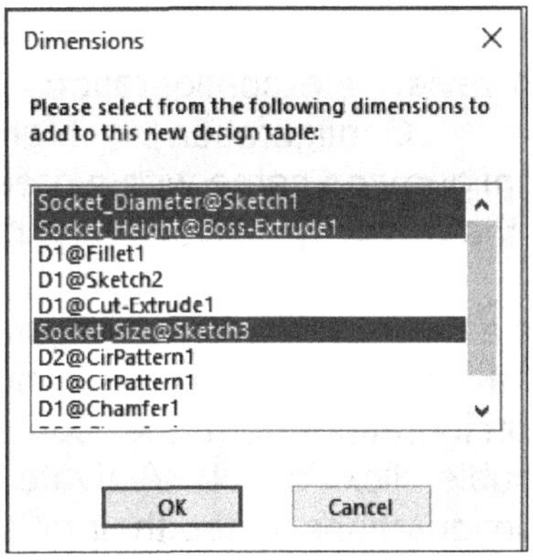

NOTE: If you attempted to add a sketch or create a feature multiple times, the **Sketch** or **Feature** name counters _may_ have a different number from the list above. If this is the case, there should be no problem; the dimension names will match your **Dimensions** list.

Immediately an Excel spreadsheet is added to the SOLIDWORKS part. The pre-selected dimensions will be added to the **Design Table** on Row 2, starting at Column B.
To add more configurations to the design table, you must enter the configuration names on Column A, starting at Row 3 going down, and fill the table with the corresponding dimension values for each configuration. Fill the Design Table as indicated.

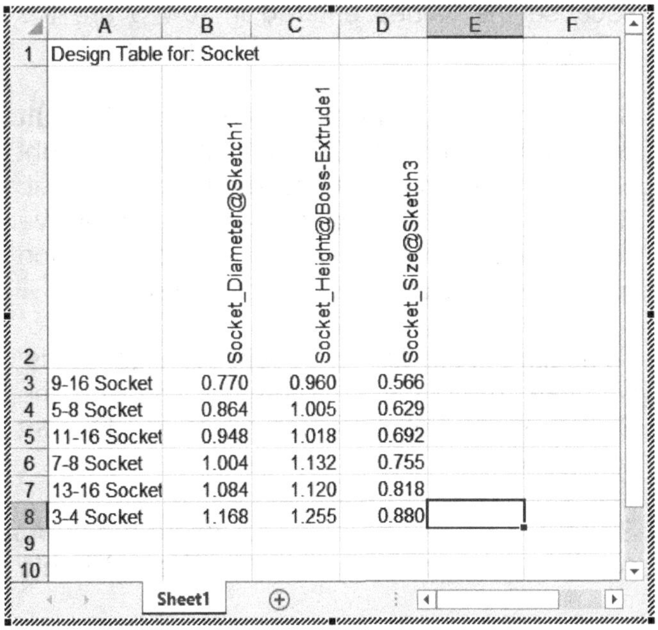

	A	B	C	D	E	F
1	Design Table for: Socket					
2		Socket_Diameter@Sketch1	Socket_Height@Boss-Extrude1	Socket_Size@Sketch3		
3	9-16 Socket	0.770	0.960	0.566		
4	5-8 Socket	0.864	1.005	0.629		
5	11-16 Socket	0.948	1.018	0.692		
6	7-8 Socket	1.004	1.132	0.755		
7	13-16 Socket	1.084	1.120	0.818		
8	3-4 Socket	1.168	1.255	0.880		
9						
10						

NOTE: Do not click outside of the spreadsheet until you are finished entering information, otherwise, the Design Table is considered complete and you will exit Excel and return to SOLIDWORKS.

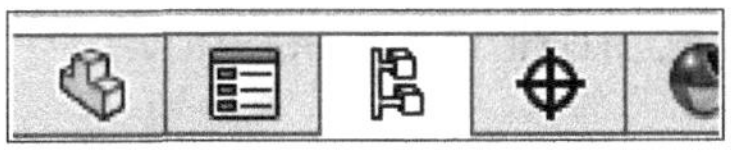

NOTE: If you need to modify your design table after completing it (or if you accidentally exited before finishing it), select the **Configuration Manager** tab at the top of the FeatureManager, expand the Tables folder, right click on **Excel Design Table** and select **Edit Table**.

When you finish entering the table information, click on the **SOLIDWORKS** graphics area to complete the Design Table and create the different configurations. The new part configurations are generated and listed as shown.

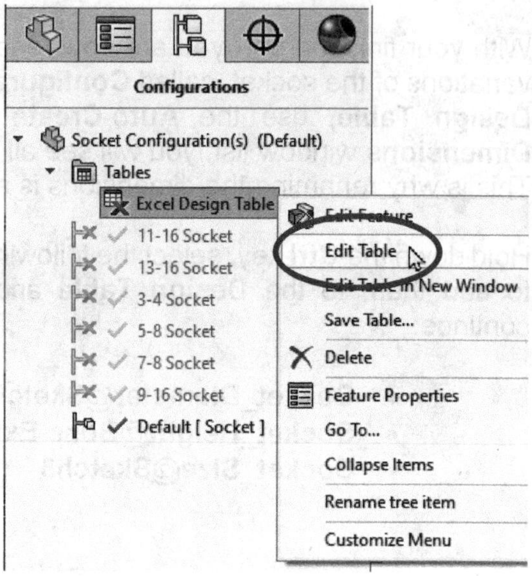

To review the configurations created select the Configuration Manager. The configuration name with a green check mark is the currently active configuration.

To activate a different configuration right click in a configuration's name and select Show Configuration from the context toolbar or double-click on it. Activate the different configurations to see their differences.

Notice the configurations driven by the **Excel** table are preceded by an **Excel** icon, and the **Default** configuration is not. Also, the **Default** configuration is the same as "**9-16 Socket**" because the values entered in row 3 are the same as your original model dimensions.

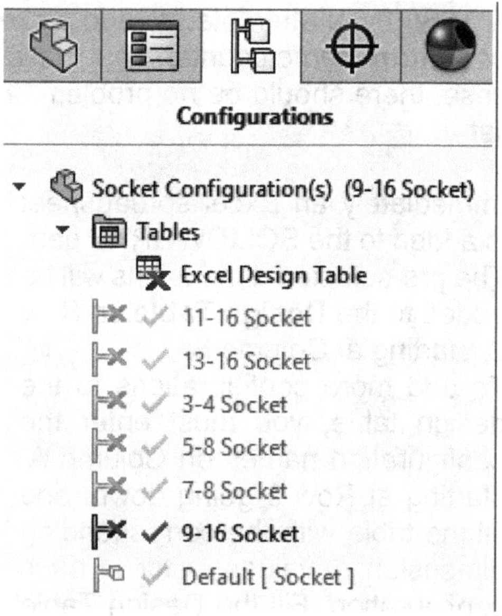

Since the "**9-16 Socket**" and the **Default** configuration have the same dimensions, right click on the **Default** configuration, and delete it. Be aware that you cannot delete the active configuration. Activate a different configuration and then delete the **Default** configuration.

To document the different Socket configurations, you can make a new assembly and add the Socket part multiple times using a different configuration for each instance.

Select the **New** command and create a new **Assembly**. When the **Begin Assembly** command is displayed, click in the **Keep Visible** pin next to the **Cancel** button.

Select the **Socket** part from the **Open Documents**: list, select a configuration from the **Configuration:** drop-down list and click in the assembly to add it. Select a different configuration, add it to the assembly and continue adding all configurations approximately as shown.

Finish the **Begin Assembly** command and save the assembly as **Socket Configurations.sldasm**.

NOTE: In an assembly, if you select a part with configurations, the context toolbar will include a drop-down list from which you can select a different configuration.

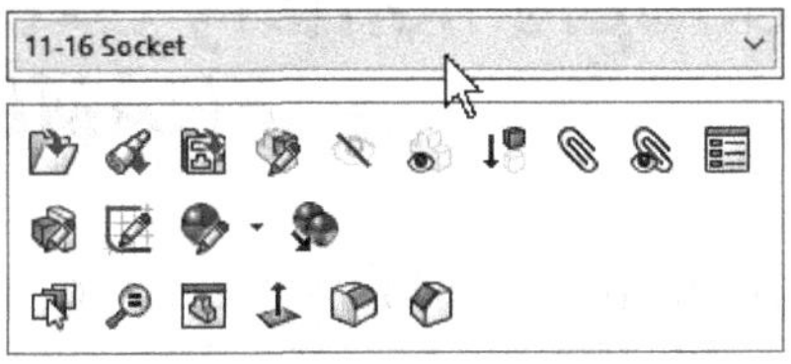

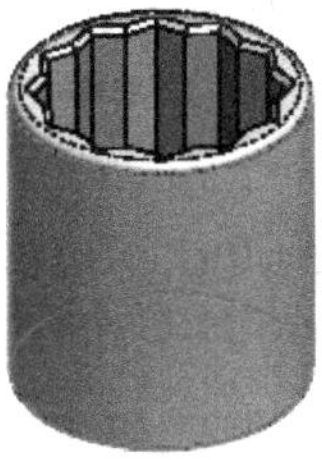

Make a new drawing using the **Inches** template and add a shaded **Trimetric** view. Add a note next to each configuration indicating its name as shown. Save the drawing as **Socket Configurations.slddrw** and print a copy for your lab instructor to finish.

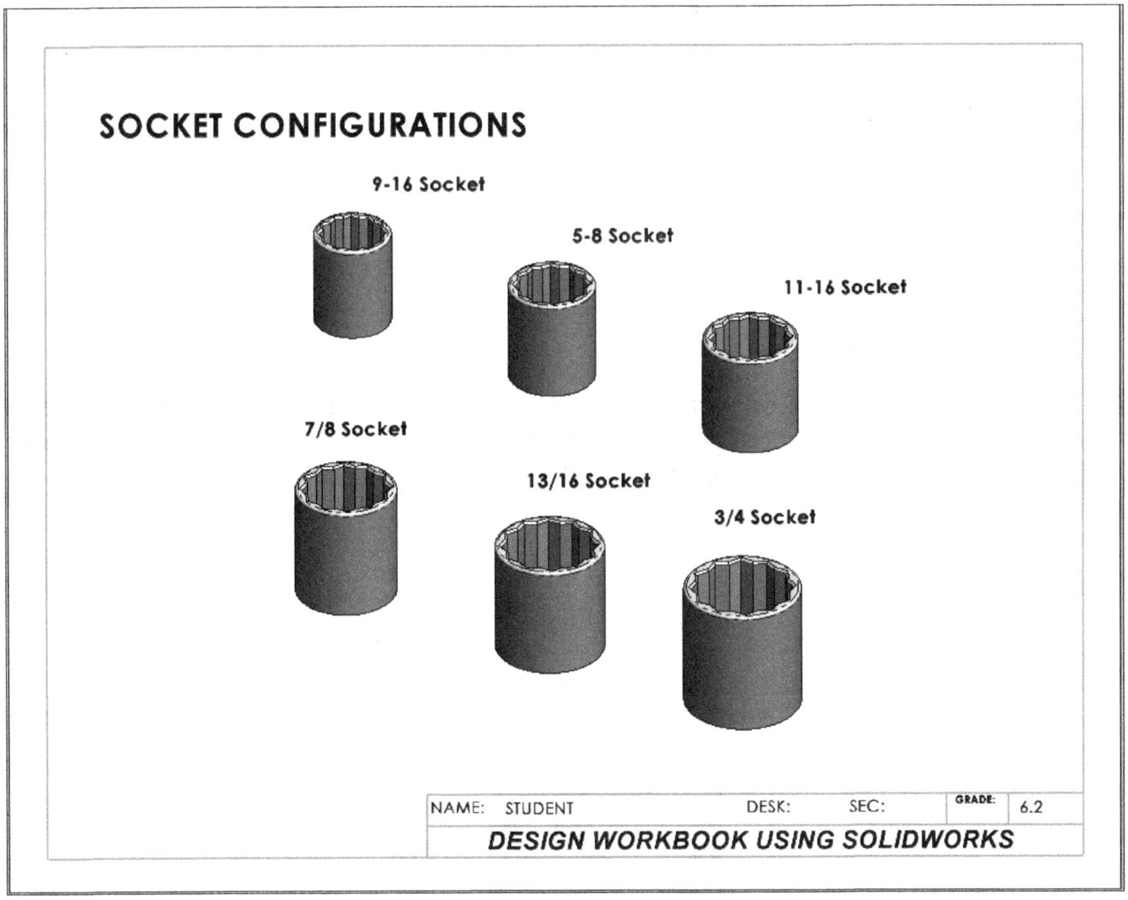

Supplementary Exercise 6-1: CASTER AND CASTER FRAME CONFIGURATIONS

Create new parts with Design Tables to generate the different configurations needed for the Caster Frames and Caster Wheels specified in the Table provided. See your instructor for specific instructions.

1. Generate an assembly drawing of the six **Caster Frames** generated with the Design Table. Use the previous Socket's drawing as an example.
2. Generate an assembly drawing of the six **Caster Wheels** generated with the Design Table. Use the previous Socket's drawing as an example.
3. In the assembly you can create Assembly Configurations. In this case, manually add the different configurations, and select the component configuration to use for each part in the different configurations. Create a configuration for each size.

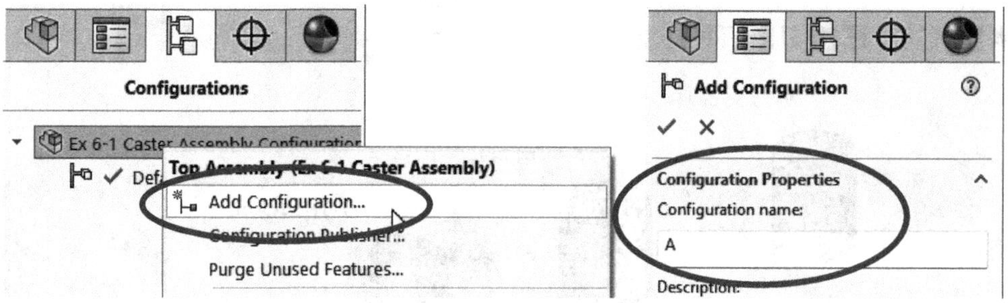

Activate the ***A*** Assembly Configuration, select the *Frame* in the screen, from the pop-up menu pick the desired configuration and repeat for the *Wheel*. Change to the next Assembly Configuration and repeat to change the component's configuration as needed.

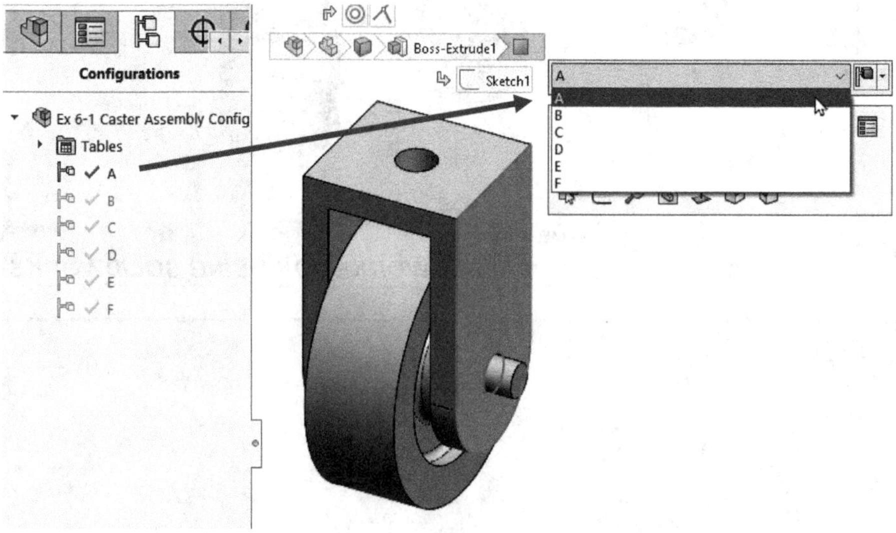

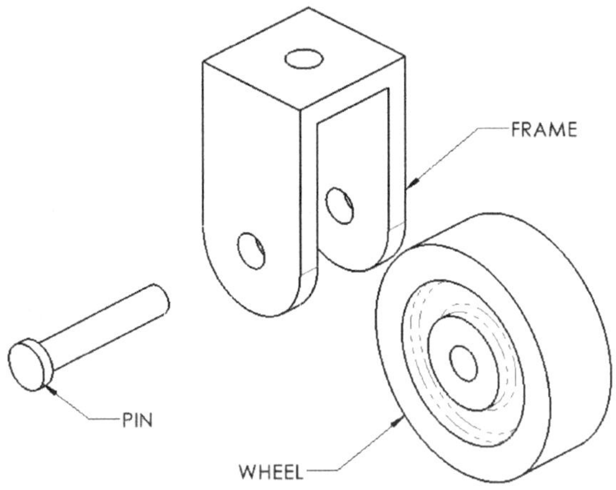

	CASTER DIMENSIONS	
CONFIGURATION	**WHEEL DIA. (D)**	**SADDLE HEIGHT (H)**
A	2.00	1.50
B	2.20	1.60
C	2.40	1.70
D	2.60	1.80
E	2.80	1.90
F	3.00	2.00

ALL DIMENSIONS ARE IN INCHES

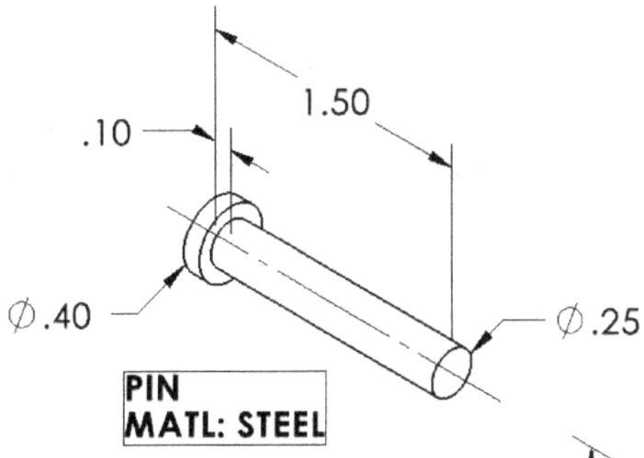

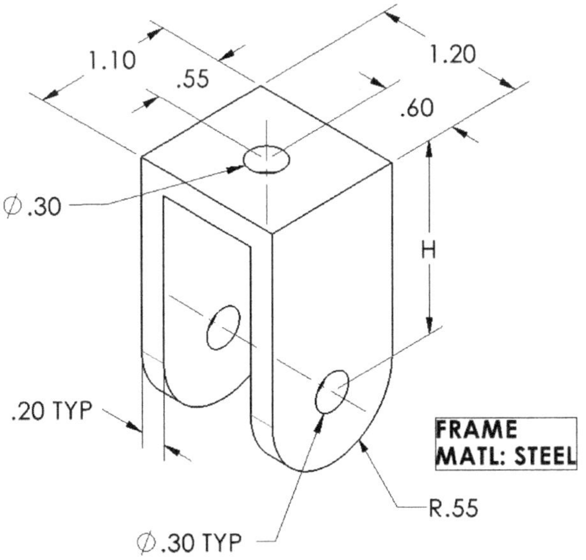

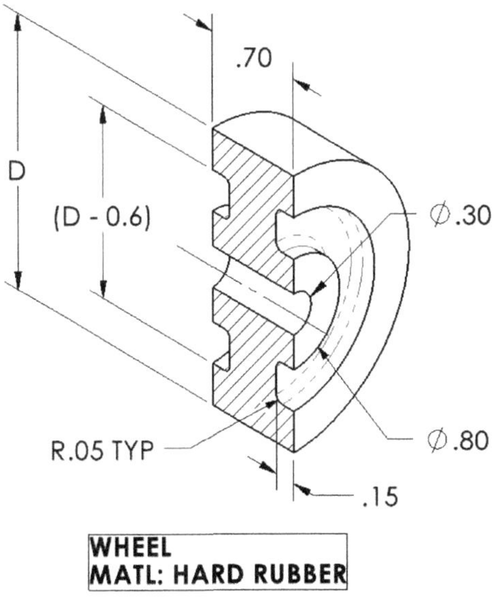

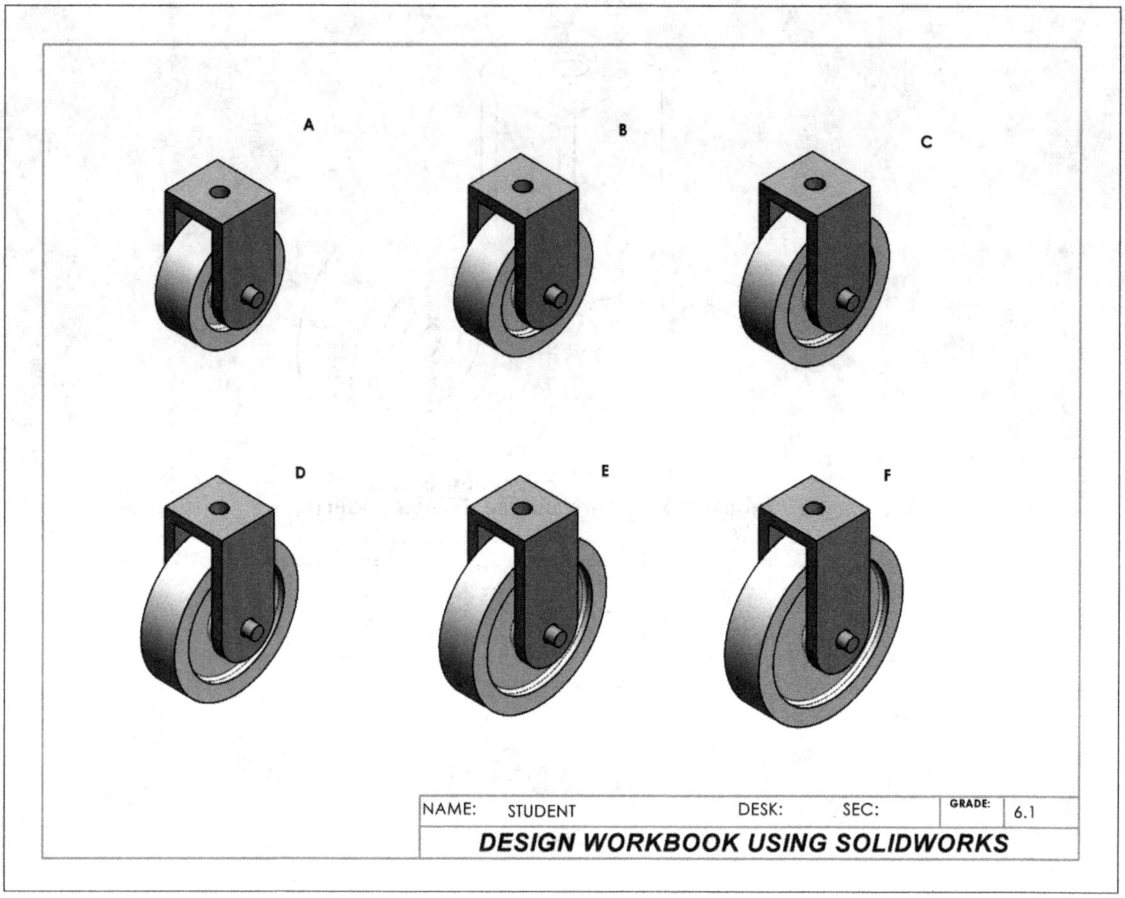
A
B
C
D
E
F
NAME: STUDENT
DESK:
SEC:
GRADE: 6.1
DESIGN WORKBOOK USING SOLIDWORKS

Supplementary Exercise 6-2: PULLEY AND PULLEY BRACKET CONFIGURATIONS

Create new parts with Design Tables to generate the configurations needed for the Caster Frames and Caster Wheels specified in the Table provided. See your instructor for specific instructions.

1. Generate an assembly drawing of the six **Pulley Brackets** generated with the Design Table. Use the previous Socket's drawing as an example.
2. Generate an assembly drawing of the six **Pulleys** generated with the Design Table. Use the previous Socket's drawing as an example.

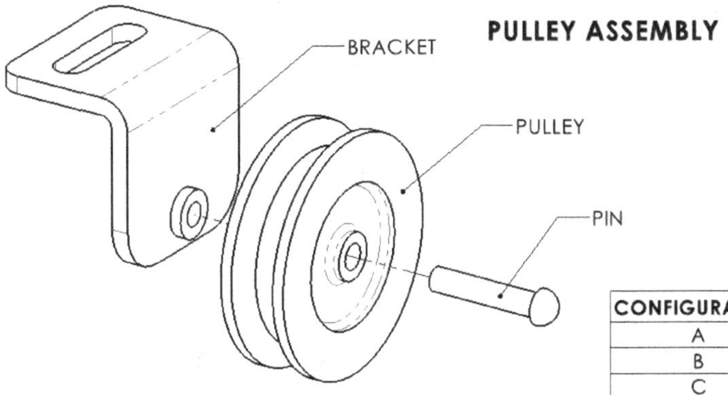

	PULLEY DIMENSIONS	
CONFIGURATION	PULLEY DIA. (D)	BRACKET HEIGHT (H)
A	2.00	1.50
B	2.20	1.60
C	2.40	1.70
D	2.60	1.80
E	2.80	1.90
F	3.00	2.00

ALL DIMENSIONS ARE IN INCHES

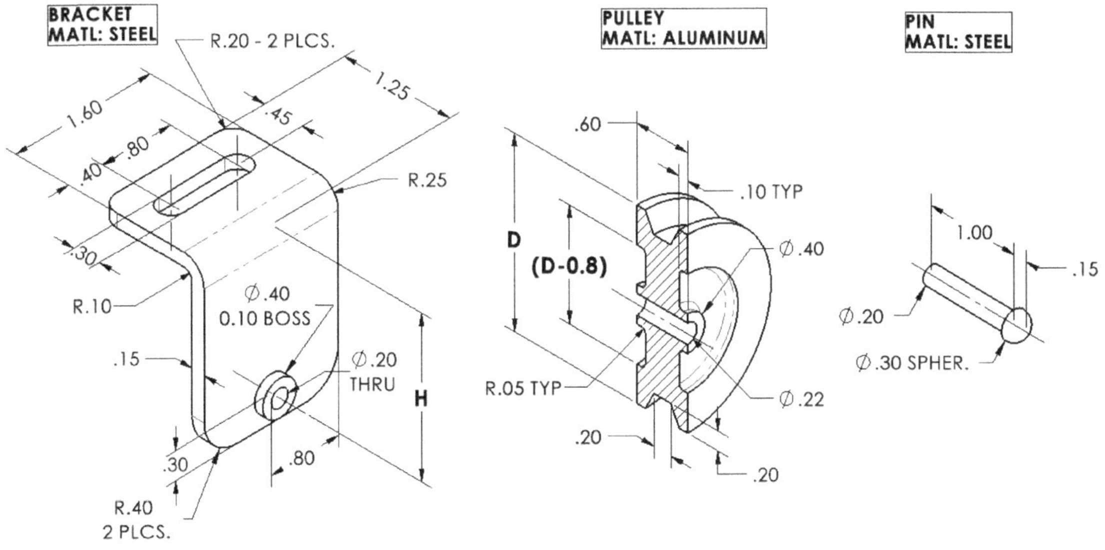

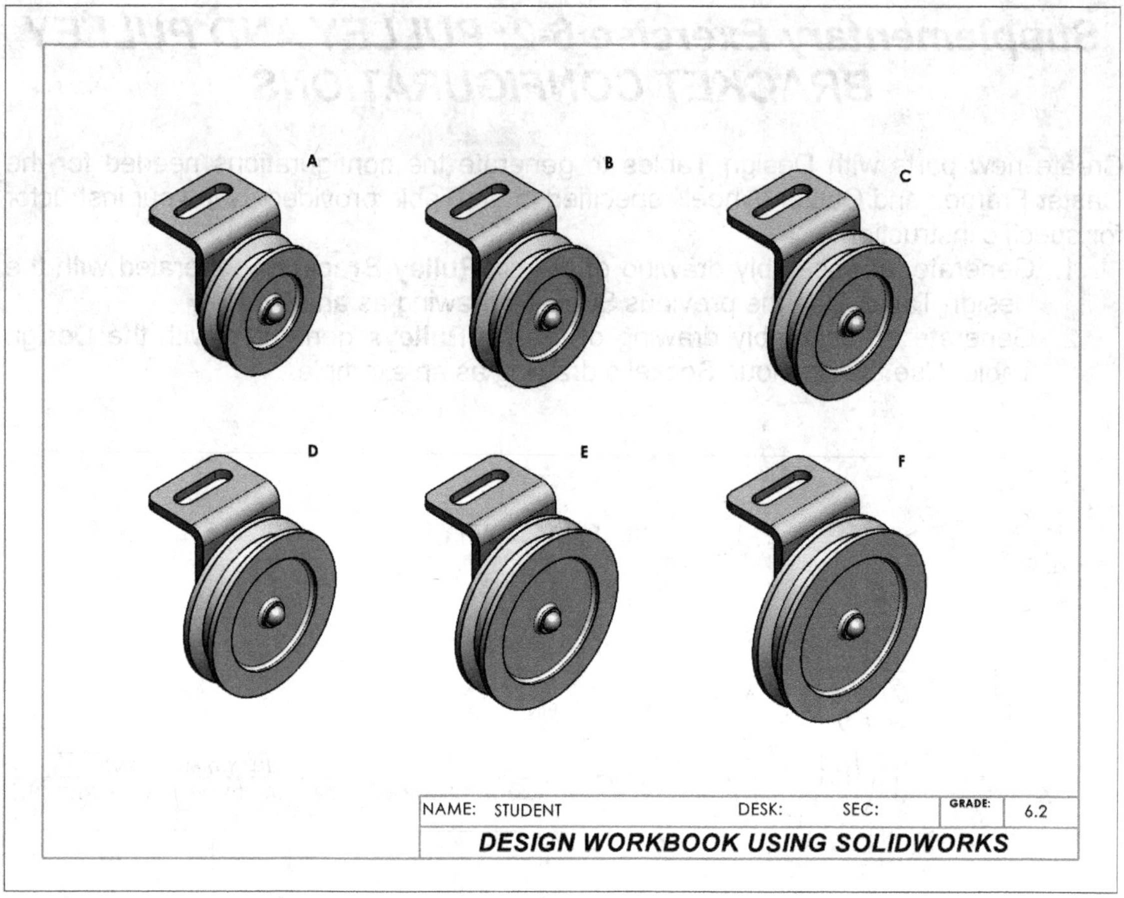
A
B
C
D
E
F
NAME: STUDENT
DESK:
SEC:
GRADE:
6.2
DESIGN WORKBOOK USING SOLIDWORKS

Design Workbook Lab 7
Static Stress Analysis

INTRODUCTION

SOLIDWORKS allows the designer to build and modify 3D solid models easily by editing their dimensions and constraints. In this lab exercise, you will build a concept for a pillow block. This model will then be analyzed using the **SOLIDWORKS Simulation** Finite Element Analysis (FEA) add-in software to find the stress distribution in the part under normal use conditions, supporting a rotating shaft.

Based on the stress analysis results, you will then make changes to the original design's geometry and/or material to improve the part's performance. After the changes are made, you will make a second stress analysis to evaluate if the changes improved the part's performance.

Finite Element Analysis using SOLIDWORKS

The purpose of performing a Finite Element Analysis early in the design process is to evaluate if your design will be capable of performing the task it was designed to do. A simulation will virtually test your design when exposed to the conditions anticipated under normal use, including external forces, pressure, temperature, vibrations, etc. Once you have the simulation results, you can make an informed decision about how to modify your design, re-run the analysis, and modify it again, until the results are satisfactory.

A huge advantage of FEA analysis is that you can significantly reduce prototype and physical testing costs.

The object of this exercise is to build a Pillow Block and run a simulation using the **SOLIDWORKS Simulation** under the anticipated loads expected.

Before you begin this exercise, make sure the simulation software is installed as an "Add-In" to SOLIDWORKS. To load the "Add-In" select the menu **Tools, Add-Ins**, and activate the **SOLIDWORKS Simulation** package.

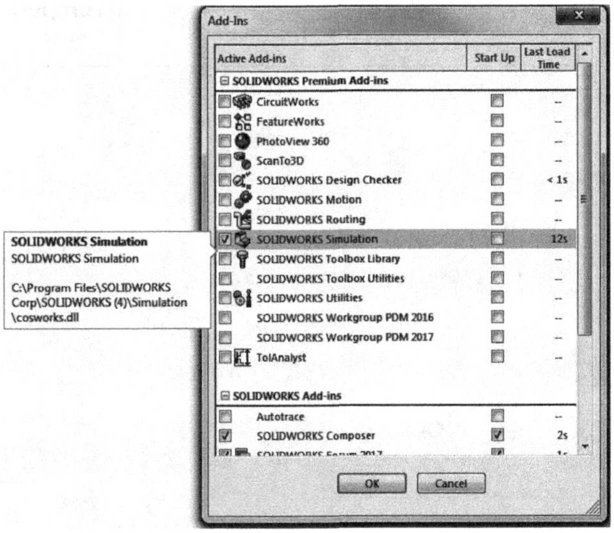

After the Add-In is loaded a new Simulation tab is added to the CommandManager, and a new toolbar is available in the menu **View, Toolbars**.

The Simulation toolbar shown includes the list of external loads you can add to your design, to approximate the actual conditions your design will be exposed to during normal operation. A description of each load command is shown.

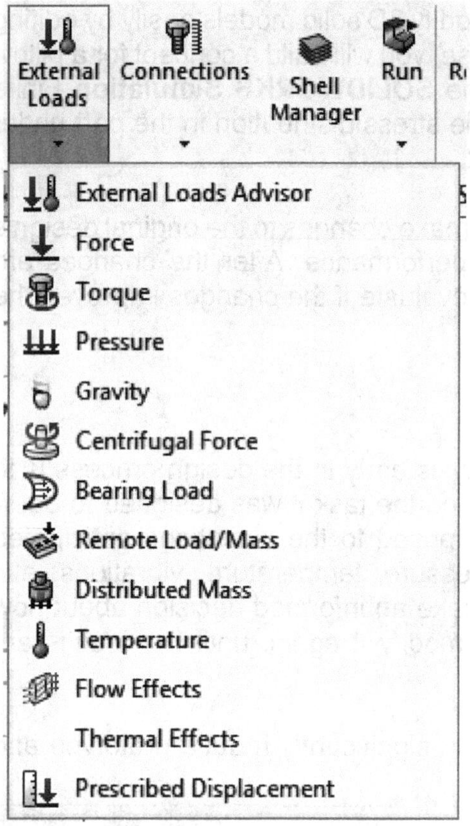

Restraints	Applies restraints to the selected entities for the current structural study to prevent the model from moving under the loads (static, frequency, or buckling).
Pressure	Applies a pressure to the selected faces for the current structural study (static, frequency, or buckling).
Force	Applies a force, torque, or moment to the selected entities in the current structural study (static, frequency, or buckling).
Gravity	Defines gravity loading for the current structural study (static, frequency, or buckling).
Centripetal Force	Applies centripetal forces for the current structural study (static, frequency, or buckling).
Remote Load	Applies a remote load, used for structural studies.
Rigid Connection	Simulates a rigid connection between selected faces in structural studies.
Bearing Load	Applies bearing loads on selected faces of different components.
Temperature	Applies temperatures on the selected entities for the current thermal or structural (static, frequency, or buckling) studies.

Exercise 7.1: Finite Element Analysis of Pillow Block

DESIGNING THE FIRST VERSION OF THE PILLOW BLOCK

The first step is to build the model of the Pillow Block. Start a new part using the **ANSI-INCHES** template, add a new sketch in the **Front Plane** and draw the profile shown.

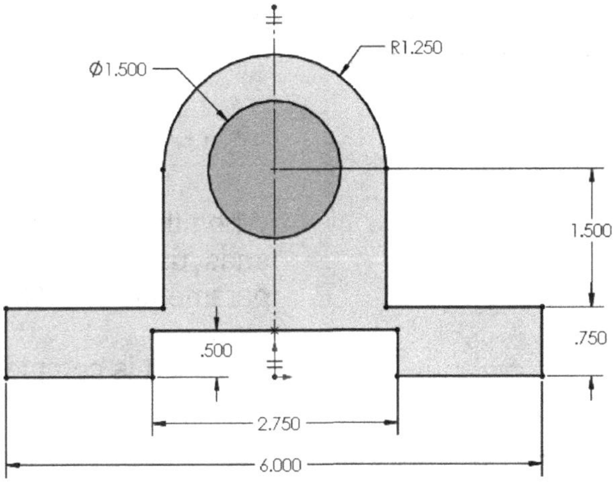

Make a **1.40"** extrusion using the **Mid Plane** end condition and click OK to continue.

Now you will make the mounting holes for the pillow block.

Change to a **Top** view, add a sketch on either flat face as shown and make an **Extruded Cut** using the **Through All** end condition.

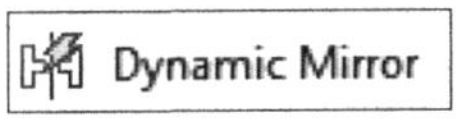

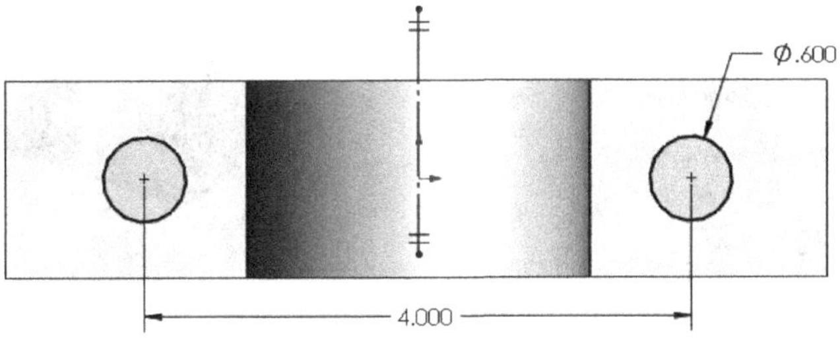

You can use the **Dynamic Mirror** about the vertical **Centerline** to make the sketch symmetric and make the center of either hole **Horizontal** with the origin to fully define your sketch.

When finished, your model should look like this. Save your part as **Pillow1.sldprt**.

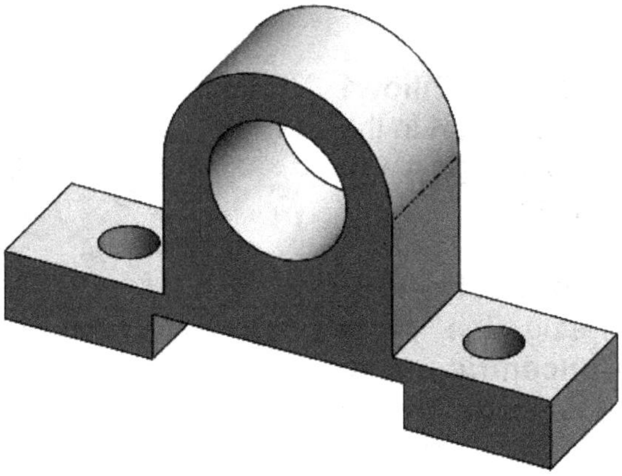

Now you will build a shaft that fits into the large hole of the Pillow Block. Make a new part using the **ANSI-INCHES** template.

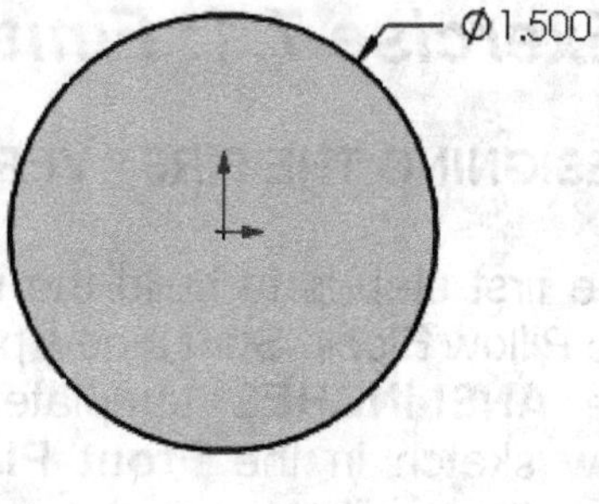

Add a new sketch in the **Front Plane**. Draw the profile shown and make a **7.00"** extrusion using the **Mid Plane** end condition.

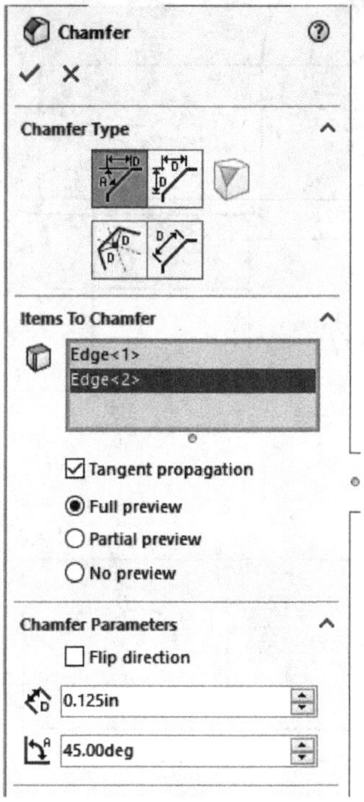

When the first extrusion is complete, add a **Chamfer** to both ends using the **Angle-Distance** option. Set the Chamfer parameters to **0.125" x 45°**.

The shaft is complete. Save your part as **Shaft.sldprt**.

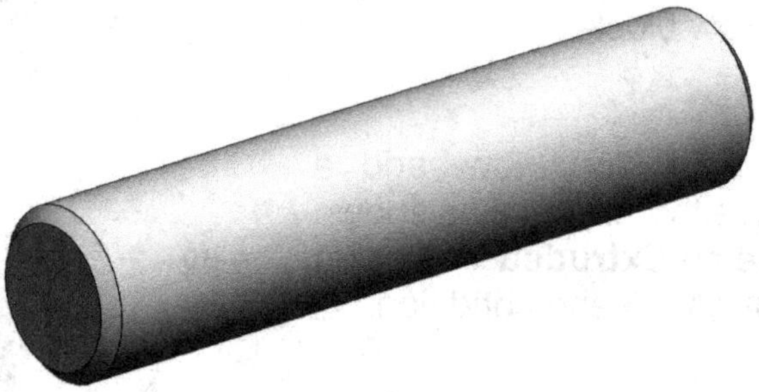

MAKE THE PILLOW BLOCK ASSEMBLY

Before you run the stress analysis, you need to make an assembly using the **Pillow1** and the **Shaft**.

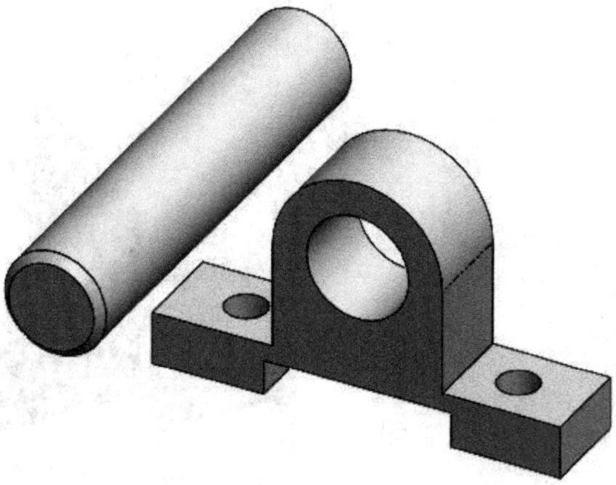

Start a new assembly and add the **Pillow1** part first. Remember to just click **OK** in the **Begin Assembly** command to align it with the assembly's default planes.

Next add the **Shaft** and locate it next to **Pillow1** as shown. Using the **Mate** command, add a **Concentric** mate between the shaft and the inside face of the **Pillow1**.

TIP: If you hold down the **Ctrl** key and select the two faces to mate, after releasing the **Ctrl** key you will see the option to add the **Concentric** mate in the context toolbar.

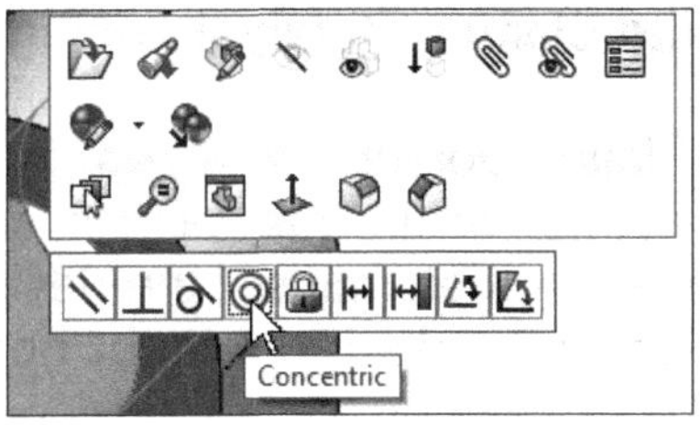

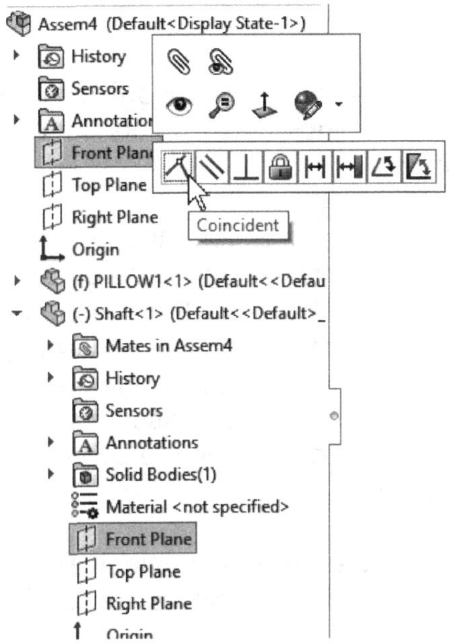

In the FeatureManager expand the **Shaft**, hold down the **Ctrl** key and select the **Front Plane** and the assembly's **Front Plane**.

Release the **Ctrl** key and click in the **Coincident** mate from the context toolbar. This way, the **Shaft** will be centered in the **Pillow1**.

After the parts are mated, save your assembly as **Pillow Assembly-1.sldasm**.

Finite Element Analysis using SOLIDWORKS SIMULATION

A Finite Element Analysis is a complex process that *approximates* a physical phenomenon, in this case a shaft rotating in a pillow block, using a mathematical process to determine if the design will meet your design specifications or not, in other words, **if the design fails or not**.

An analysis can take into consideration many different physical conditions like force, pressure, temperature, heat, vibration, etc. In your Pillow Block design, you will only make a static analysis with a vertical force on the shaft.

Word of advice: It should be noted that just because an analysis *reports* the design is safe, does not necessarily mean it will *work* as intended. An analysis only considers the external conditions entered, and nothing else. In other words, if the simulation parameters are incorrect or do not represent the actual physical conditions accurately, the results will likely be incorrect. This could be caused by an incorrect or missing force, pressure, temperature, vibration, material, speed, etc. This is the reason why an analysis needs to be performed by a qualified person who understands the conditions to be simulated and is trained to interpret the results obtained.

If the Simulation Tab does not appear in the CommandManager, select the menu **Tools**, **Add-ins,** select **SOLIDWORKS Simulation** and click **OK**. This will add a Simulation menu and tab.

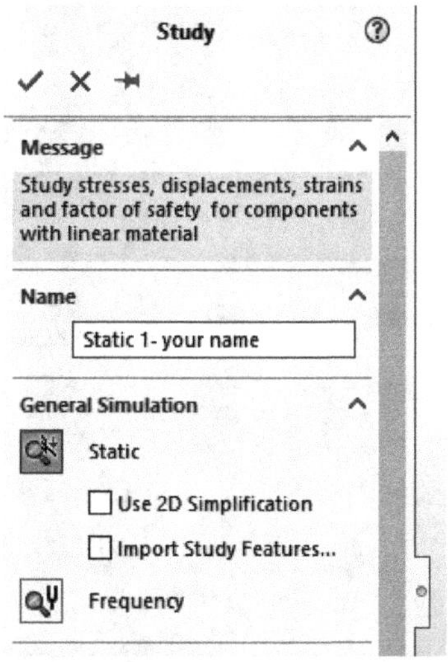

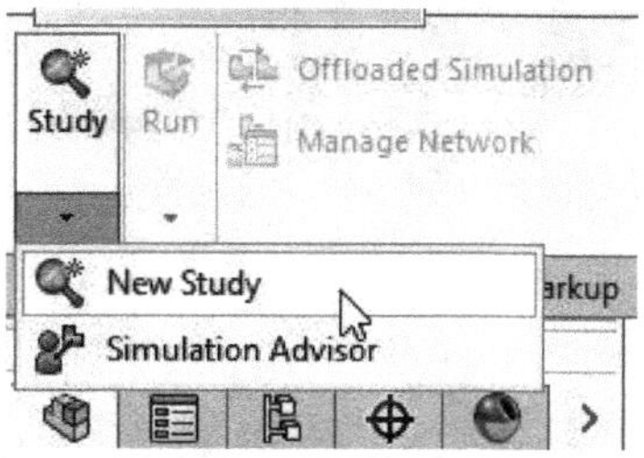

In order to make an analysis you need to define a new study. The minimum information required to run a static stress analysis is to define how the part is supported (**Fixtures**), what external forces and/or pressures are acting on the part/assembly (**External Forces**), what the component(s) are made of (**Material**), and in the case of an assembly, how the parts are connected to each other (**Connections**).

To start the analysis, select the **Simulation** tab, click on **New Study** and select the **New Study** command.

- In the Study name type "***Static 1- your-name***".
- In the type of study select "***Static***".
- And click **OK**.

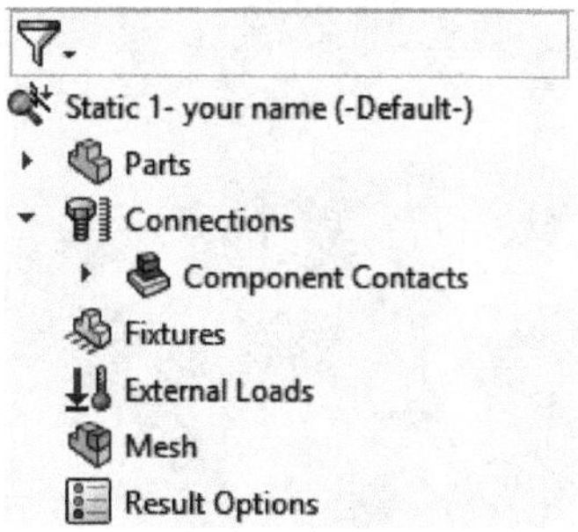

Now you can see the new study in the Study Manager Tree tab.

Assigning Material to the Parts

If you had assigned a material to a part in SOLIDWORKS, this information will be used in the analysis. Also, you can use the **SOLIDWORKS Simulation** materials library, which includes a wide variety of materials, including their physical properties needed for analysis.

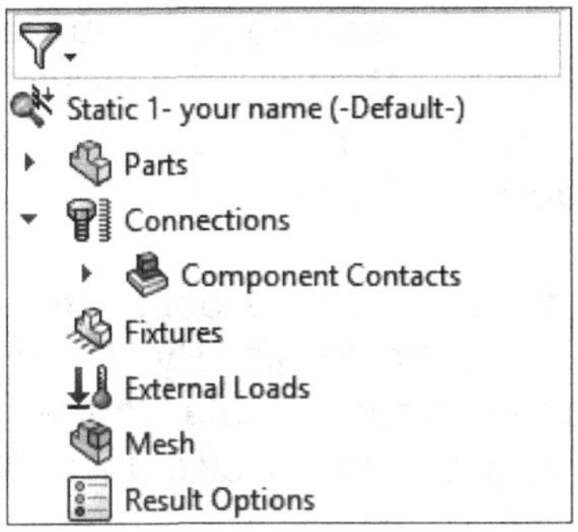

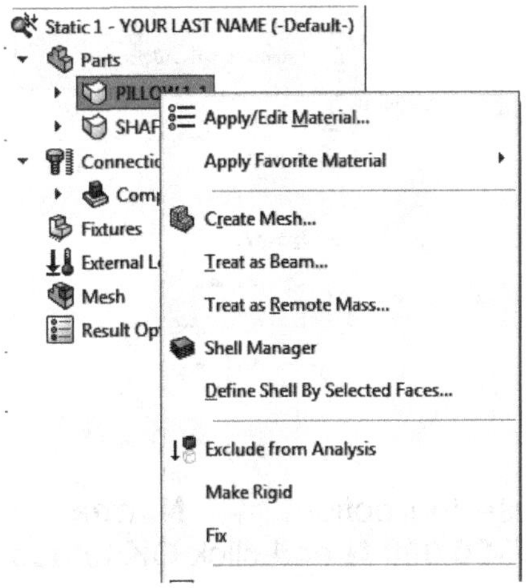

Since you have not defined a material for either part, expand the **Parts** section of the study, right click on **Pillow1** and select **Apply/Edit Material**. From the library select **Cast Alloy Steel** and click **OK**. Repeat the same process to define the **Shaft's** material as **Alloy Steel**.

A green check mark will be added to a part after the material has been defined.

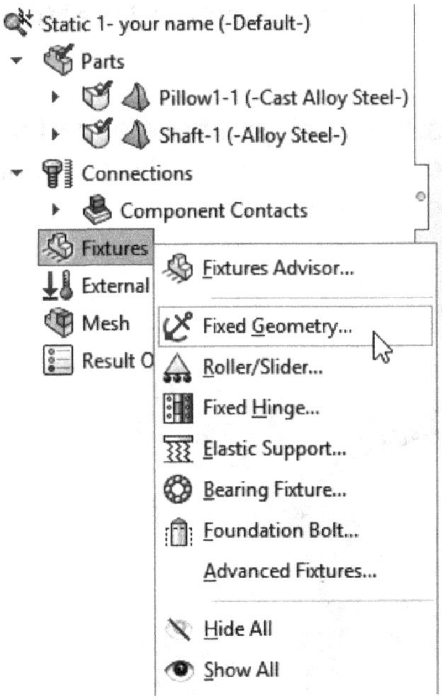

Applying Restraints

The restraints will define how the model is constrained. It is important to make sure the model cannot move, otherwise the **Static analysis** will return an error.

For this example, you will assume the bottom faces of the **Pillow Block** are fixed and cannot be moved. In the Study Manager right click in **Fixtures** and select **Fixed Geometry**. Select the two faces indicated and click **OK** to continue.

Applying the Force on the Shaft

In your **Static** analysis you will apply a force to the **Shaft** to determine the stress in the **Pillow Block**. In the Simulation Manager right click on **External Loads** and select the **Force** option.

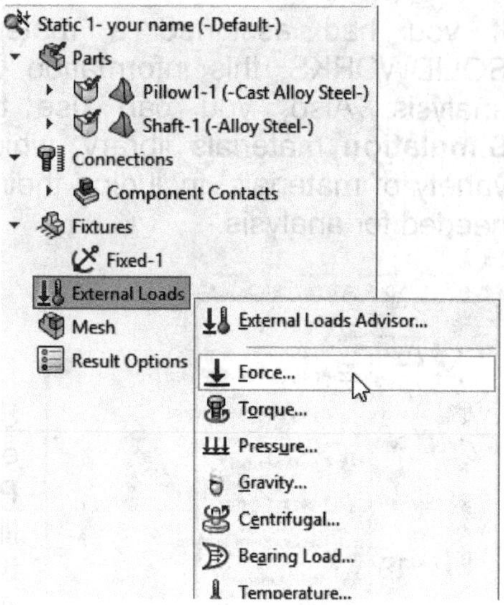

Activate the **Force** option and select the cylindrical face of the **Shaft**. By default, the direction of a force is perpendicular to the face where the force is applied. For a cylindrical face, the direction is radial.

For your analysis, the direction of the force must be downward. To change its direction, activate the option **Selected direction** and select the Top Plane from the fly-out FeatureManager. Set the units to the **SI** system.

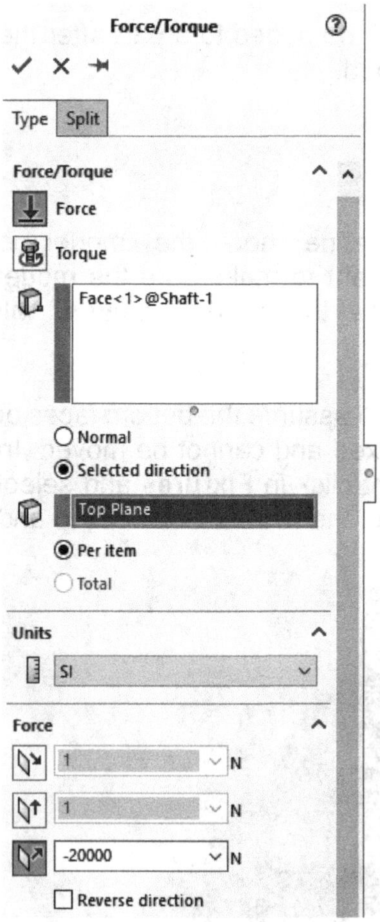

In the **Force** section use the option  **Normal to Plane**, enter a value of **-20,000 N** and click **OK** to add the force. Make sure the direction vectors are pointing down, and if needed, click in the **Reverse Direction** checkbox.

After adding the force to the shaft, you have added all the information needed to run your analysis. Your analysis study should now look like this.

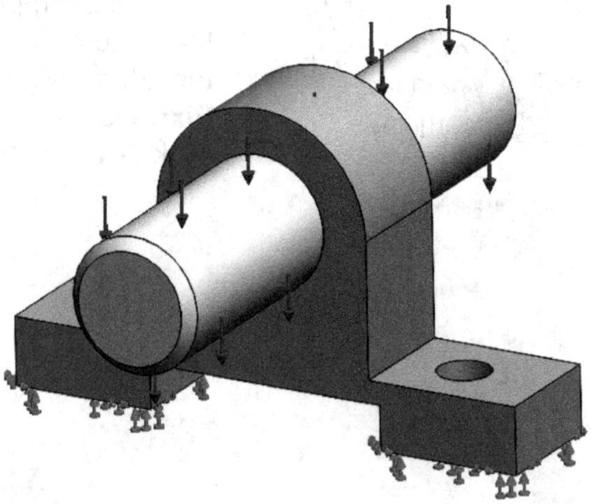

Creating the Mesh

The next step in your analysis is to generate the Mesh for the analysis. In simple terms, the model is automatically divided in thousands of small pieces or elements. An equation is generated for each element, and after the equations for all elements are solved simultaneously the stresses are calculated.

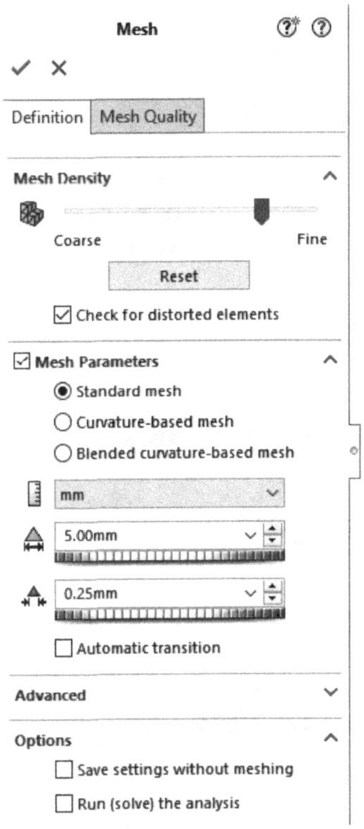

To generate the mesh, in the Simulation Manager right click on **Mesh** and select **Create Mesh.**

In the **Mesh** options set the units to **mm**, change the **Global Size** to **5 mm** and the **tolerance** to **0.25 mm**.

Keep in mind that a smaller mesh will typically yield a more accurate solution, but also will take longer to calculate. Click **OK** to generate the mesh and continue. When the mesh is complete a green checkmark is added in the Simulation Manager, and the mesh is displayed.

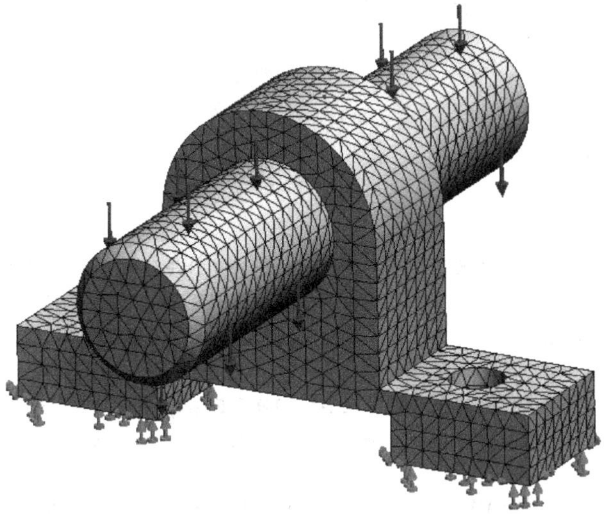

Static Analysis

After the mesh is generated and the necessary information has been defined (forces, materials, and restraints) you are ready to run the analysis. In the Simulation Manager, right click on the name of the study (**Static 1 -your name**) and select **Run** to begin the solution process.

Depending on the number of components and the complexity of the analysis, the analysis may take longer to run. In this case, the analysis is quite simple and should only take a few seconds to solve.

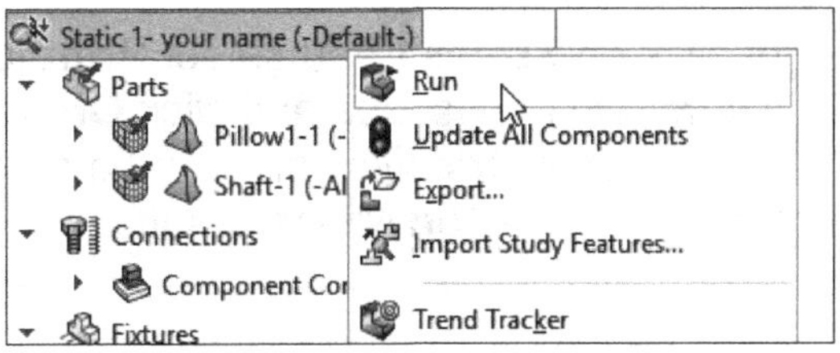

Post-processing

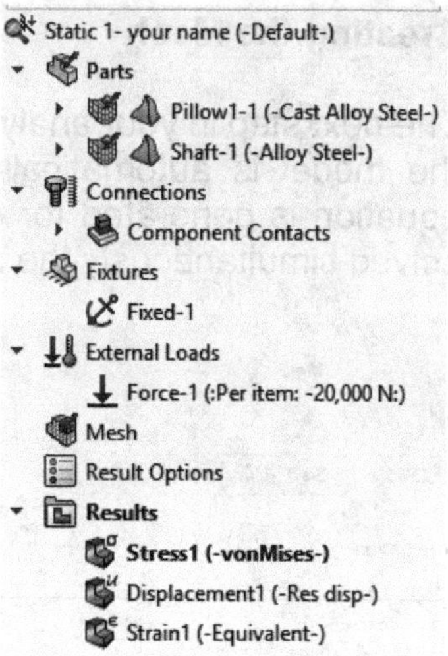

After the analysis is complete a **Results** folder is added to the Simulation Manager, including Stress, Displacement and Strain. To display the results, right click on any of the results and select **Show**. You can also change the parameters to display the results or delete them. The Stress results are shown here.

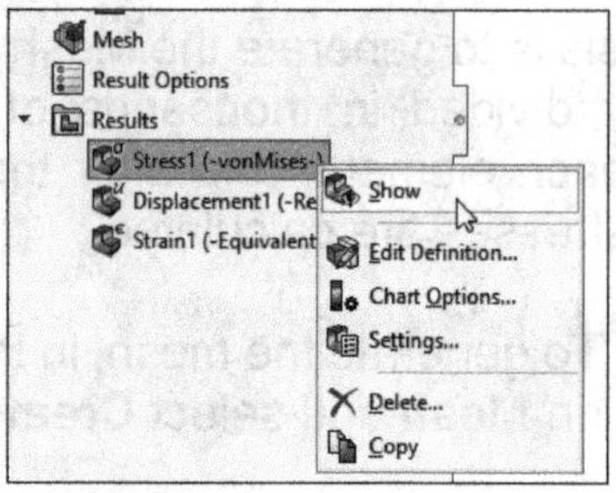

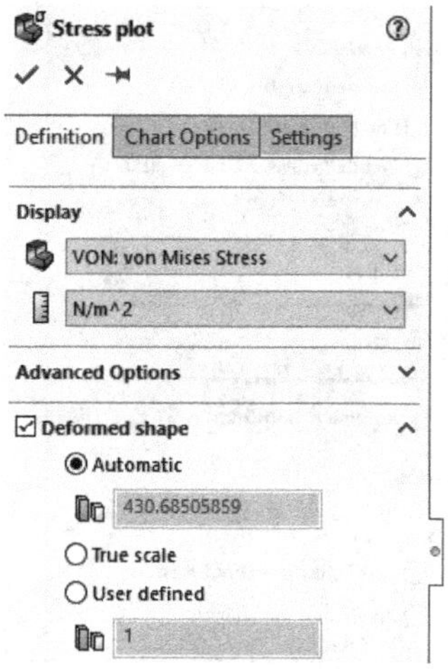

The stress results show the model with an exaggerated deformation. To change how the results are shown, right click on **Stress1** and select the option **Edit Definition**.

Here you can change the **Deformed shape's** scale, the type of results to display as well as the units of measure. Feel free to explore the different options available.

If you right click in the **Stress1** results and select the **Animate** option, the model will deform from the original shape to the final deformation using the scale used in the **Deformed shape** option, in this case 430 times the actual scale.

Using the **Section Clipping** option will allow you to show areas above or below a certain stress value.

Right click on **Stress1** and select **Print** to make a printout of the results. Change the print format to Landscape and click **OK**. Make a new drawing of the **Pillow Assembly -1** with an **Isometric** view and print a hard copy to submit to your lab instructor along with the **Stress1-Von Mises** results plot.

ANALYZING THE PILLOW BLOCK RESULTS

The results of the analysis will help you understand if your design will perform as intended. In a static stress analysis, an important parameter to consider is the material's **Yield Strength**. This is the stress at which the material starts to permanently deform. In simple terms, it means that a region with a stress **below** this level will "spring back" into the original shape, and a region with a stress **above** this level will be permanently deformed or break and would be considered to have failed.

Another factor to consider in a design is the unexpected "over-loads" that could occur due to unforeseen circumstances. A **Factor of Safety (FOS)** can be defined as the relation between the Failure Stress and the maximum Allowable Stress. For most practical purposes when using **Static Stress Analysis**, the **Yield Strength** can be considered the **Maximum Stress**.

$$FOS = \frac{Allowable\ Stress\ (Yield\ Strength)}{Maximum\ Stress}$$

Depending on the type of design, the application, environment, etc., different factors of safety values are typically used in the industry. For this lab we will assume an arbitrary **FOS = 3**. Reviewing the material properties for **Cast Alloy Steel** you will see the **Yield Strength** is approximately **241 x 10^6 N/m^2**.

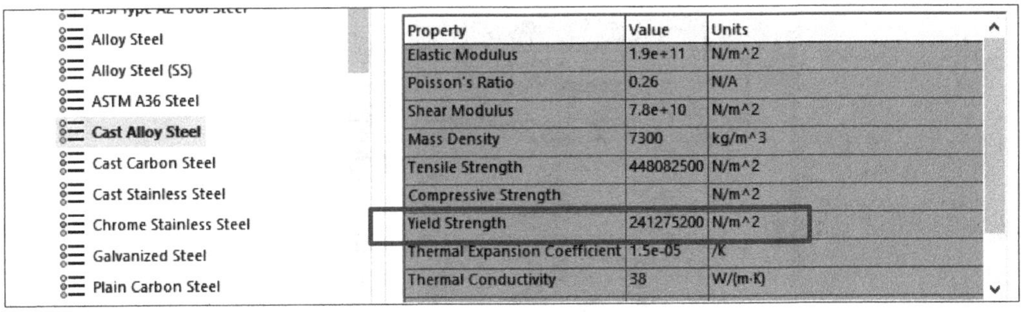

Considering this, the maximum stress allowed in the part should be:

$$Maximum\ Stress = \frac{Yield\ Strength}{FOS}$$

Maximum Stress = **(241 x 10^6 N/m^2) / 3 = <u>80.3 x 10^6 N/m^2</u>**

Reviewing the analysis Stress Plot you can see the maximum stress calculated in the part are the red colored regions around the Pillow Block's base, and the corresponding stress in the vertical scale with a value of approximately **116 x 10^6 N/m^2**, which is higher than the maximum stress allowed, calculated with the Factor of Safety = 3.

NOTE: Because of the nature of how FEA analysis calculates results, the stress values you obtain may be slightly different but generally close to these values.

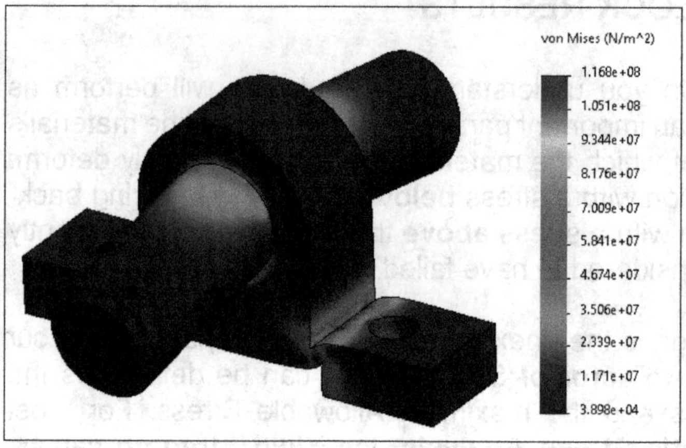

In order to reduce the stress in the part, the designer has different options, including:

- Change the geometry
- Reduce the external loads
- Change the material

Depending on the application, design restrictions, budget, etc., you could change one or more parameters. For this lab we will only change the part's geometry.

PILLOW BLOCK DESIGN CHANGES

To make the changes open the **Pillow1** part, select the Boss-Extrude1 feature and select the **Edit Sketch** command from the context toolbar. Change to a **Front** view and change the sketch dimensions as indicated. After the changes are made, Rebuild the model to continue.

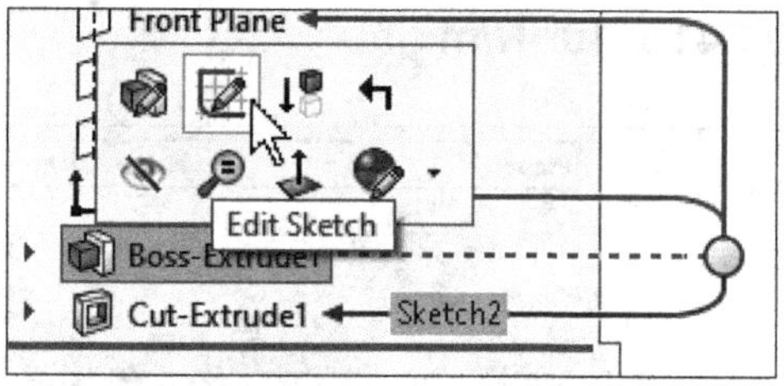

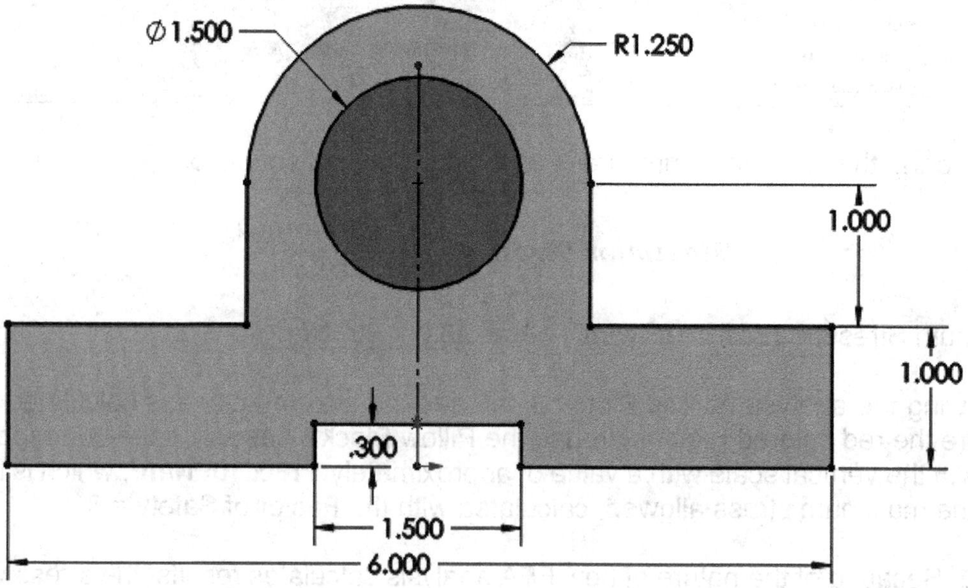

Next change to a **Top** view and **Edit** the sketch for the Cut-Extrude1 mounting holes. Change the spacing between holes as shown and **Rebuild** the part to continue.

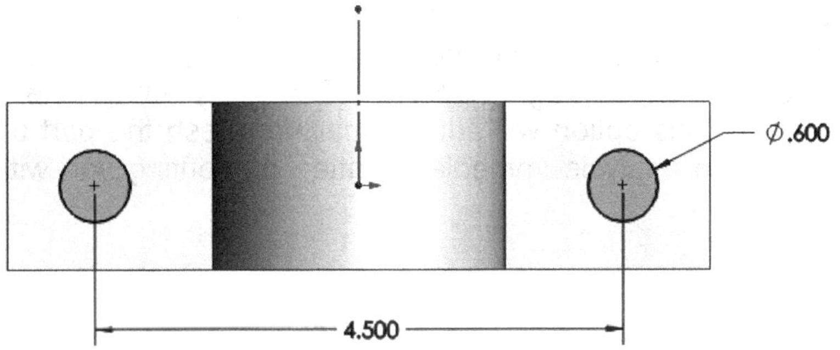

A common practice to reduce stress concentrations in a component is to round the corners. Add the fillets indicated to your model and save the changes when finished.

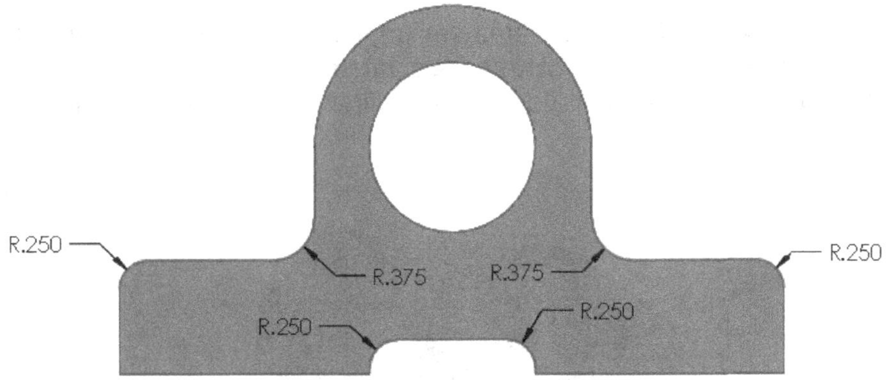

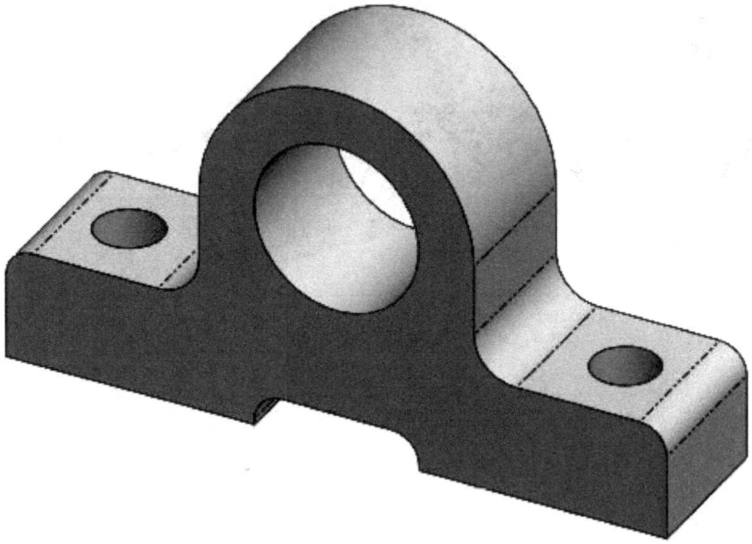

ANALYSIS OF UPDATED PILLOW BLOCK

After the **Pillow Block** has been modified, return to the assembly, and select the **Simulation** tab at the bottom to access the FEA analysis. SOLIDWORKS Simulation will recognize the geometry has been modified and alert you that the results are no longer valid. To update the study, right click in the **Static 1- your name study** and select the **Run** option. Using this option will automatically re-mesh the part using the same settings, and re-run the analysis immediately after, presenting you with the updated results.

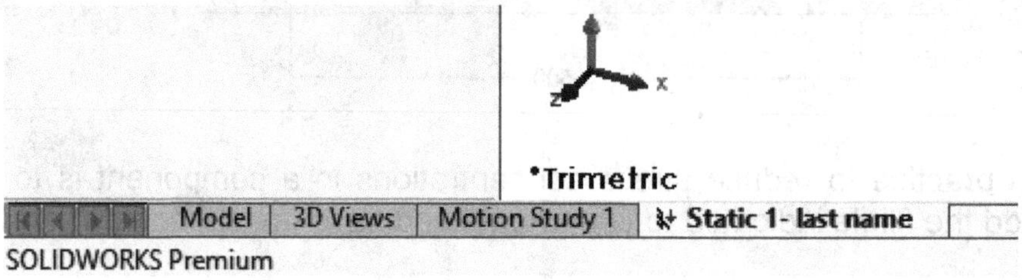

Using the analysis results for the updated design, calculate the lowest **Factor of Safety** using the maximum **Von Mises stress** calculated. Print the results of the updated **Stress1-Von Mises** and turn it in along with the first results to your lab instructor.

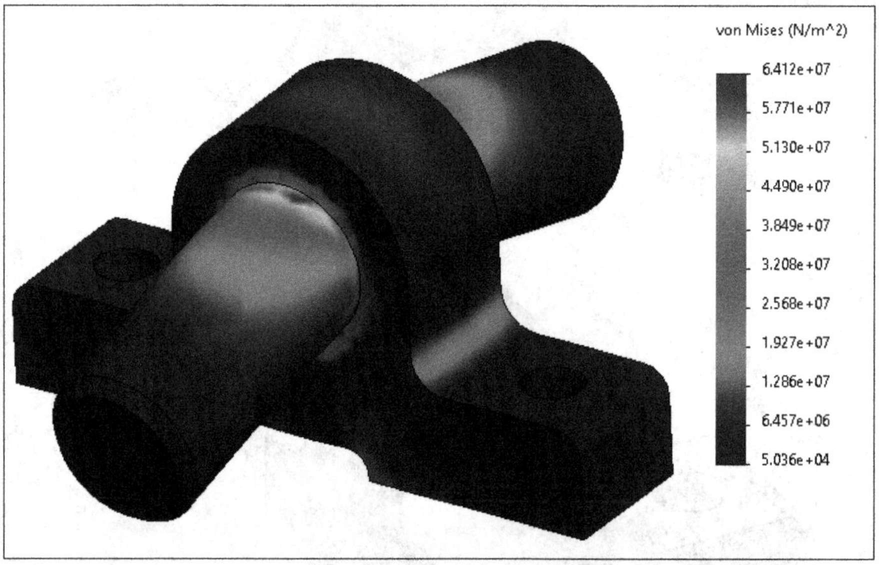

Exercise 7.2: Static Analysis of the Shaft Hanger

In this exercise you will use SimulationXpress, an analysis tool included with all versions of SOLIDWORKS, to analyze the double shaft hanger to determine if the component can support the mechanical loads it was intended to carry. SimulationXpress is limited to a single component, static stress analysis only, to determine if the component will exceed the yield stress of the material.

Open the "*Double Shaft Hanger*" part file from exercise 3-6 previously made and select the menu "Tools, Xpress Products, SimulationXpress..."

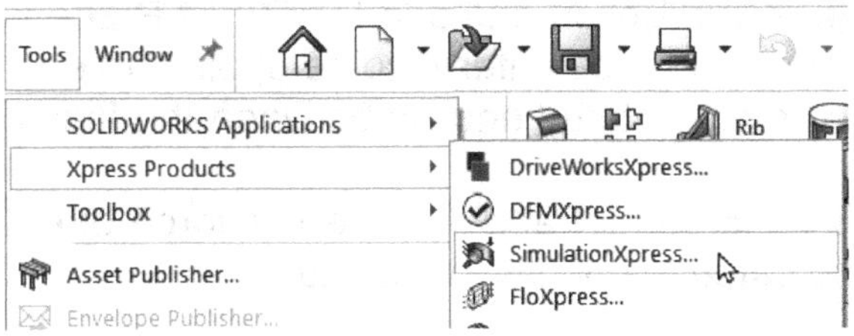

When we are first presented with the welcome screen, select Options and set the System of units for the analysis to English (IPS), and turn on the option: "Show annotation for maximum and minimum in the result plots". Click OK to accept the change and Next to continue.

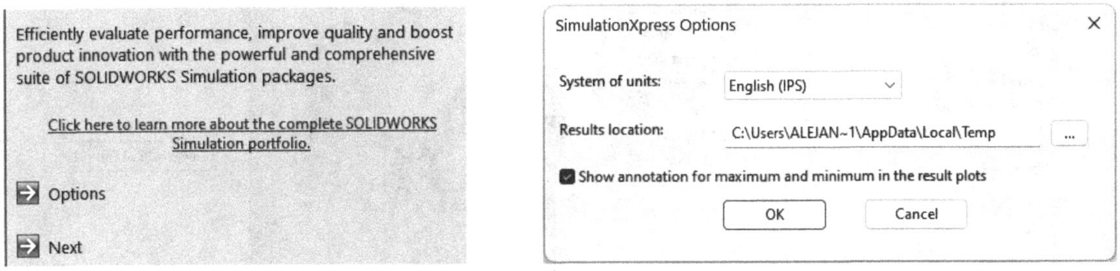

The next step is to determine how the model is supported. Select "Add a fixture" and add the four mounting holes in the base to the Fixed Geometry selection box. The selected faces will become immovable for analysis. Click OK to continue.

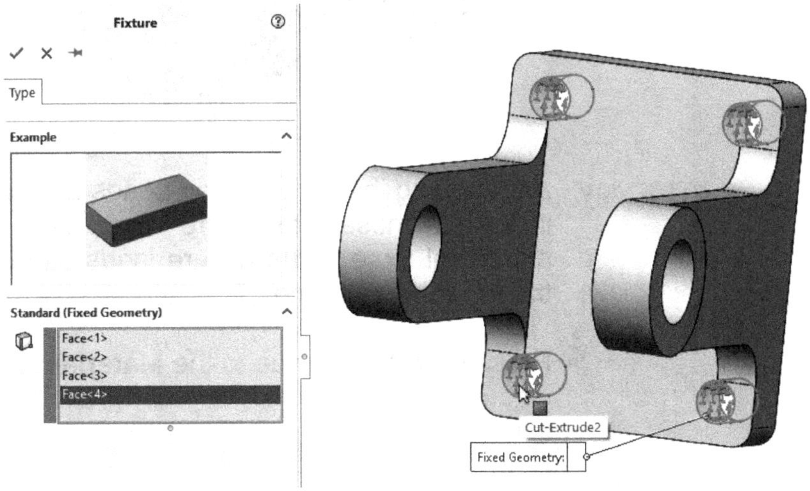

After the fixtures have been defined, click **Next** to define the loads acting upon the model, and select **Add a force**.

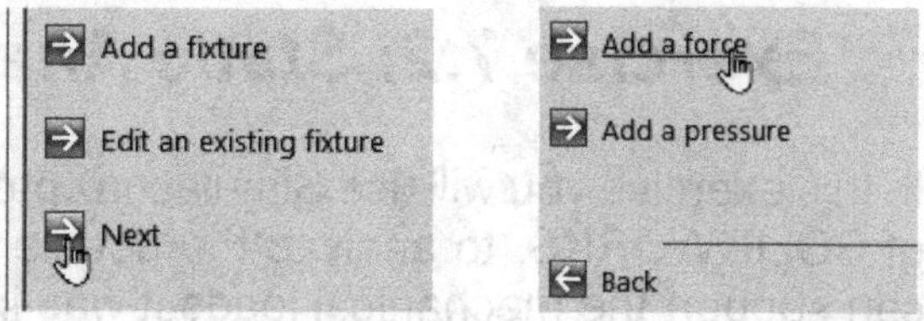

In the **Force** properties, add both round holes to the selection list. Activate the **Selected direction** option and add the **Right Plane** with the **Per item** option activated. This will cause the force entered to be applied to each face. Using the **Total** option would divide the force entered equally between the faces selected.

Enter 5,000 lbf and reverse the force direction to point perpendicular in the direction indicated. Click OK to continue.

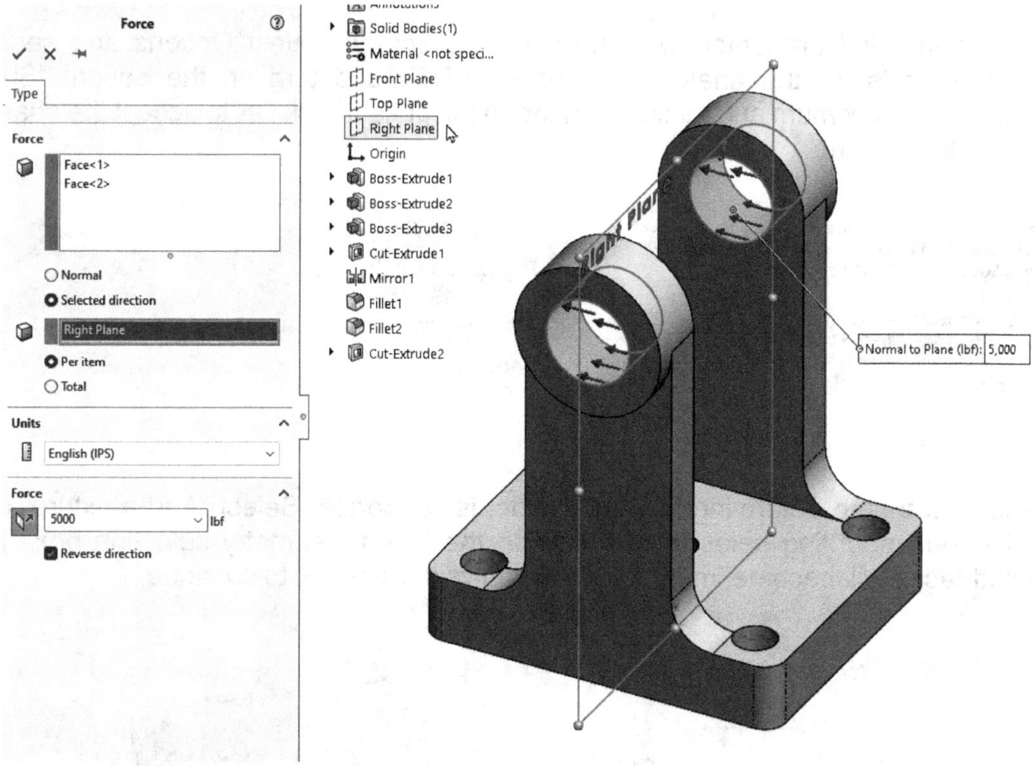

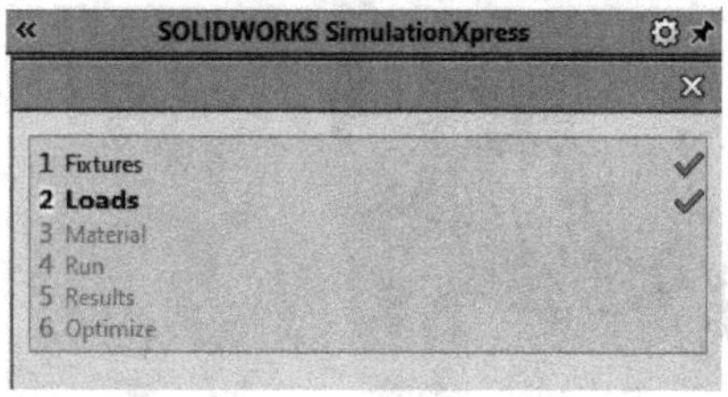

After adding the Fixtures and Loads SimulationXpress indicates those steps are complete. If needed, additional force and pressure loads can be added to the model at this time.

Press **Next** to continue to the **Material** selection step.

If the model to be analyzed has a material defined, it will be marked as complete and listed. This is where you can define the material of the part to be analyzed or choose a different one.

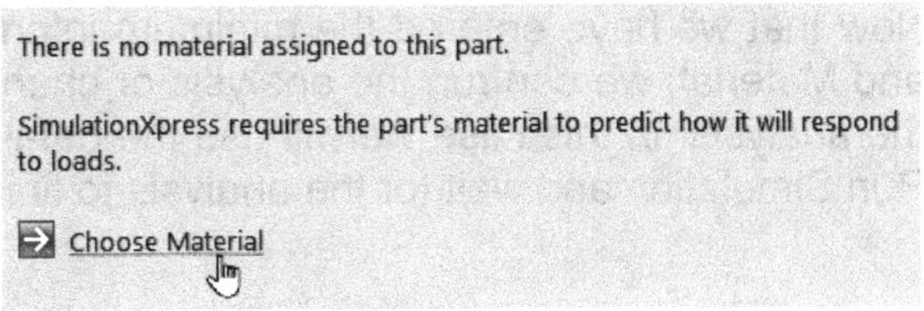

Click **Choose Material**, and from the **SOLIDWORKS Materials, Steel** library select **AISI 1015 Steel, Cold Drawn (SS)**. Click **Apply** and close to continue.

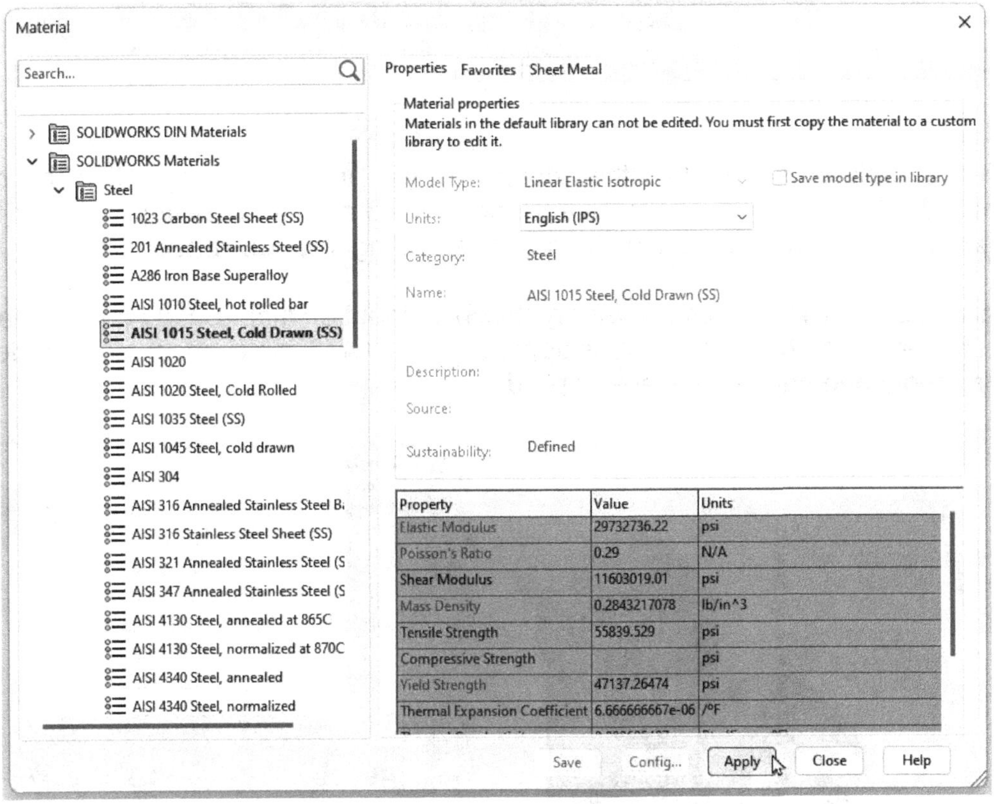

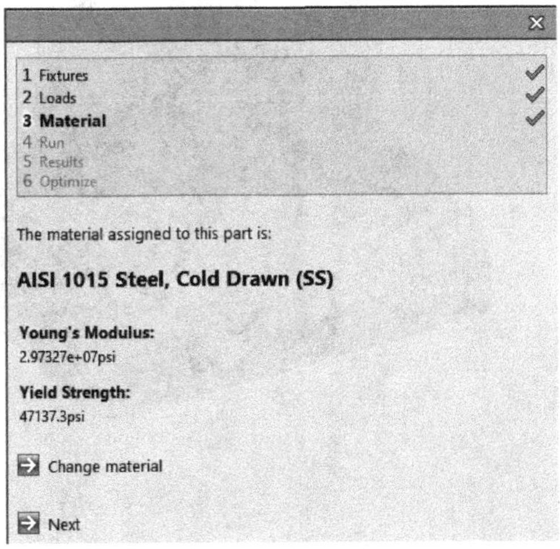

The selected material will be listed including its Young's Modulus and Yield Strength. Click Next to continue.

Now that we have entered the minimum information for an analysis (Supports, Loads and Material) we can run the analysis or change the mesh parameters before running the analysis. In this case we will use the default mesh parameters and continue. Click Run Simulation and wait for the analysis to finish. It should be but a couple of seconds.

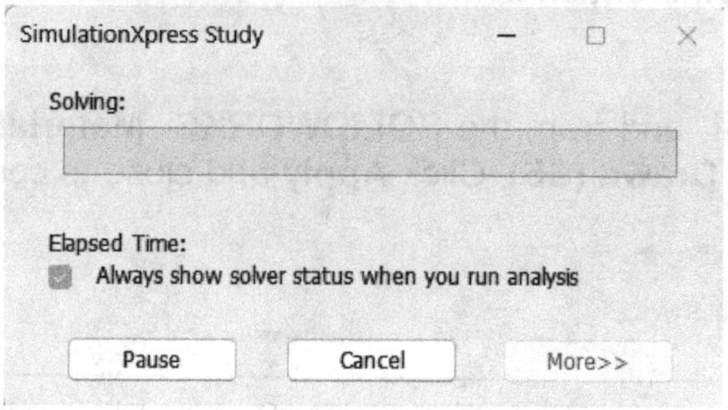

When the analysis is complete the component will animate exaggerating the part's deformation. If the deformation is as expected based on the loads applied, click **Stop animation** and continue to review the results selecting **Yes, continue**.

You are now shown the Factor of Safety (FOS) distribution map. Based on the analysis conditions entered, the lowest FOS is approximately 0.478, which means our model will fail with about one half of the loads applied.

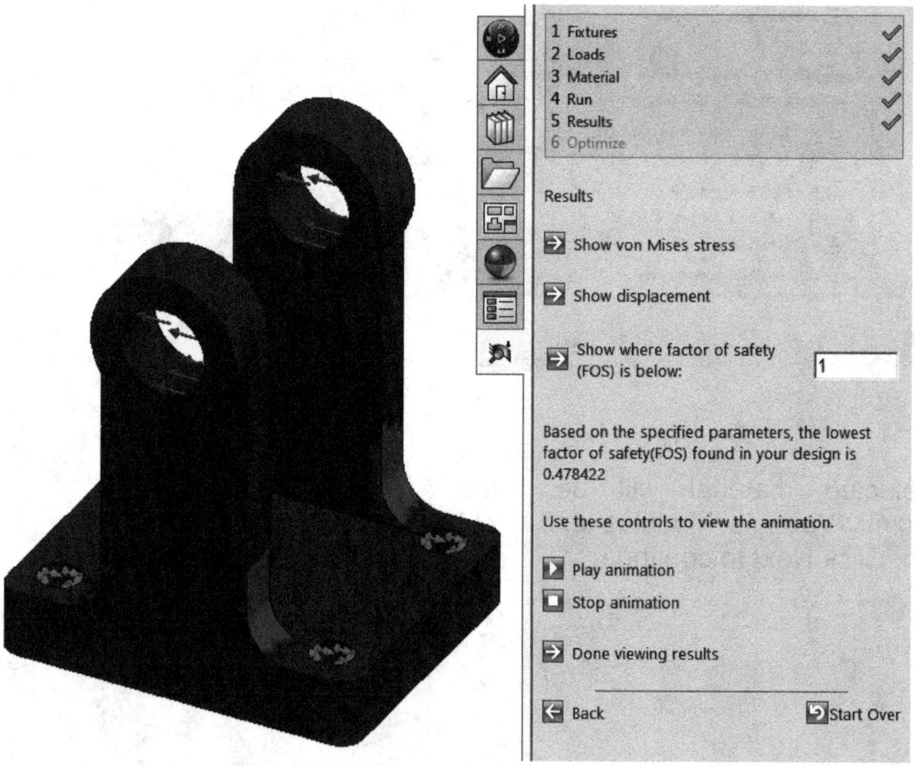

Next select the **Show von Mises stress** results to evaluate the stress distribution. In the stress scale you can see an arrow pointing to the Yield Strength (approx. 47,000 PSI), which means everything above this line will fail.

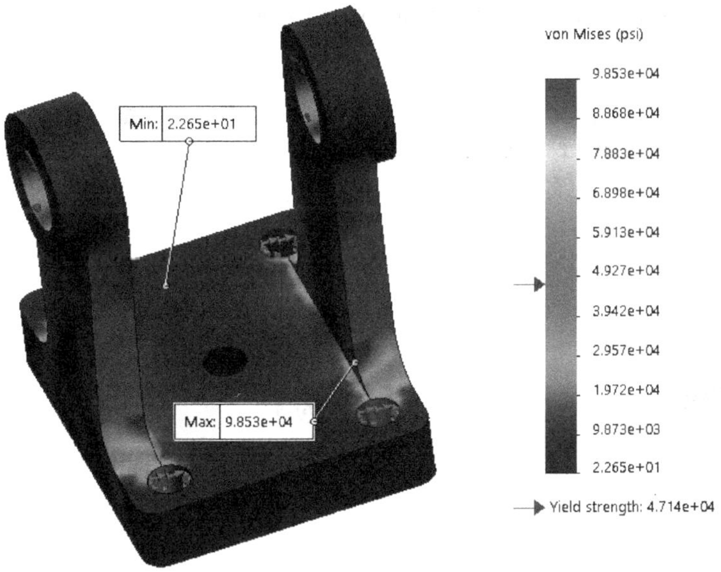

Selecting Show displacement will indicate the deformation in inches. As expected, the top part of the model will displace the most, approximately 0.022"

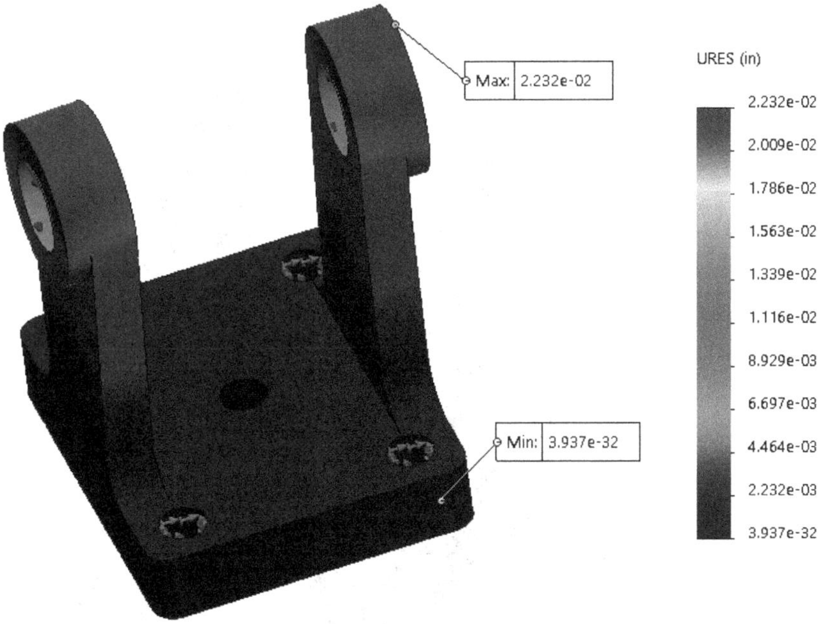

To fix our design to prevent a failure, we can reduce the loads acting on the part, modify the geometry or change the material. In this exercise you will change the model's geometry, change the material it is built with and modify the forces acting on it. Close the **SimulationXpress** tab in the upper right corner. When asked to save the analysis results select **Yes**.

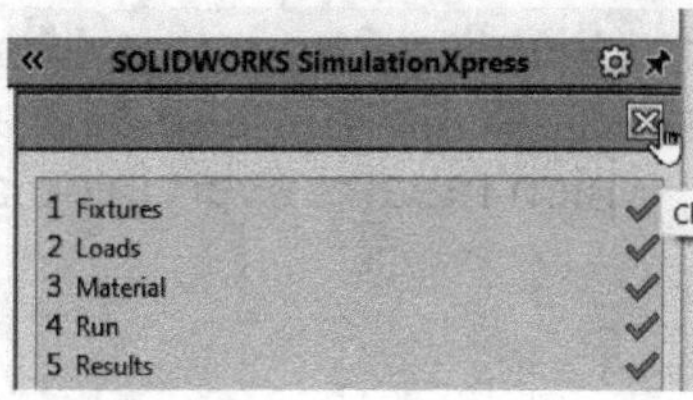

Change the width of the vertical arms from **0.500"** to **0.625"** and rebuild the model. If the second arm was mirrored, it will update automatically.

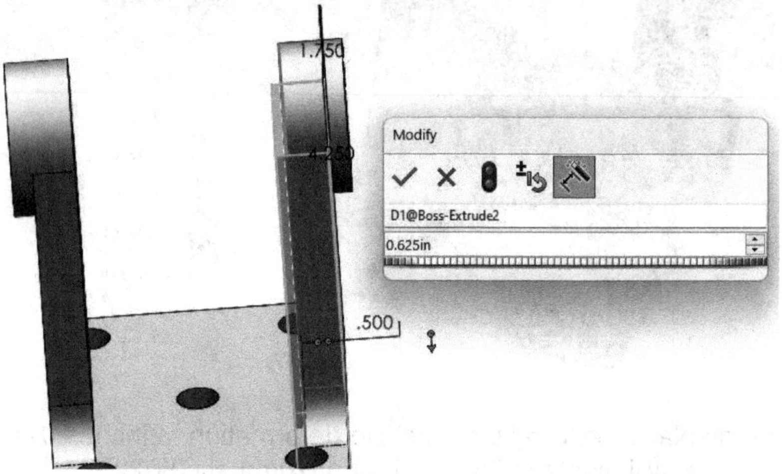

Add a **0.500"** fillet to the inside edges of the vertical arms to help reduce the stress concentration.

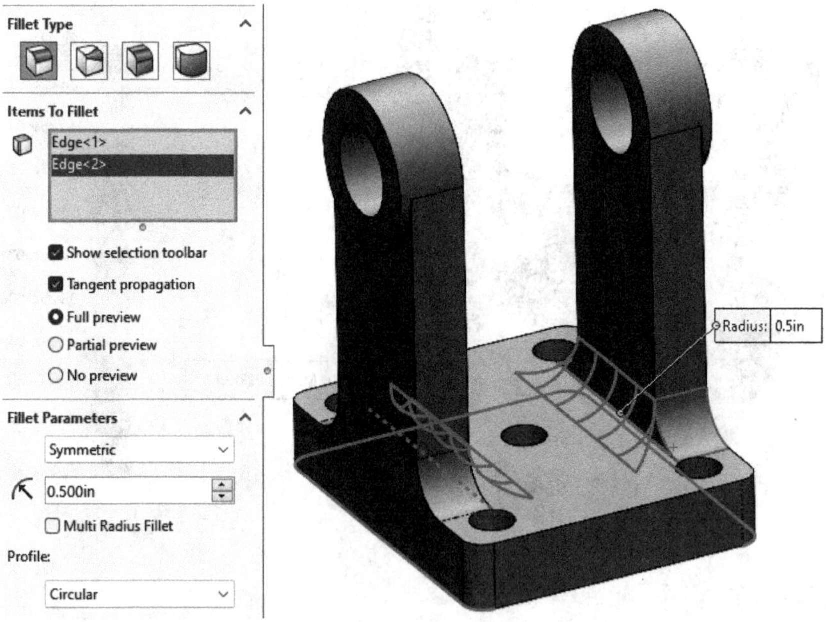

Rebuild the model and load SimulationXpress again. The simulation study will show a warning sign at the top of the study, alerting you that the results are no longer valid because the model and/or the conditions have changed. To modify the component's material right click on the mesh icon and select **Apply/Edit Material**.

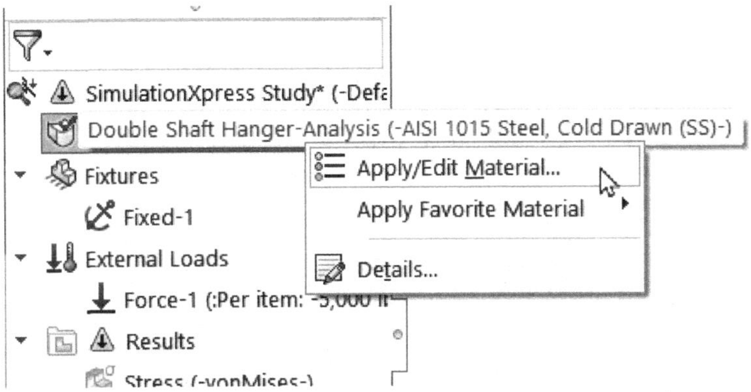

Select **Alloy Steel** which has approximately double the Yield Strength of AISI 1015 steel. Click **Apply** and **Close** to continue.

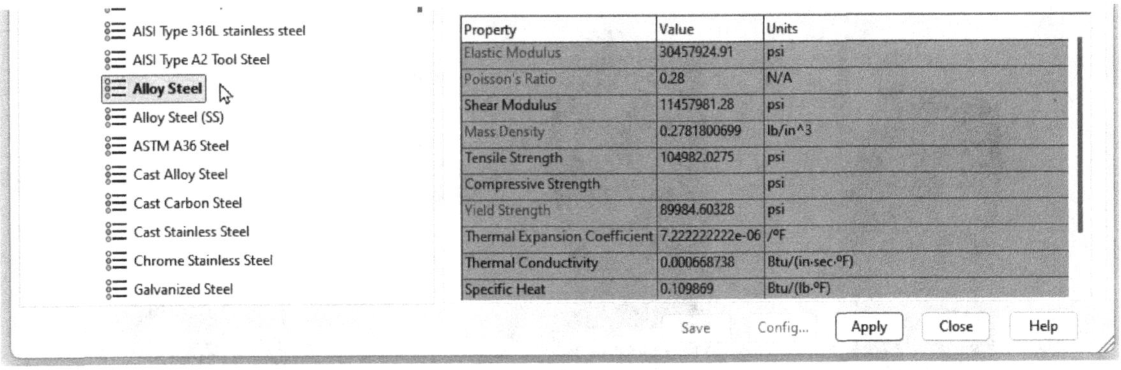

Right click on **Force-1** in the study manager and select **Edit Definition**.

Use the option **Total** instead of **Per item** to apply the load to all faces selected and change the **Force** value to **4,000 lbf**.

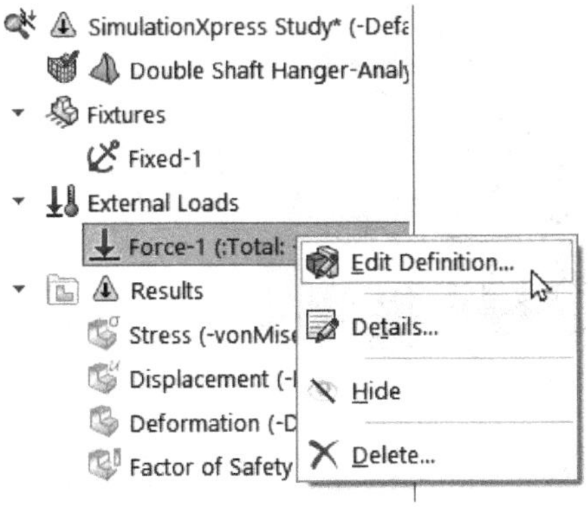

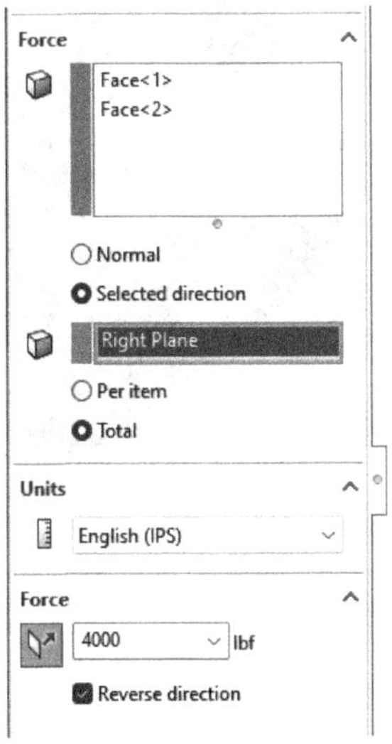

To update the analysis with the new geometry and material, right click at the top of the study and select **Run** from the context menu.

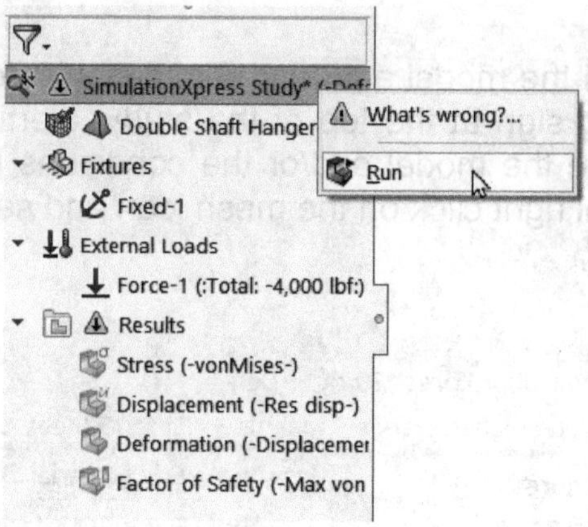

Now we can see the lowest **Factor of Safety** is almost 2.5, and the maximum stress is approximately 36,000 PSI, far below the 90,000 PSI Yield strength of the material, making our design safer under the simulated conditions.

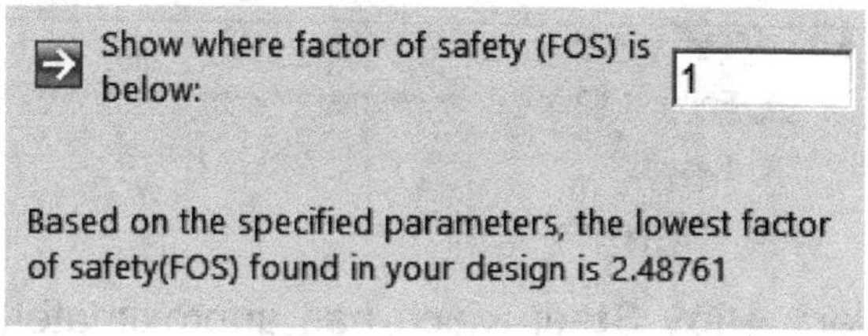

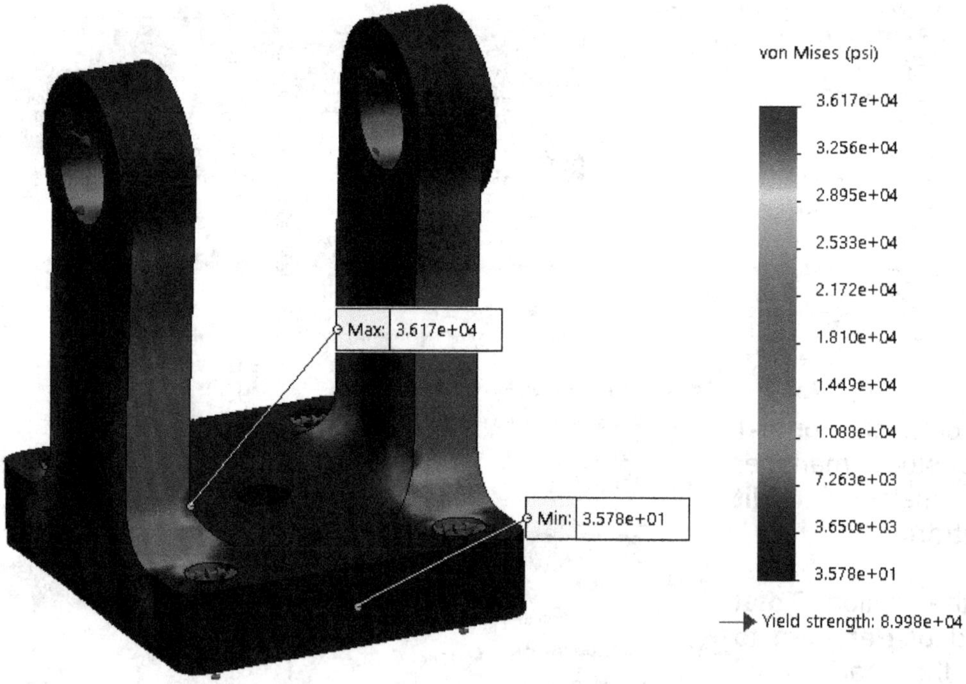

Design Workbook Lab 8
Animation, Detailing and Rapid Prototyping

In this Lab you will learn how to create an Exploded view of an assembly, used mainly for illustration, assembly instructions or documentation, and how to create a video animation of it. You will also learn how to add different drawing views of a part to a detail drawing, used to document how a part will be manufactured, and finally, you will cover the basic principles of Rapid Prototyping, additive manufacturing, and how to make an .STL file to create a physical prototype of your designs.

ASSEMBLY EXPLODED VIEW

In a previous lab you learned how to assemble individual components, and now you will learn how to create an **Exploded View**. With an assembly open, select the **Exploded View** command from the Assembly tab, or from the menu **Insert, Exploded View**.

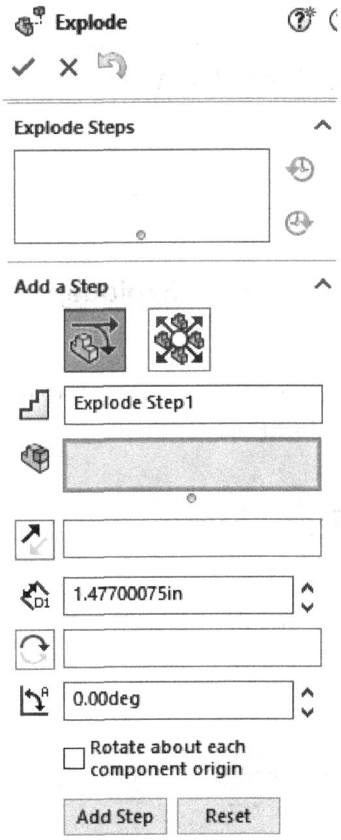

When the **Exploded View** command is loaded, you can select the components to explode, the distance, direction and other options for each step of the assembly's exploded view.

To add a translation or rotation step, activate the **Standard Step** option and select the part to explode. Immediately after selecting the part, you will see three axes to define the exploded step direction. You can:

- Click and drag the direction arrow in the screen, or
- Select the direction arrow, enter the exploded step distance, and click on **Add Step**.

If you decide to drag the direction arrow, a ruler will be automatically displayed to help you define the exploded step distance. After the direction arrow is released a new **Explode Step** is added to the Explode Steps list.

Once the step is complete, click the **Done** button to finish the step and continue with the next one.

Continue adding more steps and click **OK** to finish when done.

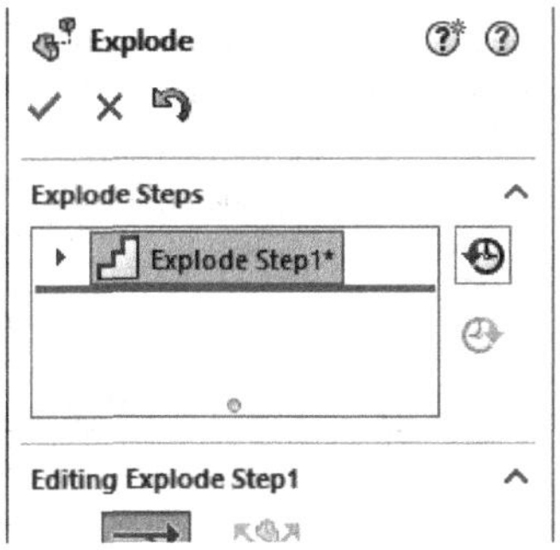

An assembly with three exploded steps is shown here.

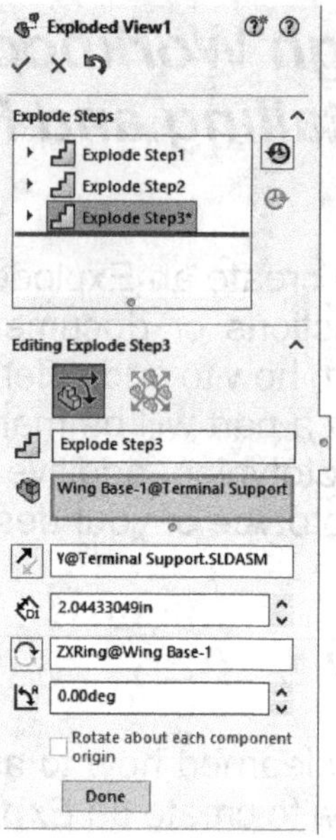

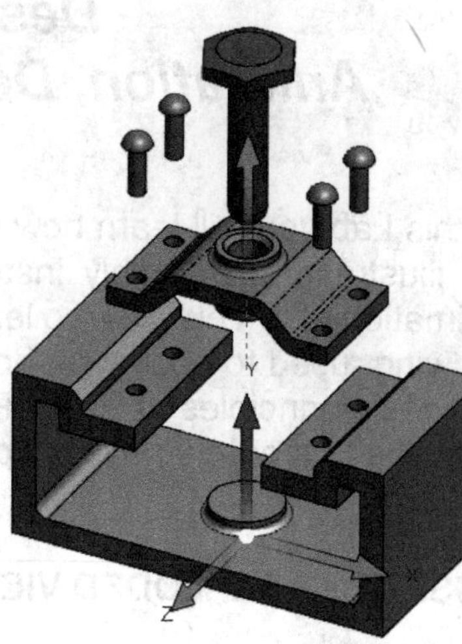

After the assembly's exploded view is finished, select the **Configuration Manager** tab, expand the **Default** (or the currently active configuration), and expand the **Exploded View1** section.

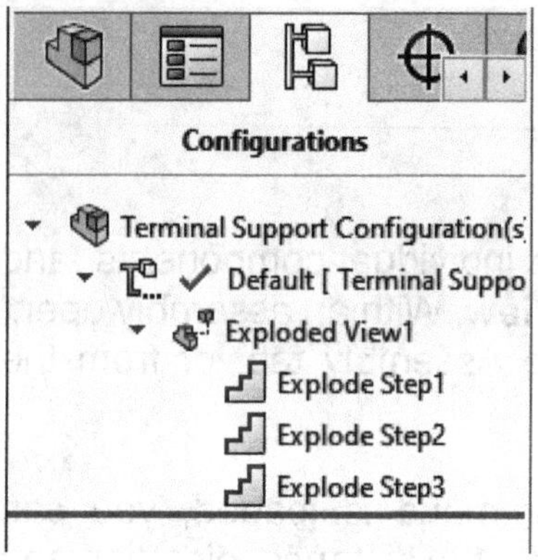

If you select an exploded step, you can see the explode view direction arrow, where you can click and drag it to adjust the step distance if needed.

If you want to edit the entire exploded view, right click on the **Exploded View1** and select **Edit Feature** from the context menu.

To collapse the exploded view, right click on it and select **Collapse** from the context menu, or double click on it.

To explode it again right click on the collapsed exploded view and select **Explode**, or double click on it.

You can also animate the explode and collapse from this context menu, where you will see an animation controller to play, pause, stop and save the animation.

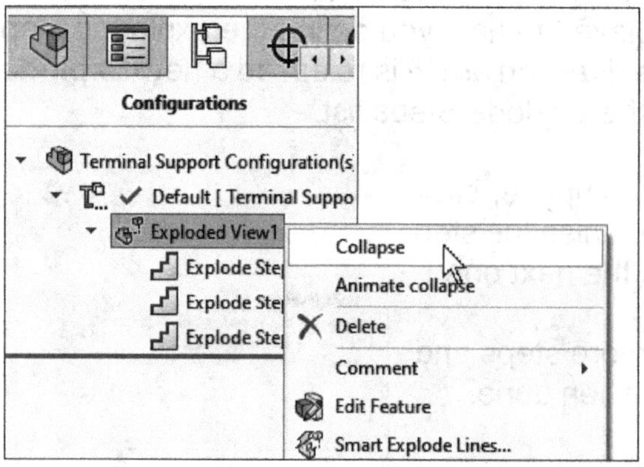

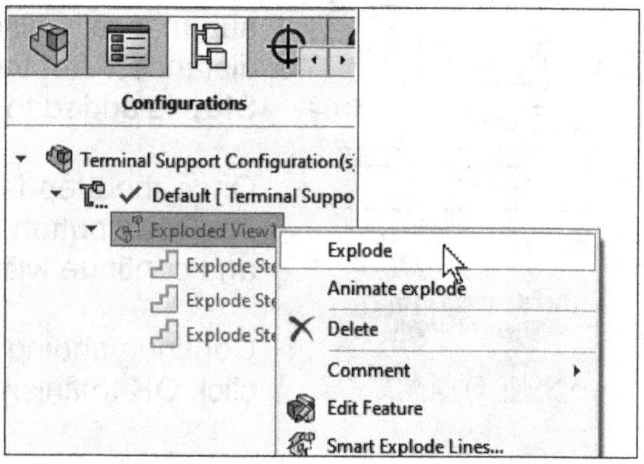

ANIMATION WIZARD

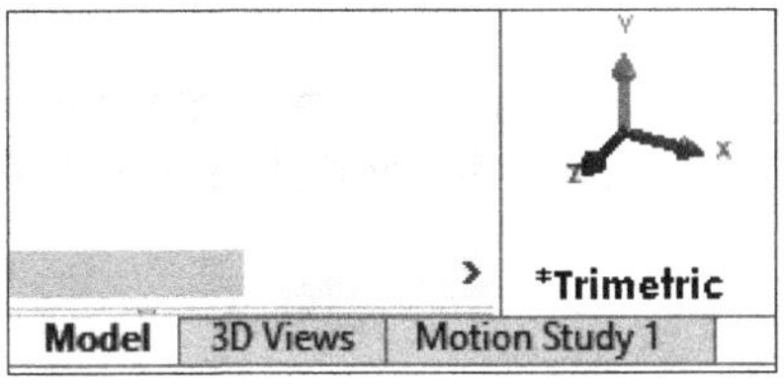

The **Animation Wizard** is an easy-to-use tool to create an exploded view animation. You can access it by activating the **Motion Study1** tab on the bottom left side of the screen, or you can add a new study by selecting the **New Motion Study** command from the Assembly tab.

In the Motion Study toolbar select the **Animation Wizard** command. If you have created an exploded view to your assembly, additionally to **Rotate model**, you will have the option to animate the **Explode** or **Collapse** the exploded view, and later save it to a video file.

The difference between the option in the context menu previously shown and the **Animation Wizard** is that in the latter you can define the length of the animation, the timing of the explosion steps and change the order of the different steps.

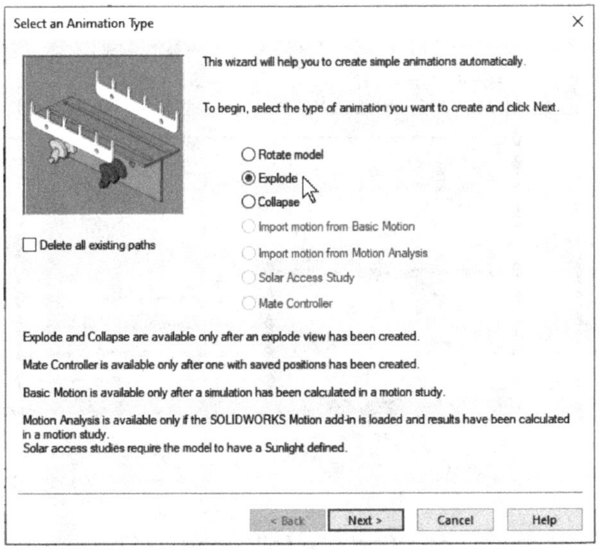

When you select the **Explode** or **Collapse** option and click **Next**, you will be asked to enter the duration of the animation.

Selecting the **Rotate Model** option, you will have the option to select the axis to rotate around and the number of rotations.

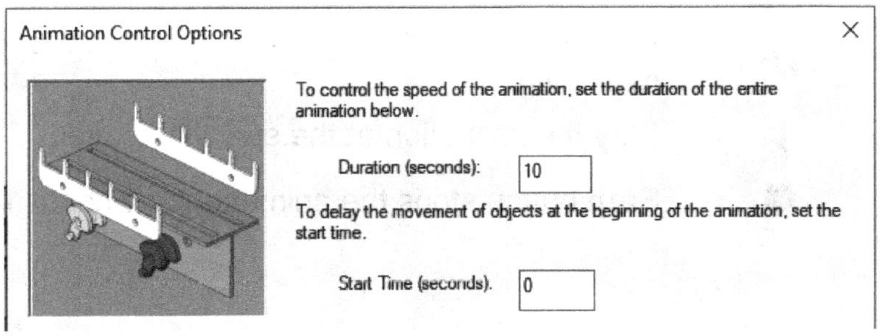

ANIMATION CONTROLLER AND MOTION MANAGER

After creating an animation with the Animation Wizard, the MotionManager is automatically populated with the steps and their timing.

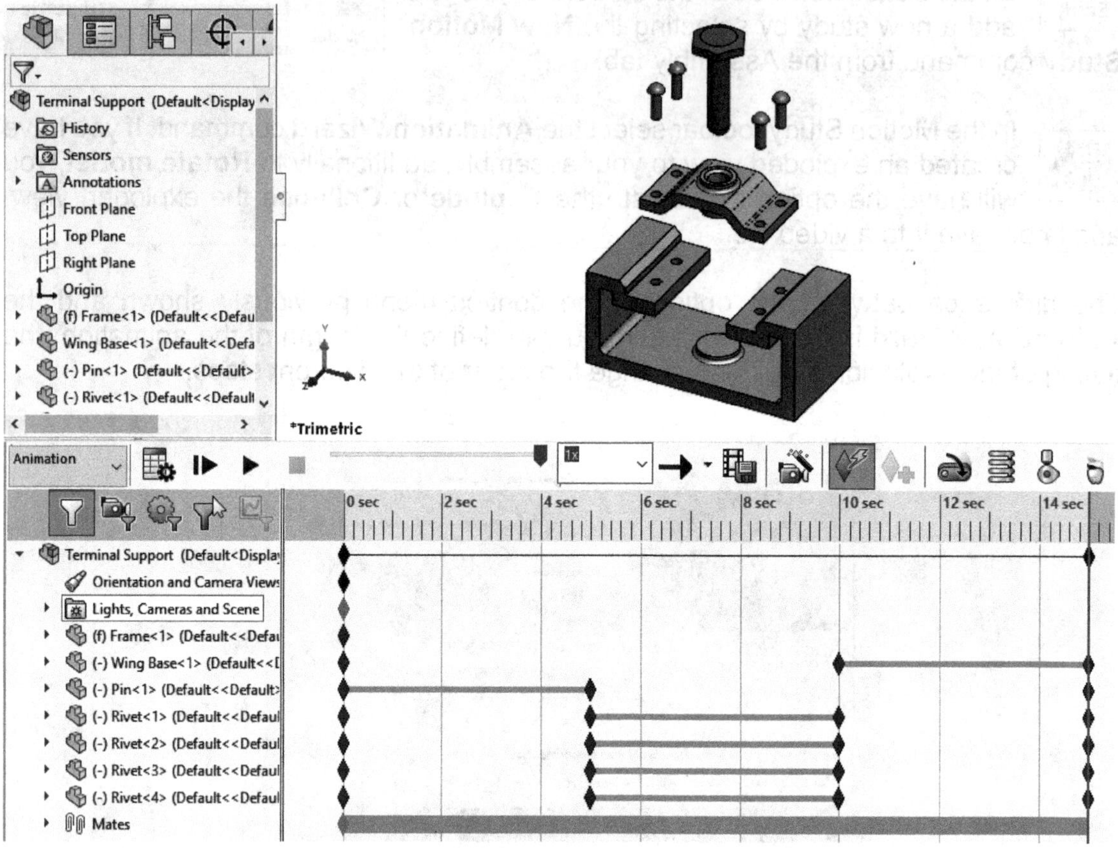

In the Animation Controller you can find the following options.

Calculate the animation from the start.

Play from Start stops the animation and starts to play from the beginning.

Play the animation at the specified speed.

Stop button stops the animation and sends it back to the start position.

There are three "**Playback Modes**" on the controller.

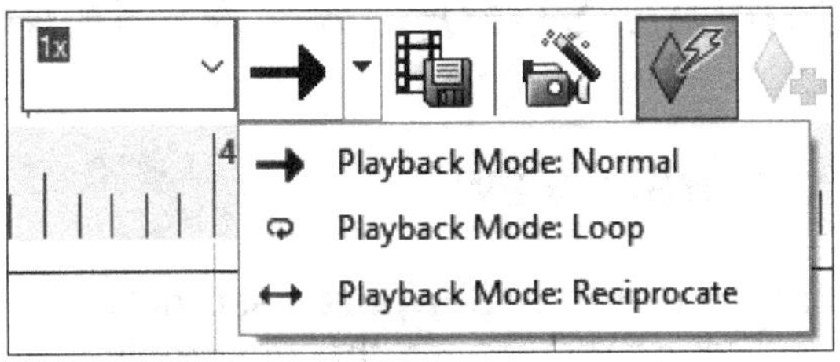

- **Normal Mode** button sets the animation to run once from start to end at the specified speed multiplier.
- **Loop Mode** button sets the animation to run in a continuous loop.
- **Reciprocate Mode** button sets the animation to run forward and then backwards in a repeating loop.

Save as AVI File button will open the "Save Animation to File" menu and allows you to save the animation as an .AVI file that can then be played on a media player.

Animation Wizard launches the "Animation Wizard" tool covered previously.

INTRODUCTION TO PHYSICAL SIMULATION

Physical Simulation allows you to simulate the effects of motors, springs, and gravity on your assemblies. Physical Simulation combines simulation elements with SOLIDWORKS tools such as mates and Physical Dynamics to move components in your assembly.

The options available included for simulation are:

Linear Motor or Rotary Motor

Linear or Torsional Spring

Contact between components

Gravity effects along a selected axis and value

Exercise 8.1: Exploded Animation of the TERMINAL SUPPORT ASSEMBLY

CREATE EXPLODED ASSEMBLY VIEW

Open the previously made **Terminal Support** assembly. First you need to add an Exploded View to the assembly. Select the menu **Insert, Exploded View**, or click on the **Exploded View** command from the Assembly tab.

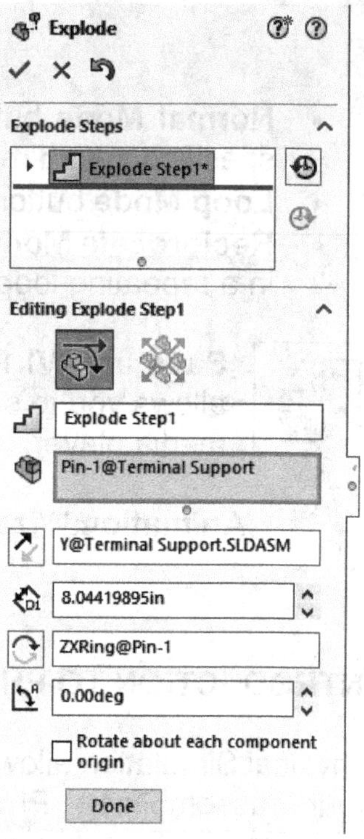

The first part to explode is the **Pin**. Select it in the assembly, click and drag the **Y direction arrow** up, and using the reference ruler move it approximately a distance of **8.00”**. To complete the explode step click on the **Done** button. *Explode Step1* is added to the Explode Steps list at the top as shown.

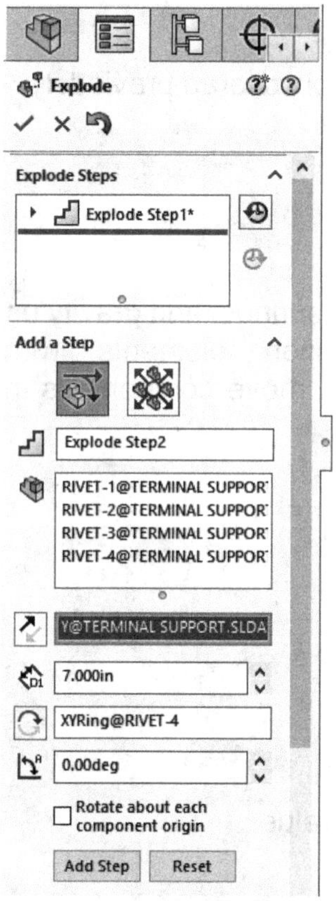

In the next step you will explode the four **Rivets** at the same time, but in this case, you will define the exact distance to explode them. Select the four **Rivets** on the screen, and then click in the **Y arrow** to define the explode direction.

Enter **7.0”** in the Explode Distance box and, if needed, toggle the **Reverse Direction** button to make the Rivets go up. To complete the step, click on **Add Step** to continue. Note *Explode Step2* is added to the Explode Steps list.

The correct explode settings for the Rivets are shown here.

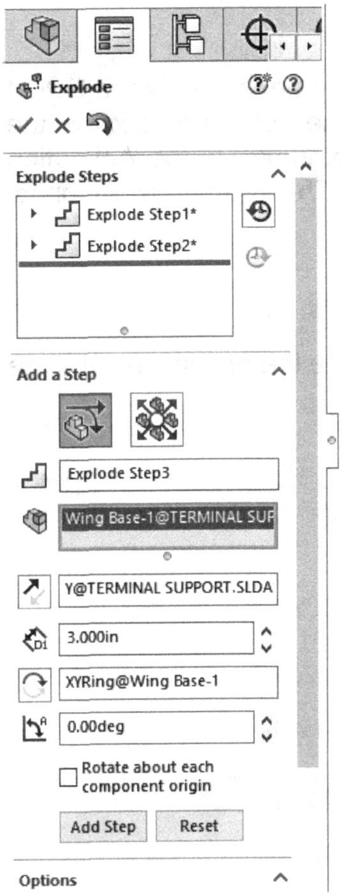

The final step is to explode the **Wing Base**. Select it in the assembly, click on the **Y direction arrow**, enter a distance of **3.00"** and click on **Add Step** to continue. *Explode Step3* is added to the Explode Steps list. The explode settings for the Wing Base are shown here.

After the three exploded steps are added to the assembly, click **OK** to finish the **Explode** command.

The exploded Terminal Support assembly is shown next.

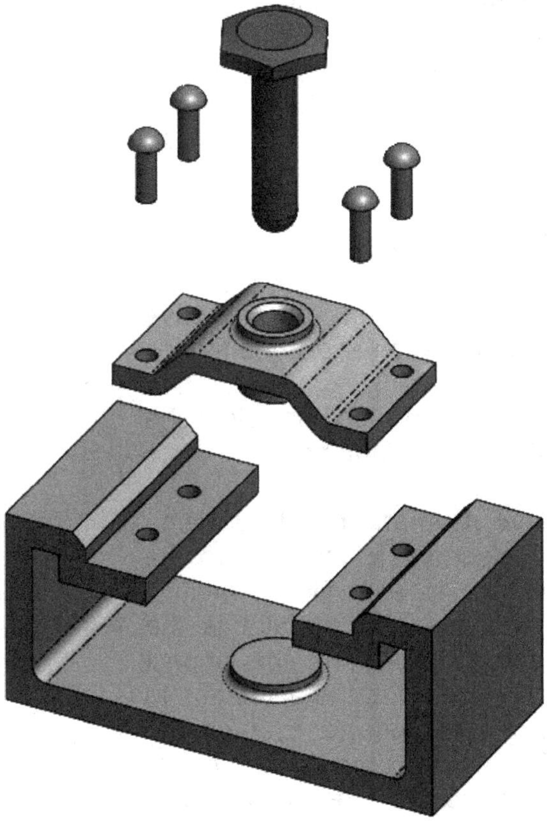

ASSEMBLY ANIMATION

In this step you will create an animation of the assembly's exploded view. Select the **Motion Study1** tab on the lower left side of the screen to access the MotionManager. Click on the **Animation Wizard** command in the Animation toolbar, select the **Explode** option, and click on Next. Set the **Duration** to **15** seconds, leave the **Start time** at **0** and click on **Finish** to continue.

PLAYING THE ANIMATION

In the Animation Controller toolbar press the **Play** button to see the animation. Change to the different play options including the **Normal**, **Loop** and **Reciprocate** modes to see the effect on the animation.

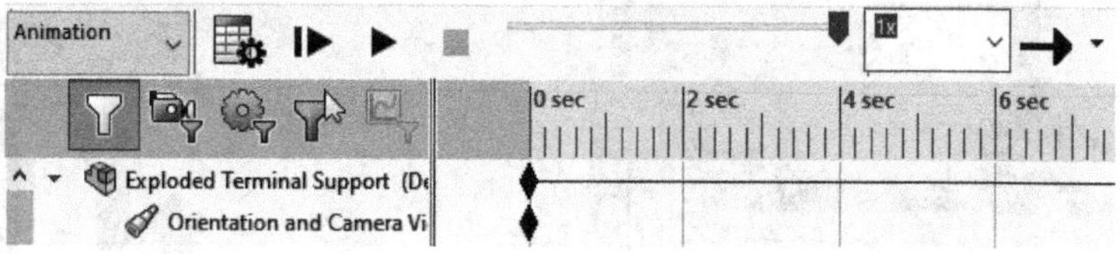

SAVING THE ANIMATION

In the next step you will save your animation to a video file.

From the Animation toolbar select the **Save Animation** command.

In the **Save Animation to File** dialog you can select the folder to save the animation, the type of file (default is **.AVI**), the image size (default is the current screen size) and the aspect ratio for the video. Set the animation to **15 Frames per Second**, change the video format to **.mp4** for better compatibility across different devices, and click the **Save** button to continue.

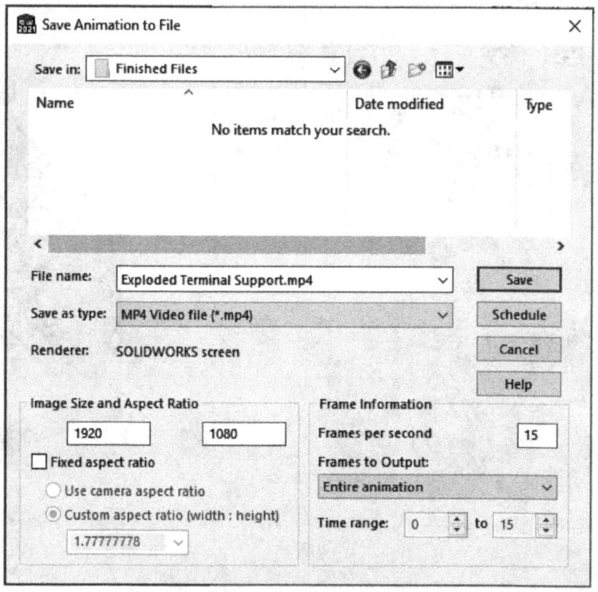

If asked to recalculate the animation to update the results, click **Yes** to continue. To finish this part of your lab, use the **Save As** command to save your exploded assembly file as **Exploded Terminal Support**.

When saving a video using the **.AVI** video format, you will be asked to select the **Video Compression** codec to use. Select the default **Microsoft Video 1** and click **OK**.

Note: You may have to adjust your model viewport to fit the whole animation on your screen as it is being captured.

ANIMATION MOTION ELEMENTS

In the Motion Manager toolbar, you can optionally add **Linear** and **Rotary motors**, **Linear** or **Torsional Springs**, define **Contact** between different components and **Gravity** effects acting on the assembly.

It must be noted that the Motion Elements can only be added to components that have the necessary degrees of freedom. In other words, if the Mates added to the assembly prevent the motion of a component, adding a motion element to it will not make it move.

In this step you will add a **Rotary Motor** to the Pin. Start by adding a new Motion Study to your assembly. Click on the **New Motion Study** command from the assembly tab.

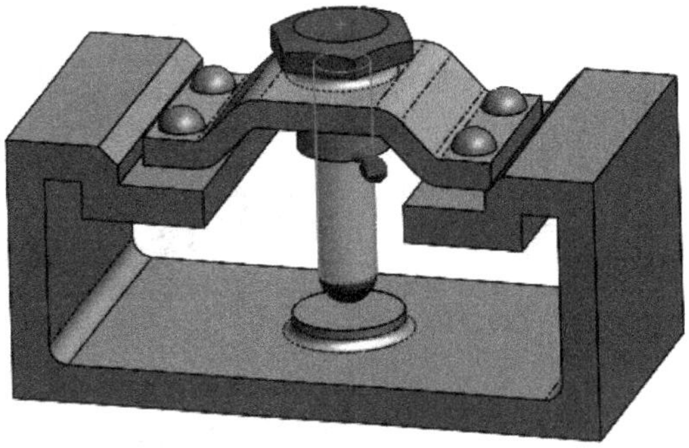

In the animation toolbar click on the **Motor** command and select the **Rotary** option. Select the cylindrical surface of the **Pin** to define the component and direction of the motor.

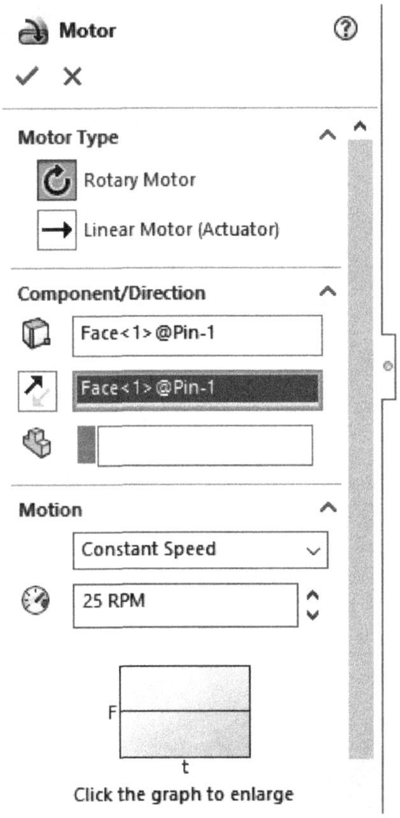

In the Motion section set the **Function** to **Constant Speed**, set the speed to **25 RPM** and click **OK**.

After adding the **Rotary Motor** select the **Calculate Simulation** button and wait for the simulation to be completed. By default, your animation will last 5 seconds. If instructed by your lab instructor, show your animation by pressing the **Play** and **Stop** buttons in the Animation toolbar.

After calculating the animation, you can replay it using the **Normal**, **Loop** or **Reciprocate** options.

The last step to finish your lab is to make an assembly drawing with an exploded view. Add a shaded Isometric and print a copy for your lab instructor, as shown.

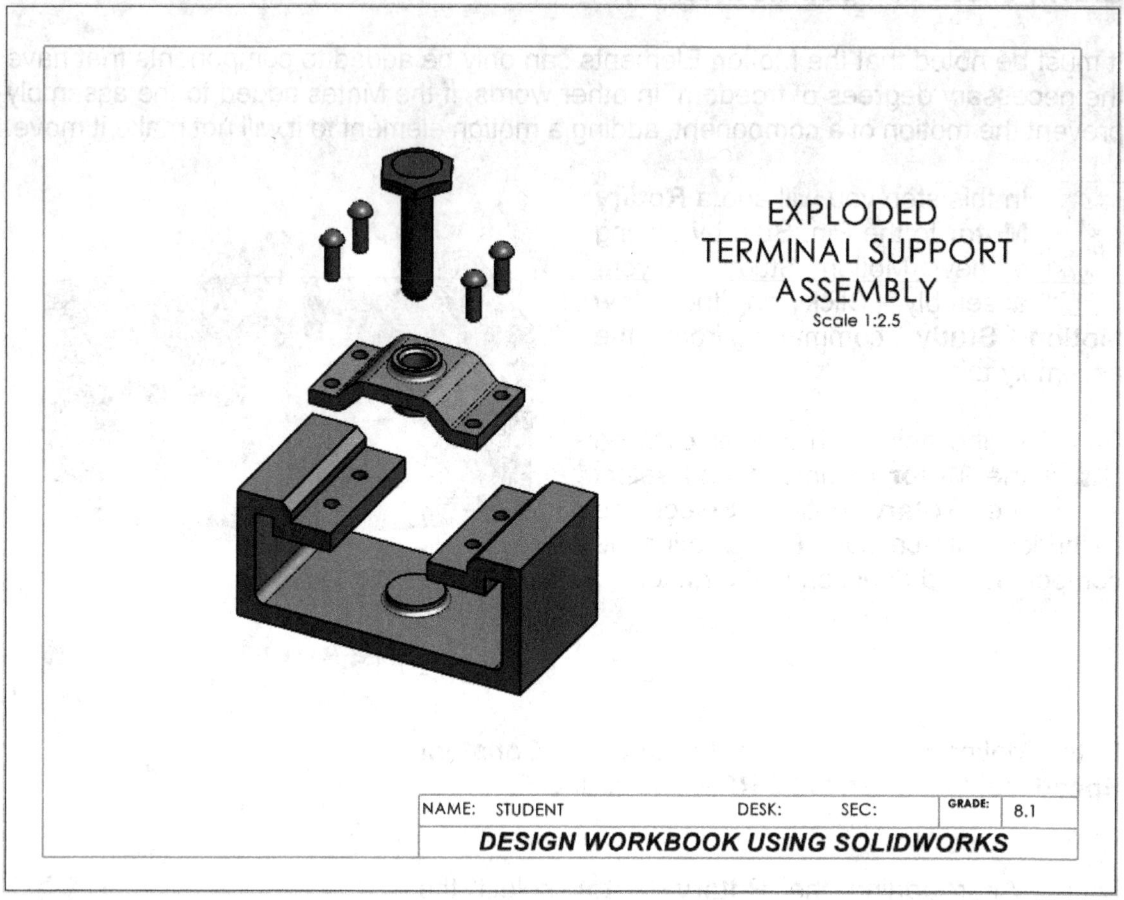

Exercise 8.2: Exploded Animation of the PULLEY ASSEMBLY

ASSEMBLY EXPLODED VIEW

In this lab you will make an exploded view of the previously made **Swivel Eye Block** assembly, make an animation, and save it as a video.

Open the **Swivel Eye Block** assembly and click on the **Exploded View** command from the Assembly tab.

The first component to explode is the **Big Rivet**. Select it in the screen, set the **Explode Direction** in the **Z** direction and enter a distance of **6.0"**.

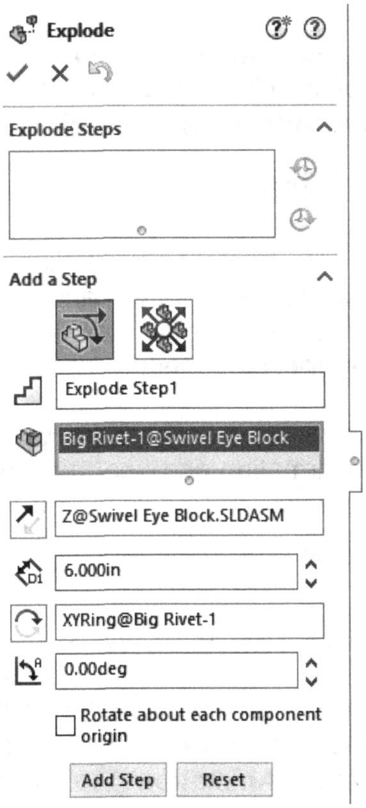

If needed, change the explode direction to make the **Big Rivet** go to the left, and click **Add Step** to continue to the next step, as shown.

Using the same steps, explode the **Pulley** in the **Y** direction down **6.00"**, change the direction to go down and click on **Add Step**.

To finish the exploded view, add the rest of the exploded steps as follows:

- Both **Small Rivets**, **6.00"** to the left
- Front **Base Plate**, **2.00"** to the left
- Back **Base Plate**, **3.00"** to the right
- **Pulley**, **3.00"** going down.
- **Eye Hook**, **3.00"** going up.

In this exploded view the **Spacer** will remain in place and not move. When the exploded steps are completed click **OK** to finish the **Explode** command. Your exploded assembly should look as follows.

EXPLODED VIEW ANIMATION

Now you will create an animation using the **Animation Wizard**. Activate the **Motion Study1** tab on the bottom left of the screen, and from the Animation toolbar click on the **Animation Wizard** command.

In the **Animation type** select the **Explode** option, click **Next** and set the **Duration** to **15 seconds**. Finally click **Finish** to create it.

PLAYING THE ANIMATION

In the Animation controller press **Calculate** or **Play** to generate the animation and play it, and evaluate the effect of the **Normal**, **Loop** and **Reciprocate** mode.

Show the animation to your lab instructor for confirmation and use the **Save As** command to save your assembly as **Exploded Pulley Assembly.sldasm**.

SAVE THE ANIMATION

From the Animation Controller select the **Save Animation** command. When the **Save Animation to File** dialog is presented, use the **.mp4** save type, set the animation to **15 seconds,** and click **Save** as shown.

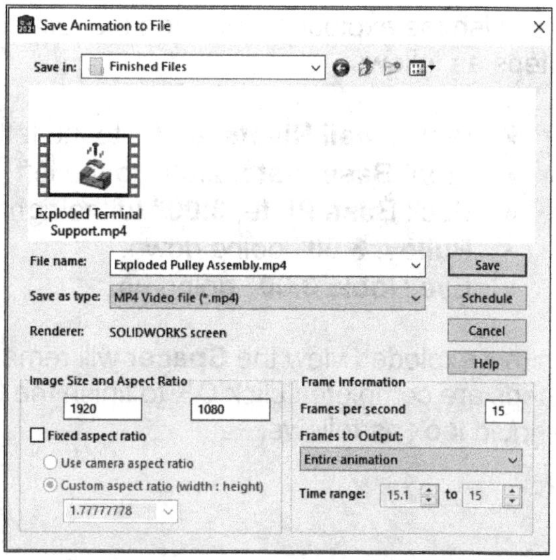

Use the default video compression setting with the default options and click **OK** to save it.

The animation will be played on screen at the same time it is being saved. Remember to adjust your screen size before saving the animation.

ANIMATION OF PULLEY ASSEMBLY MOTION

In this step you will make an animation of the pulley using a rotating motor. Open the **Swivel Eye Block** assembly, and with an isometric view orientation, select the front Base Plate and from the context toolbar select the command to make **Change Transparency**. The reason to make the part transparent is to be able to see the pulley rotate.

Since the **Pulley** is a revolved feature and it does not have any feature that can make its rotation visible, you will add a texture to better visualize the motion.

In the **Task Pane** select the **Appearances, Scenes and Decals** tab as shown. Click in the Appearances section, click and drag the **Texture** as indicated and drop it on the **Pulley**. In the context toolbar shown immediately after dropping the texture, select the **Assembly** option.

This way the texture will only be added to the **Pulley** in the assembly, and not in the part file. Think of it as painting the part *after* it is assembled, instead of painting it *before* adding it to the assembly.

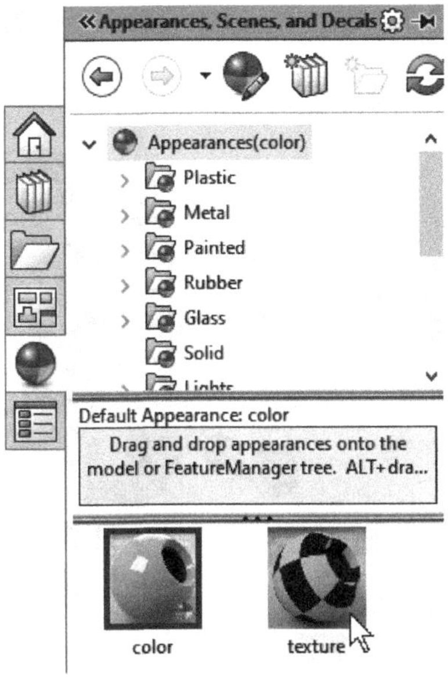

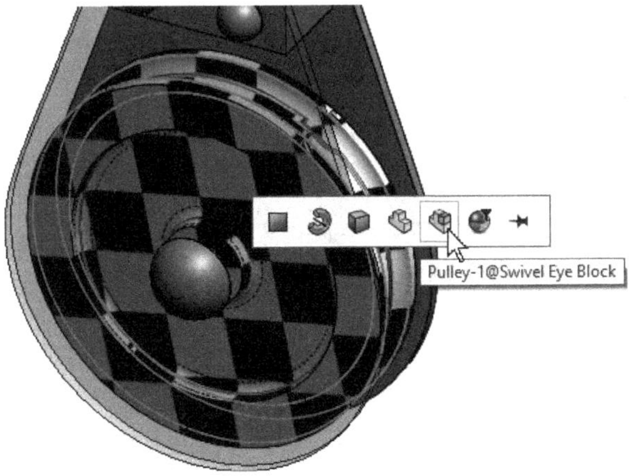

Now select the **Motion Study1** tab on the bottom left of the screen and click on the **Motor** command from the Animation toolbar. Select the **Rotary Motor** option and click on a circular edge (or cylindrical face) of the **Pulley** to attach the motor, and then select an outside edge to define the rotation direction. Using the **Constant Speed** motion set the speed to **20 RPM** and click **OK** to add the motor.

Now click on the **Calculate** or **Play** command from the Animation toolbar to calculate the animation. Play the animation to show it to your lab instructor.

To finish this lab, make a new drawing of the Exploded Pulley Assembly, including a shaded Isometric view and print a copy for your lab instructor.

Save your drawing as **Exploded Pulley Assembly.slddrw**.

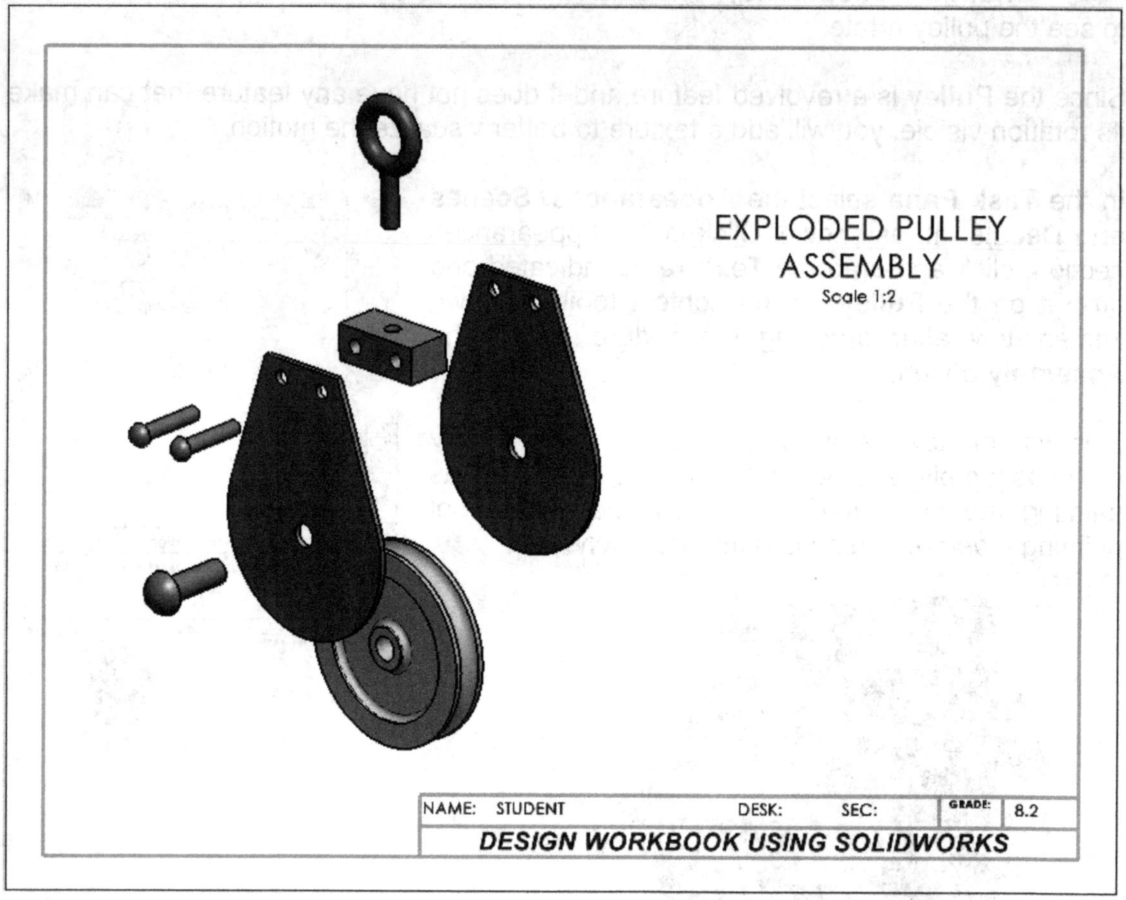

Exercise 8.3: Creating Drawing Views for Manufacturing

In this lab you will learn how to add multiple model views to a drawing, used to correctly document a part's features and dimensions for manufacturing. In this lab you will be working in the detail drawing environment.

Open the **Toe Clamp** part previously made in **Exercise 3.4**, make a new drawing using the **Inches** drawing template, and the **Model View** command is automatically opened. Since the **Toe Clamp** part was already open it is listed in the Open Documents list. Click on the **Next** button to continue.

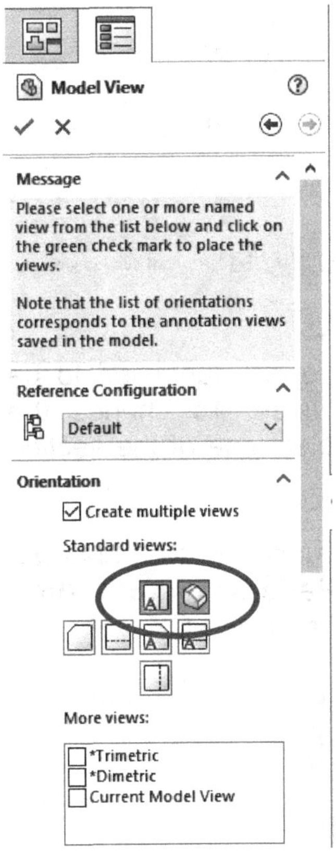

In the next page check the **Create Multiple Views** checkbox, select the **Top** and **Isometric** views, and click **OK** to add the views to the drawing sheet.

Now select the Isometric view, and change the **Display Style** to **Shaded with Edges**, set the scale to **User Defined** and enter **1:3**. Drag to locate the isometric view on the upper right side of the drawing.

Select the **Top** view, change the **Display Style** to **Hidden Lines Visible** and move it up in the drawing sheet.

From the **Annotation tab** select the **Center Mark** command. In the **Slot Center Marks** section select the **Slot Ends** option. Click in one of the slot's arcs and click **OK** to continue.

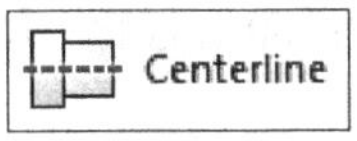

The next step is to add the centerlines. Click on the **Centerline** command, in the **Auto Insert** section check the **Selected View** option and click to select the **Top** view to add the centerline. Click **OK** to finish.

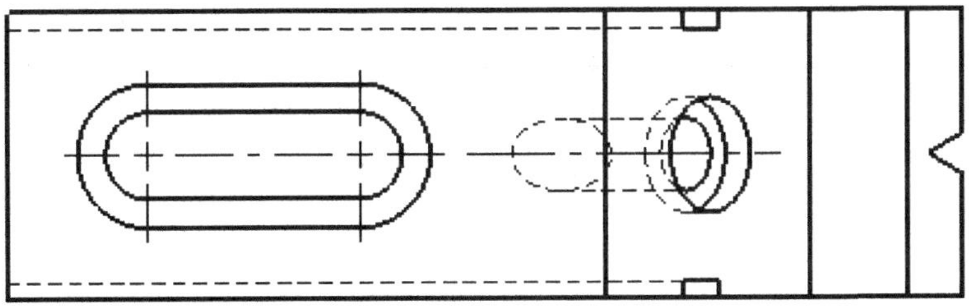

Section View

Now you will add a section view of the **Top** view. A section view is created by drawing a section line through a model to show internal details.

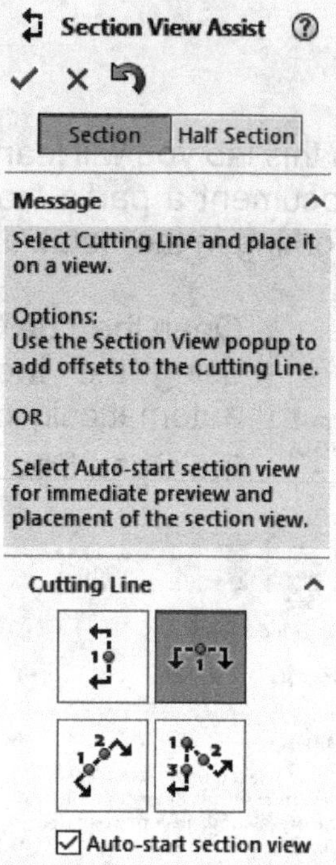

From the **Drawing** tab, select the **Section View** command. In the properties select the Horizontal Section option as shown.

Now move your cursor to the **Top** view and click on either arc of the slot to locate the section line. Once located, click OK in the floating toolbar to make the section.

As you move the cursor, a preview of the section will follow it. Move the section view below the **Top** view and click to locate it.

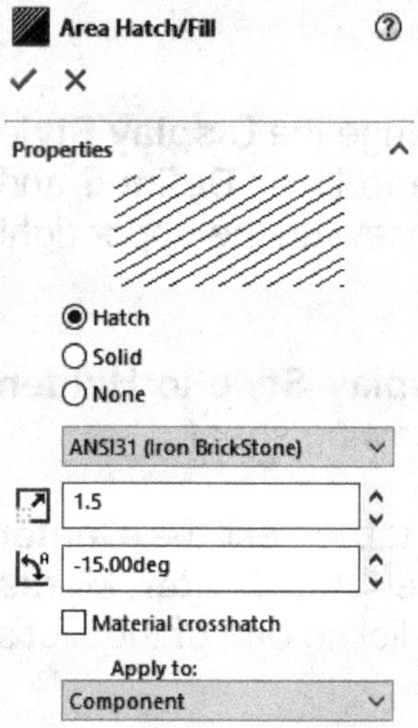

Right click on the **Section A-A** view, and from the context menu select **Tangent Edge**, **Tangent Edges Removed**. To complete the section view, select the **Centerline** command, and add the centerlines to the view as we did in the **Top** view.

The cross hatching of section lines is not supposed to be parallel or perpendicular to model lines. To change the orientation of the hatching, click to select one of the section view's faces.

In the **Area Hatch/Fill** properties uncheck the **Material crosshatch** option, change the **Hatch Pattern** to **ANSI31 (Iron BrickStone)**, set the scale to **1.5** and the Pattern Angle to **-15°** and click **OK**.

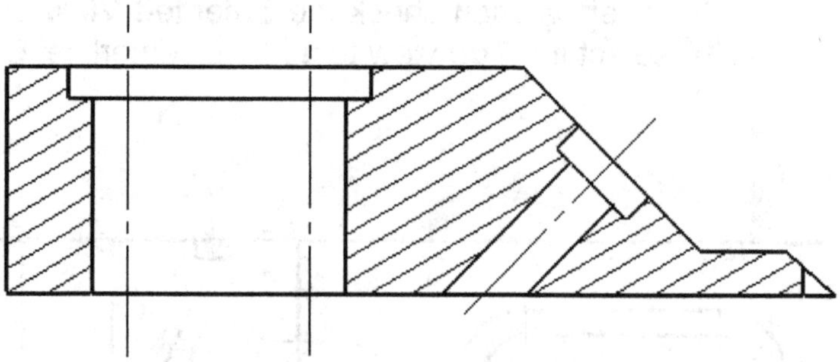

SECTION A-A

Projected View

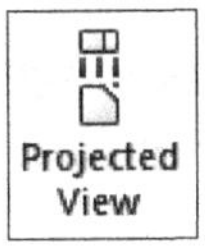

The next view that you are going to add is a projected view to the right of the Section view. From the **Drawing** tab select the **Projected View** command. When asked to select a view from which to project, select the Section view. The projected **Right** view will be pre-viewed after you move the cursor to the right side. Locate the view next to the Section view and click **OK**.

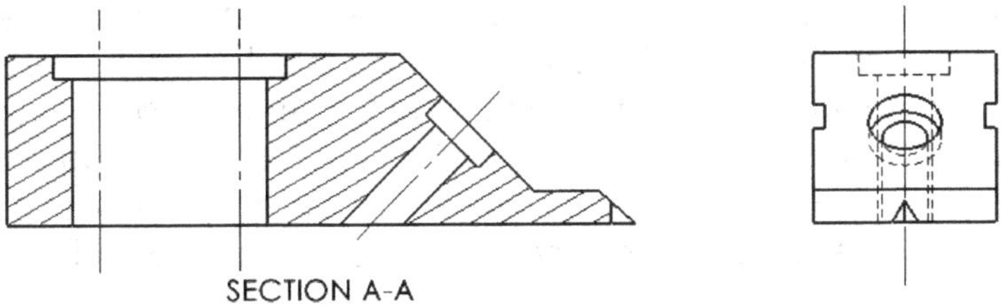

Since the **Right** view was projected from a view with **Hidden Lines Removed**, it will have the same **Display Style**. Select the **Right** view and change it to **Hidden Lines Visible**. To complete the view, add the **Centerlines** as before.

Auxiliary View

After inspecting the drawing views added so far, you will notice that the counterbore hole in the inclined face is not parallel to the page in any view. In order to correctly dimension the features in this face, you need to add an **Auxiliary** view. From the Drawing tab, select **Auxiliary View**. When asked to select a reference edge, select the inclined edge in the section view to make the auxiliary view parallel to this edge. After selecting the edge move your mouse pointer to locate the **Auxiliary** view. After locating the **Auxiliary** view, an arrow indicating the view's label and orientation is added. As you did before, add the **Centerlines** and change the view's display style to **Hidden Lines Visible**. You can also drag the label arrow to resize and locate as needed.

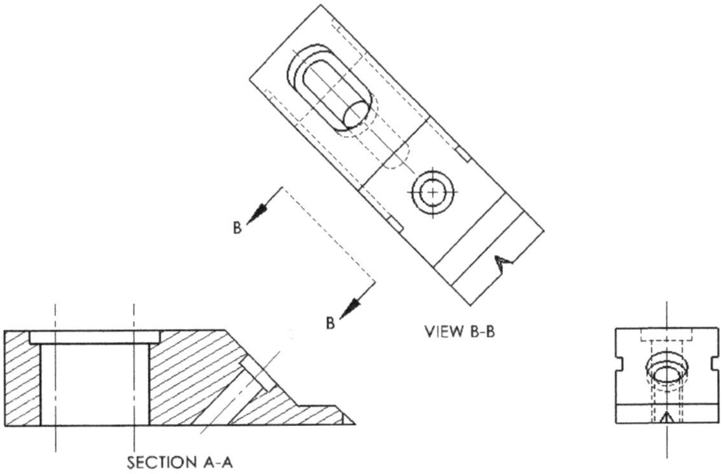

Detail View

When a model has small features that cannot be effectively dimensioned in the standard views, you need to create a **Detail** view, which enlarges a selected area to clearly show and dimension small features. From the Drawing tab select the **Detail** View command.

The sketch **Circle** tool is automatically selected, and you are asked to draw a circle around the area to be enlarged. Draw the circle around the "V" cutout at the end of the **Top** view. When the circle is finished, locate the **Detail** view between the **Top** and **Section** views. A note with the view's label and scale is automatically added. If the **Detail** view is too large or too small, you can select the **Detail** view and change its scale in the view's properties. If needed, you can also drag the detail circle to enlarge or reduce the area of the detail view. Your completed drawing should look as shown next.

Save your finished drawing as **TOE CLAMP VIEWS.slddrw** and print a copy to be turned in to your lab instructor.

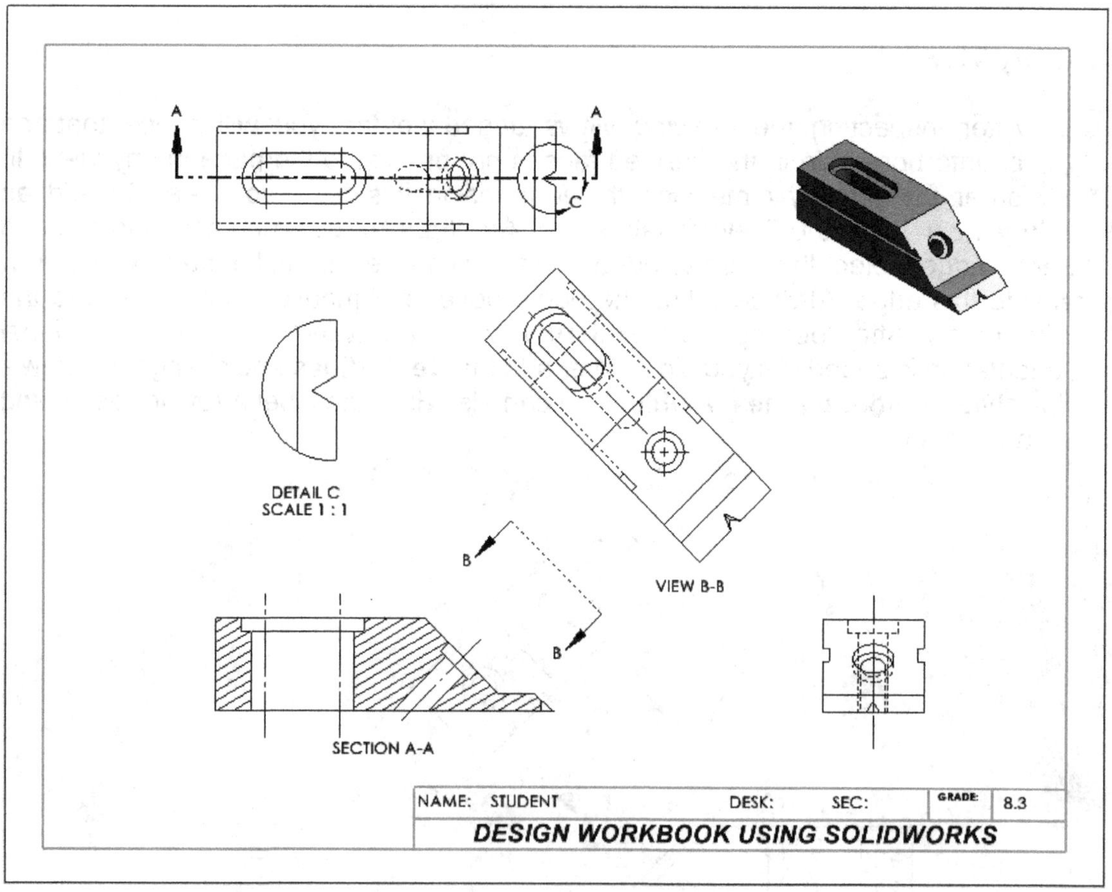

Exercise 8.4: Rapid Prototyping of a Solid Model Part

In this Exercise you will build a model assigned by your instructor and make the **.STL** file needed to make a physical prototype using a 3D printer. The **.STL** file extension is the standard input file format used in 3D printers and other rapid prototyping machines. **.STL** is an acronym for **Stereo Lithography**, the early process used to create rapid prototypes; however, more modern systems can also read the **.STEP** file format.

SAVE THE SOLID MODEL AS AN .STL FILE

After you build your assigned model or open an existing one, select the menu **File, Save As**, and from the File Type option list select **STL** and save your file.

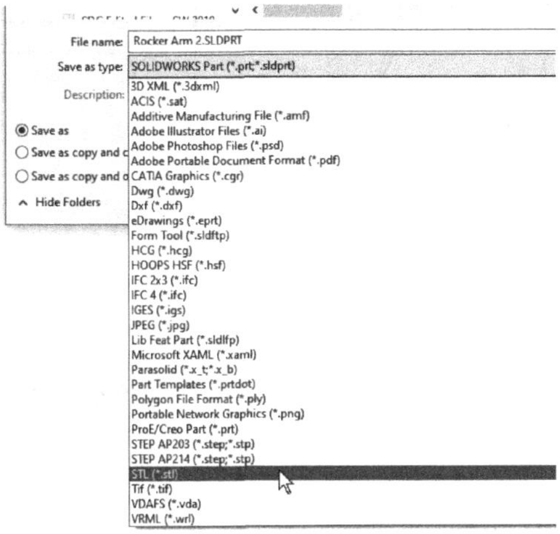

Before you can make your physical 3D printed part, depending on the type of equipment you are using, you will have to open the machine's software and prepare the file that will be sent to your specific machine.

In most 3D printers, commonly known as additive manufacturing, a stock of plastic material in the form of a thin filament is fed through a heated nozzle to melt it. The molten plastic is then deposited one thin layer at a time until the part is complete. The software to prepare the information used by the 3D printer to define the motion of the nozzle to "print" each layer is commonly known as "slicer" software, which reads the **.STL** file and 'slices' the model to define the trajectory of the nozzle to build each slice of the model.

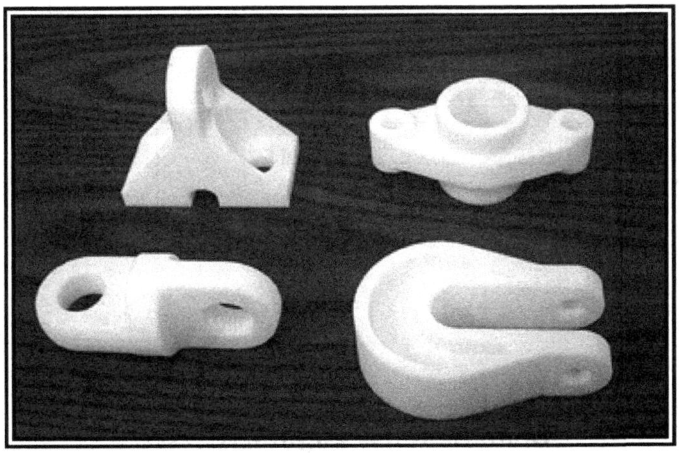

Rapid Prototype Models of Parts in Assignments 8.4.1 to 8.4.4.

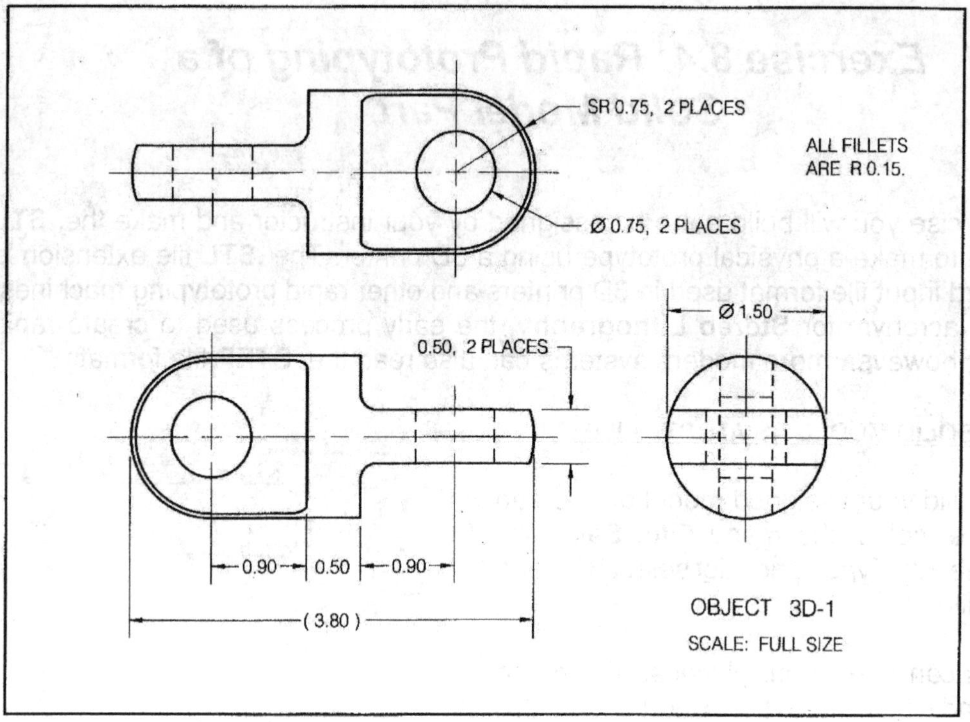

Solid Model Assignment 8.4.1

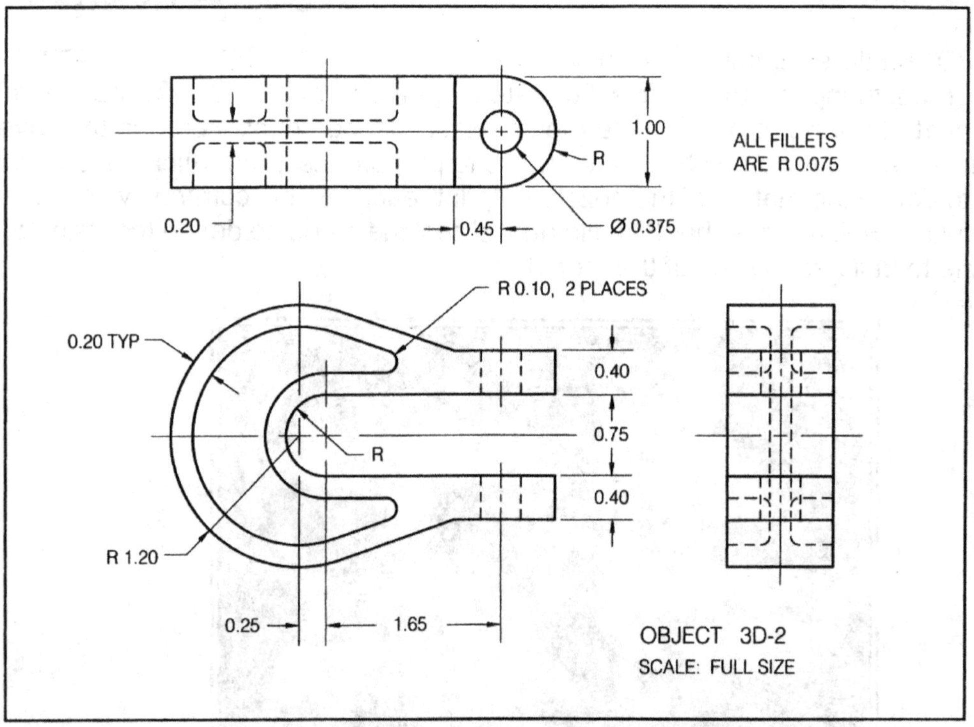

Solid Model Assignment 8.4.2

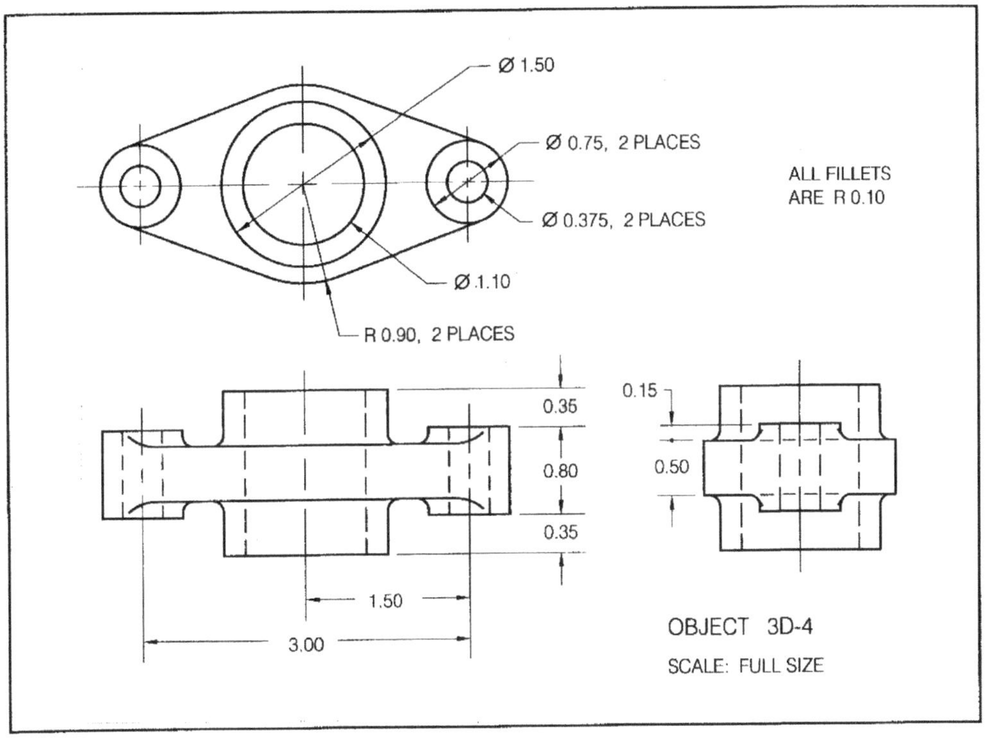

Solid Model Assignment 8.4.3

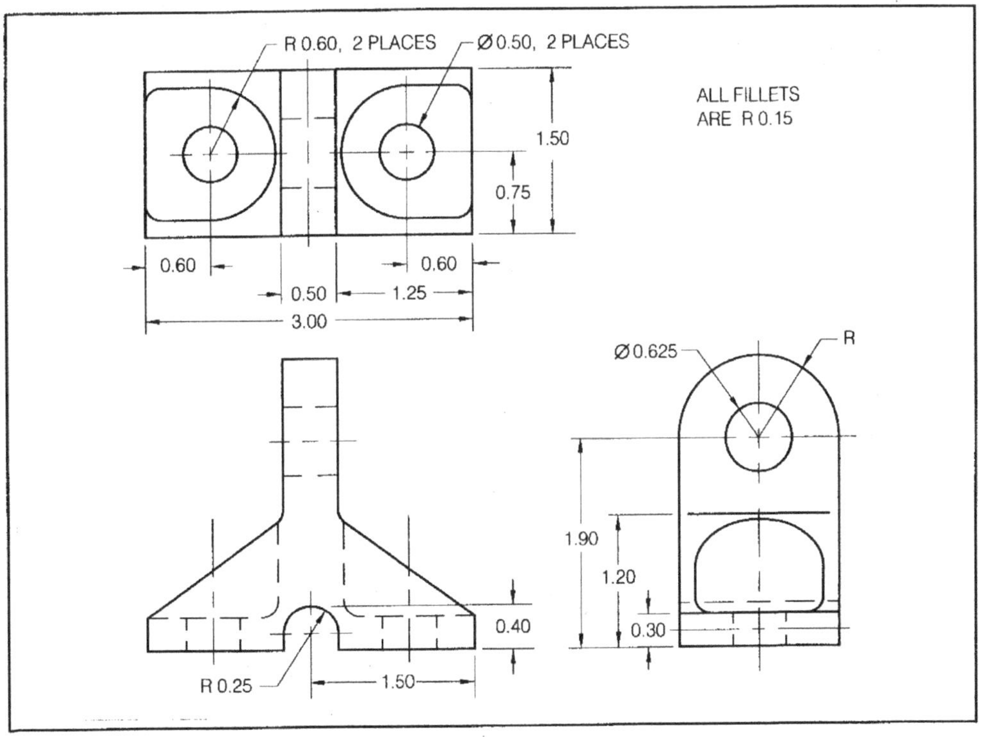

Solid Model Assignment 8.4.4

NOTES:

Design Workbook Lab 9
Section Views in 2D and 3D

In this lab you will learn how to make section views of solid models for inspection and visualization purposes, as well as how to use different options for drawing section views.

3D SECTION VIEWS

To visualize a 3D model's internal features, select the menu **View, Display, Section View**, or the **Section View** command from the View toolbar.

Your model will be cut using the last settings used, or if you have not used the **Section View** command in the current session, the **Front Plane** will be used as a default section plane, and the options will be displayed as shown below.

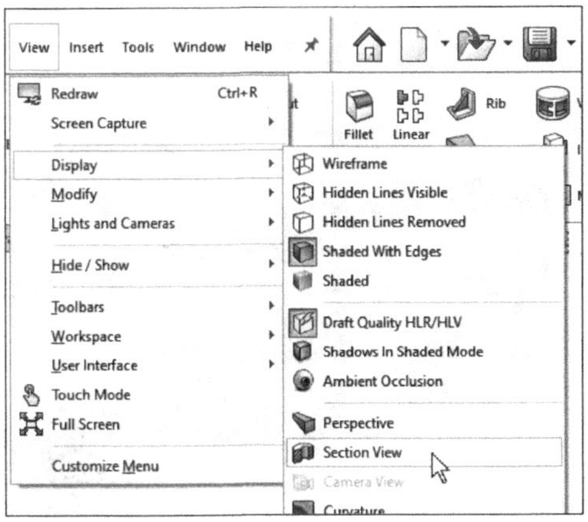

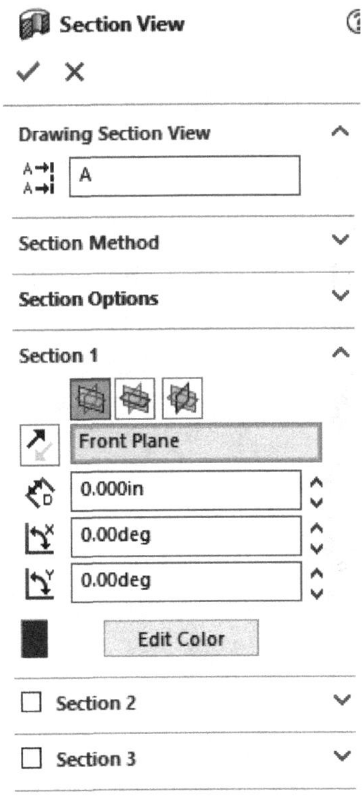

In the **Section View** options you can modify the section plane's orientation, the view's direction, the **distance** from the origin, and **rotation** about the **X** and **Y** axis as needed to inspect internal features of your model.

Additionally, you can click-and-drag the section plane manipulator arrows directly in the screen.

If needed, you can use up to three different section planes by activating the **Section 2** and **Section 3** options.

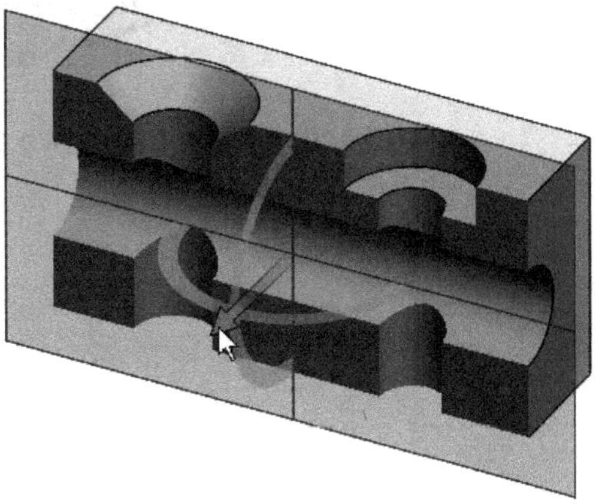

You will see a dynamic preview of the section view as you make changes. If you click OK, the section view will remain visible and the Section View command will be activated.

Remember the model has not been cut; it is only a temporary view used to measure and inspect hidden or otherwise difficult to see features of your design. To exit the Section View, select the menu **View**, **Display** and deselect the **Section View** command, or click on the **Section View** command again to restore the original 3D model view.

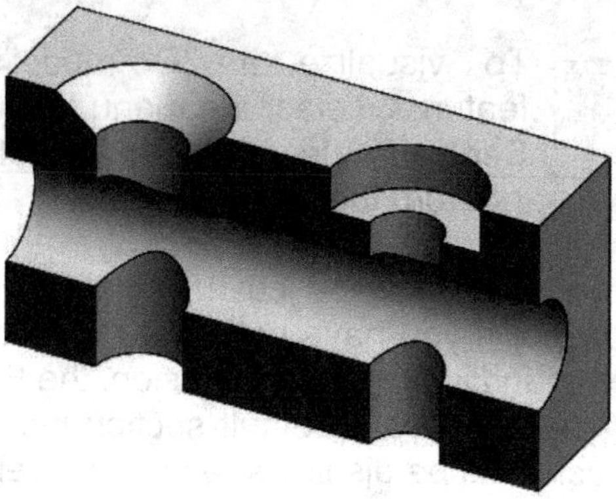

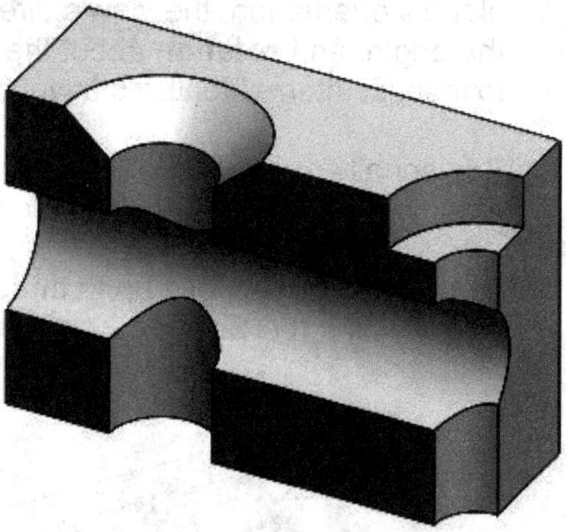

Section View using two section planes

2D DRAWING SECTION VIEWS

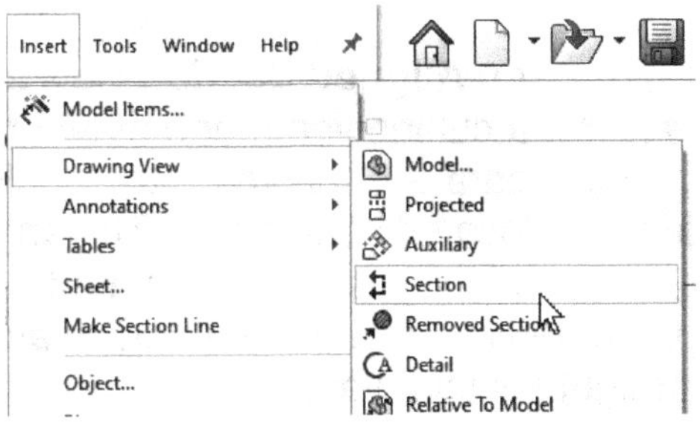

In the previous lab you learned how to add a section view in a drawing. In this lab you will learn additional options and types of drawing section views, as well as some advanced options.

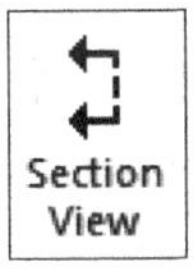

After you select the **Section View** command from the Drawing tab, you can see the different options available, depending on the model and type of the section view needed. The section line arrows point in the direction of view. To reverse the Section View's direction, you can select the view and click on the **Flip Direction** button in the view's properties.

A **Vertical Section** view adds a vertical line to the view to be sectioned. After the section's line is located, you will see a preview of the section, and click to locate the new view.

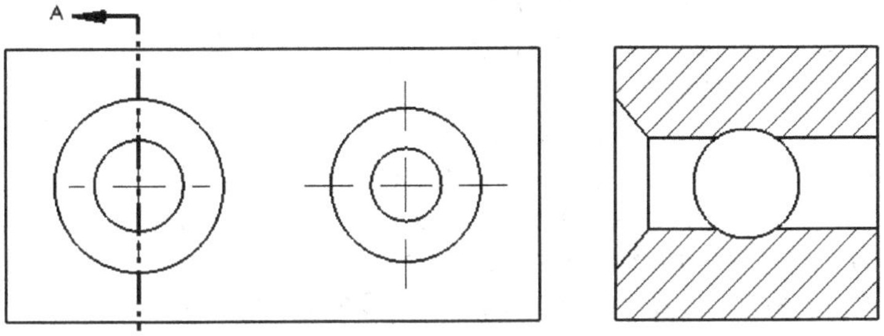

SECTION A-A

A **Horizontal Section** view adds a horizonal line to the view to be sectioned. After locating the section line, the new section view is created.

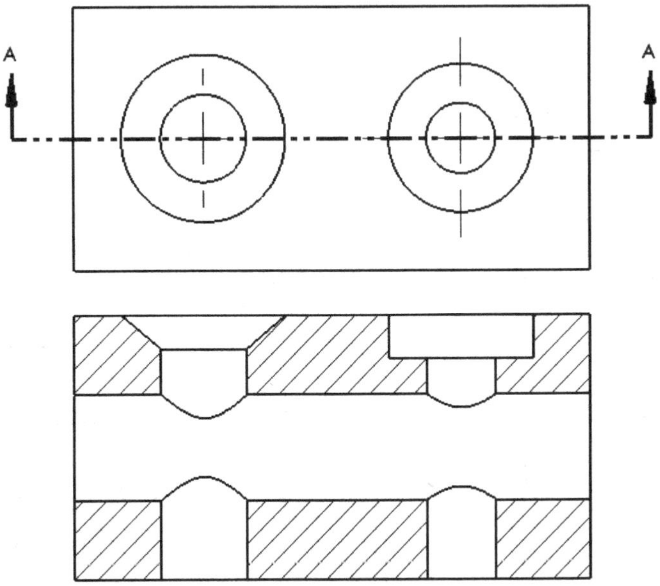

SECTION A-A

An **Aligned Section** view adds a cutting line made of two non-parallel planes to include features located at an angle. The angled section plane is then rotated and presented in the same orientation as the straight section plane.

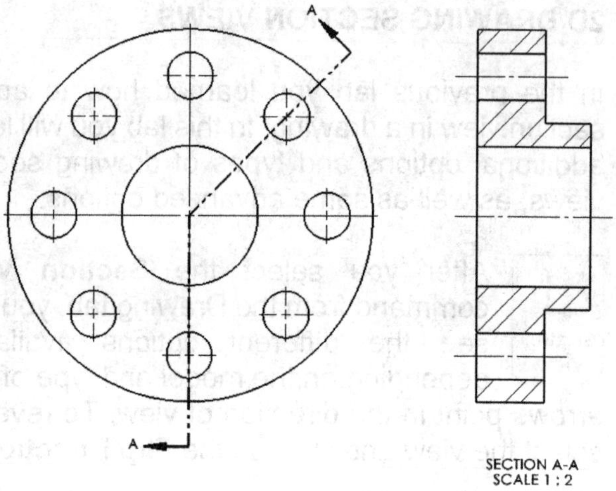

A different type of view is the **Broken-Out Section**, which allows you to make a cutout of a main view to see internal features in the same view, without creating a new view. After selecting the **Broken-out Section** you are asked to draw a closed spline curve to define the area to cut out.

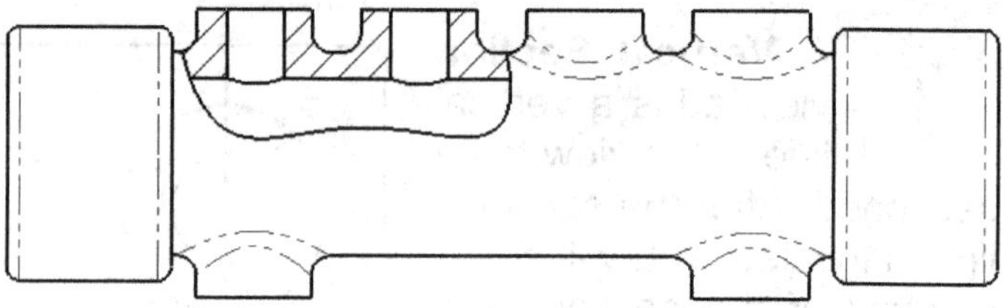

Another option is to add an **Offset Section** view. First you need to draw the connected section lines, and then click on the Section Line command. You will be asked to make a **Foreshortened** section normal to the cutting line, or a **Standard** section line, in which the cutting line is unfolded.

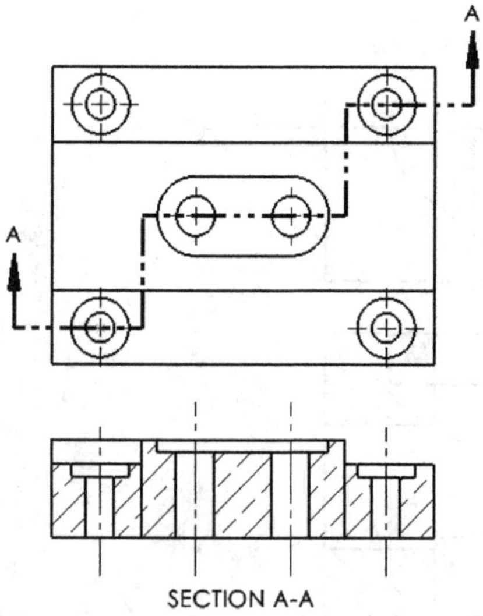

Foreshortened Section View

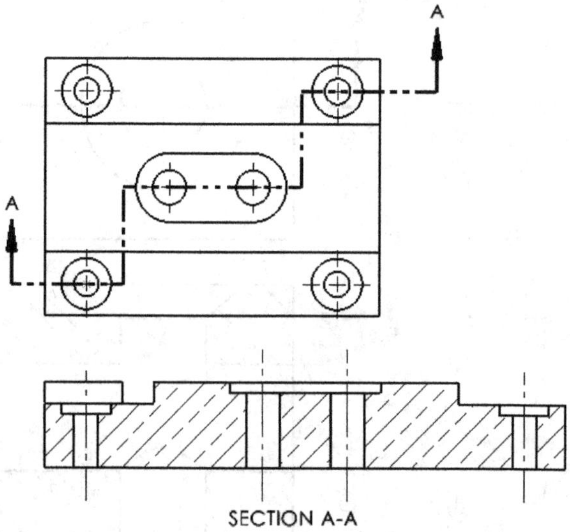

Standard Section View

Exercise 9.1: Rod Base Section View

In Exercise 9.1 you will build the **Rod Base** part using commands learned in previous labs. You will make a 3D section view of the model to visualize the internal features of the model, and then create a 2D drawing with a Top, Isometric and a Section view. To get started, you will build the solid model first.

MAKE THE ROD BASE

Start a new part using the template in **Inches**, add a new sketch in the **Front Plane** and draw the profile shown. When finished, use the **Revolved Base** command, and make a **360°** revolution.

To finish the part, add a **0.125" x 45° Chamfer** to the top inner edge, and the three **0.0625"** fillets indicated.

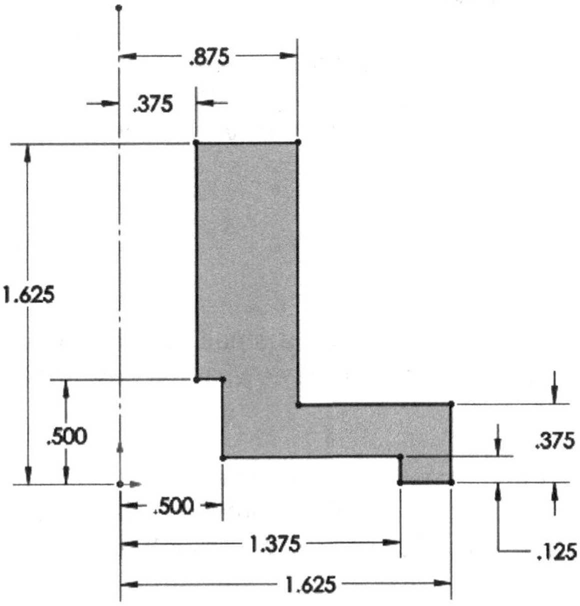

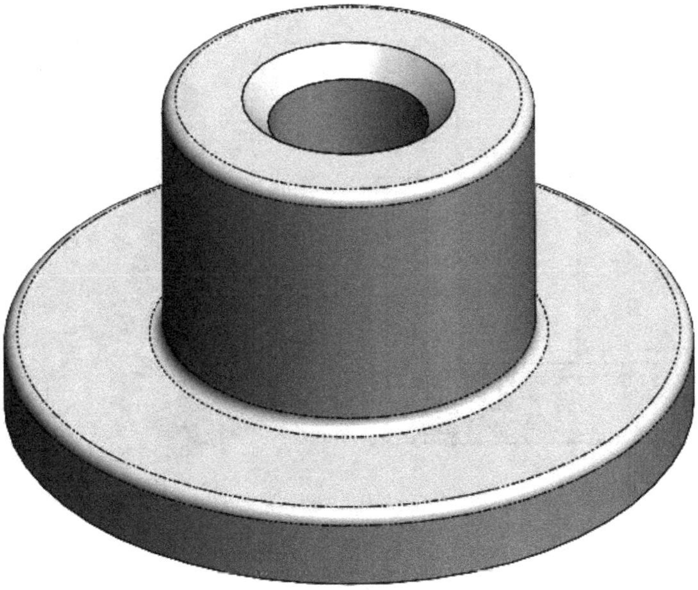

Change to a **Top** view, add a sketch on the flat face and draw the profile indicated.

Add a **2.25"** diameter circle in the origin. Change it to **Construction** geometry and use it to locate the **0.25"** hole and make a Through All **Extruded Cut**.

Next, use the **Circular Pattern** command to make **4** equally spaced copies of the hole.

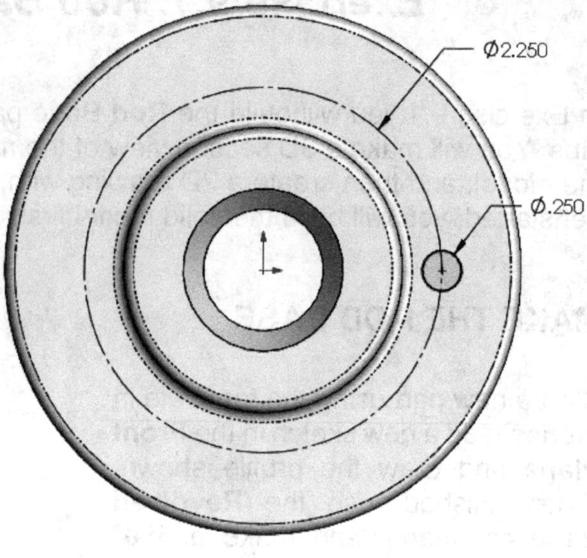

The last feature will be a small hole that goes through one side of the top cylindrical face. Change to a **Right** view, add a new sketch in the **Right Plane** and draw the profile shown next.

Draw and dimension the circle, and to fully define your sketch add a **Vertical** relation between the center of the circle and the Origin. Change to an Isometric view and make an **Extruded Cut** using the Through All end condition towards the right side of the part.

Set the part's material to **Manganese Bronze** from the **Copper Alloys** library and save your file as **Rod Base.sldprt**.

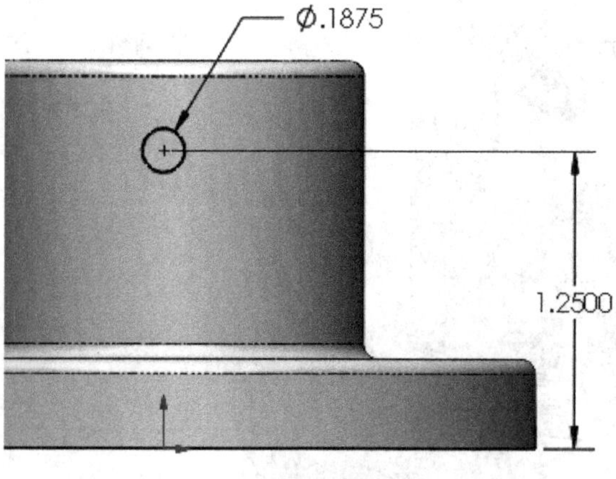

3D SECTION VIEW OF THE ROD BASE

You will now make a 3D section view of the **Rod Base**. Select the menu **View, Display, Section View**, or the **Section View** command from the **View** toolbar.

The Front Plane is the default option for Section 1. Leave the Offset Distance at **0.00”** and click OK to continue.

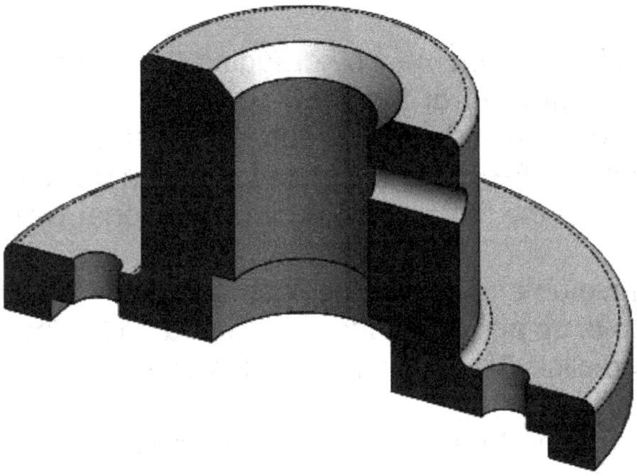

If needed, click on the **Reverse Direction** button to change the section view’s orientation. Your sectioned model will look as shown.

When ready, click on the **Section View** command again to turn it off and save your model before making the detail drawing.

ROD BASE DETAIL DRAWING

Using the **Inches** template make a new drawing. Before adding drawing views, there are some system settings that can be changed to control the default options of new drawing views. Select the menu **Tools, Options, System Options**, and go to the **Drawings, Display Style** section, as shown.

In the **Display Style** section set the default to **Hidden lines visible**, and in the **Tangent Edges** section select the **Removed** option. Click **OK** to continue.

Remember, the settings listed in the **System Options** tab control the behavior of all documents, and the settings listed under the **Document Properties** tab only affect the current document.

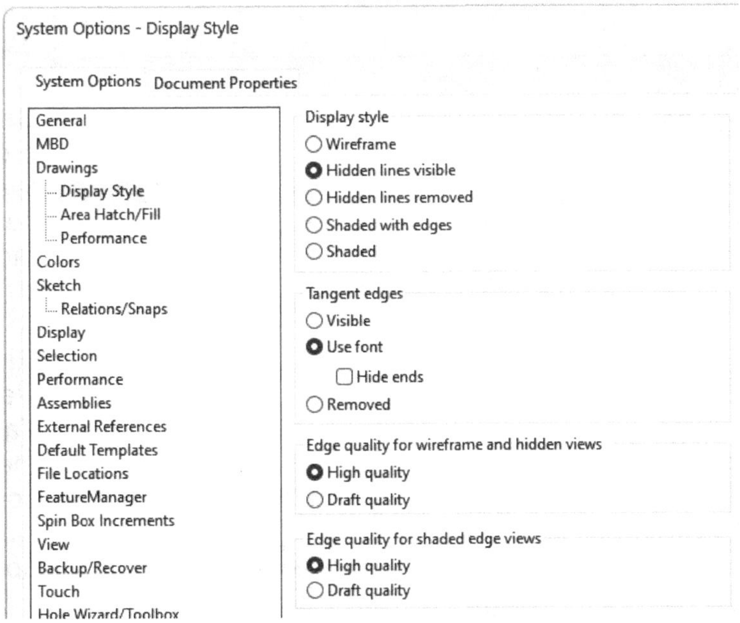

After setting the system options go back to the drawing sheet and select the **View Palette** tab in the **Task Pane**. In the **View Palette** you can see a preview of each view orientation of the part currently open in **SOLIDWORKS**, or you can select a different part from the drop-down list at the top of the **View Palette**.

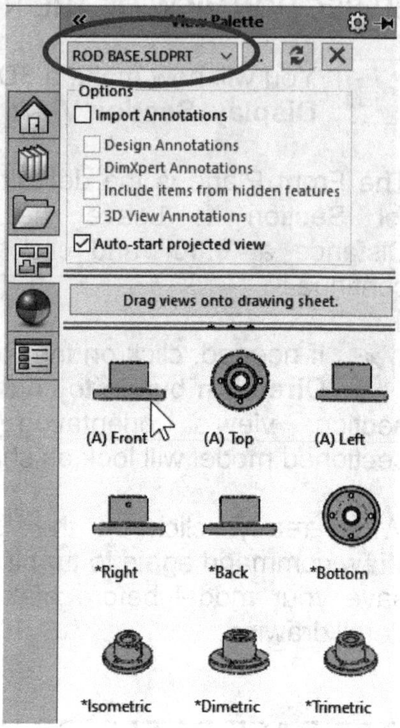

Add the **Front** view by dragging the **Front** preview to the lower left of the drawing sheet; since the "Auto-Start projected view" option is activated, immediately after the **Front** view is added, you can move the cursor around the first view to project additional views.

Move up and click in the sheet to add a **Top** view, then move up and right, and click to add an **Isometric** view, and click **OK** to finish projecting views.

Since the Front view will be replaced by a section of the **Top** view, select the **Front** view, and **delete** it. If needed, right click on **Sheet 1** in the FeatureManager and select **Sheet Properties**, change the sheet's scale to **1:1** and click **OK**. This option will change the scale of all views in the drawing sheet.

Note: For a view's scale to be controlled by the sheet's scale, select a view, and make sure the option **Use sheet scale** is selected in the view's properties.

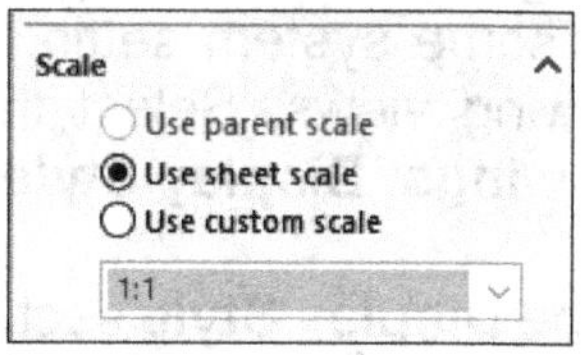

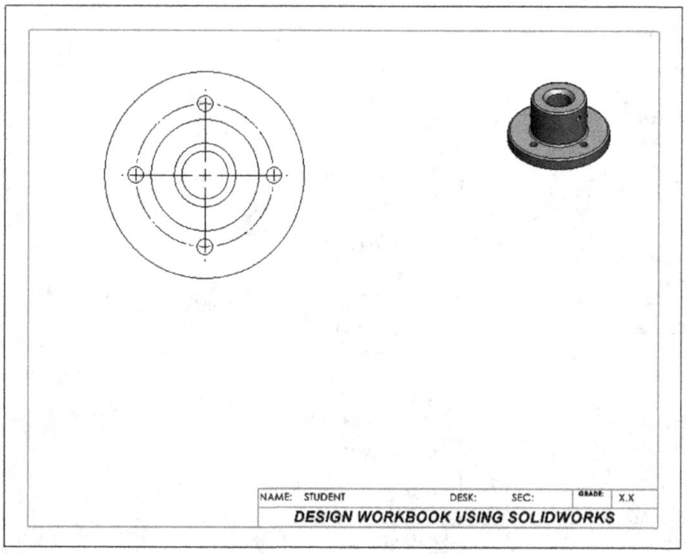

Select the **Isometric View**; in its properties make sure the **scale** is set to **1:2,** and Display Style is set to **Shaded with edges** mode.

Right Click in the isometric view, select **Tangent Edge** and **Tangent Edges Visible**. Move the view to the upper right-hand corner of the drawing sheet. Now move the image to the upper left of the drawing sheet.

To add a section view, select the **Section View** command from the Drawing tab. In the Section View properties select the **Horizontal** section line option. Move your cursor to the center of the **Top View**; when the **Section Line** snaps to the center, click to locate the section line, move your cursor down to where the Front view used to be, and click to finish the command.

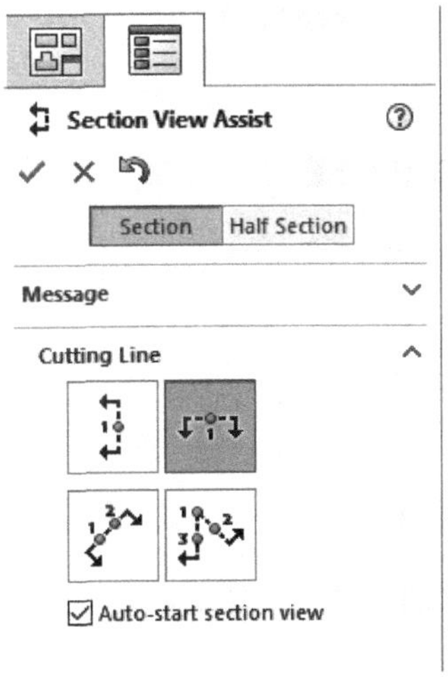

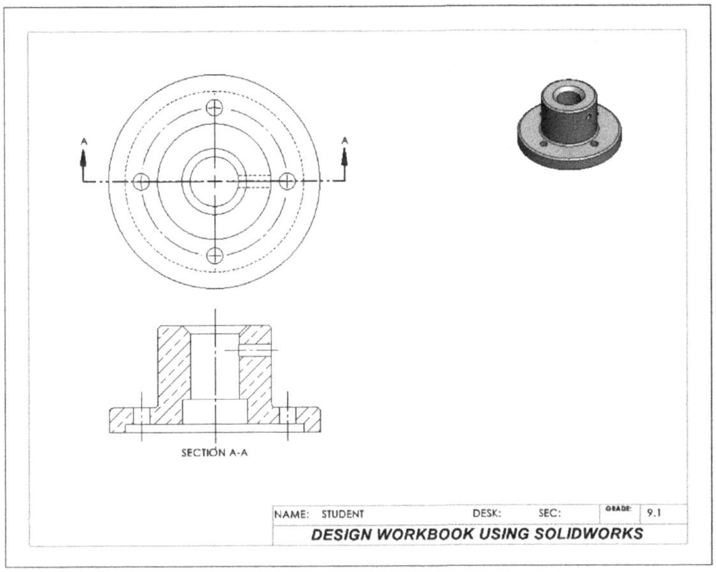

Notice the section view is aligned vertically with the top view, and parallel to the section line. The section line is automatically labeled **A-A,** and the section view is also labeled **Section A-A** as indicated.

If the section line arrows are pointing down, select the section view and click in the **Flip Direction** button. Also, if the section view has the **Hidden Lines Visible** display style, change it to **Hidden Lines Removed**. Finally, select the **Centerline** command from the Annotations tab, in the Auto Insert section activate the option **Select View** and select the section view to add centerlines.

Now you need to add a Right view. From the Drawing tab, select the **Projected view** command. Select the section view, move to the right to project the **Right** view, and click to locate the view on the sheet. By projecting the view, the **Right** view will be aligned horizontally with the section view and inherit its display style properties.

Change the **Right** view's display style to **Hidden Lines Visible**, add the view's centerlines and add a note to indicate the part's name as **ROD BASE**, the scale and material. Your finished drawing will look like the following.

When you are finished save your drawing as **ROD BASE.slddrw**, and print a hard copy to submit to your instructor.

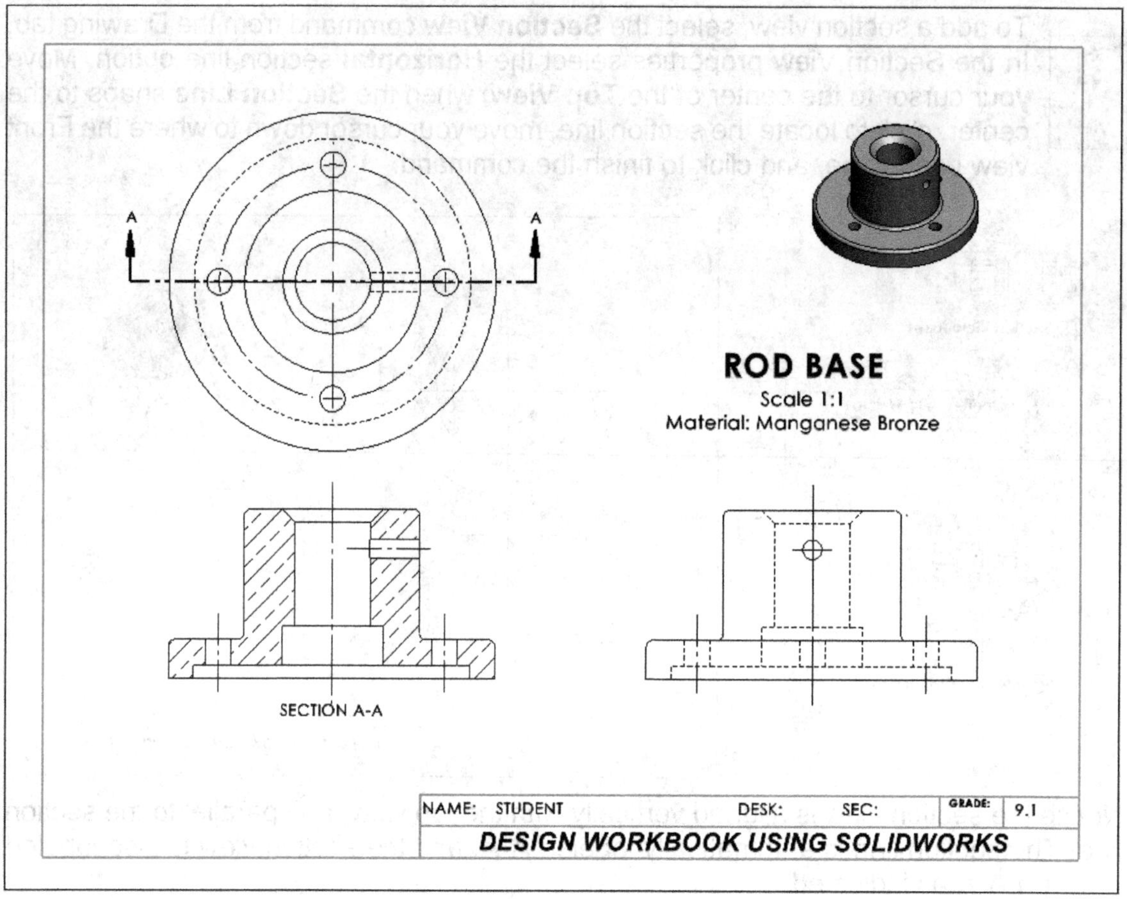
A
A
ROD BASE
Scale 1:1
Material: Manganese Bronze
SECTION A-A
NAME: STUDENT
DESK:
SEC:
GRADE:
9.1
DESIGN WORKBOOK USING SOLIDWORKS

Exercise 9.2: Tension Cable Bracket SECTION VIEW

In Exercise 9.2 you will build the **Tension Cable Bracket**. You will make a 3D section view of the model to visualize the internal features of the model, and then create a 2D drawing with a Top, Isometric and a Section view. To get started, you will build the solid model first.

BUILDING THE TENSION CABLE BRACKET

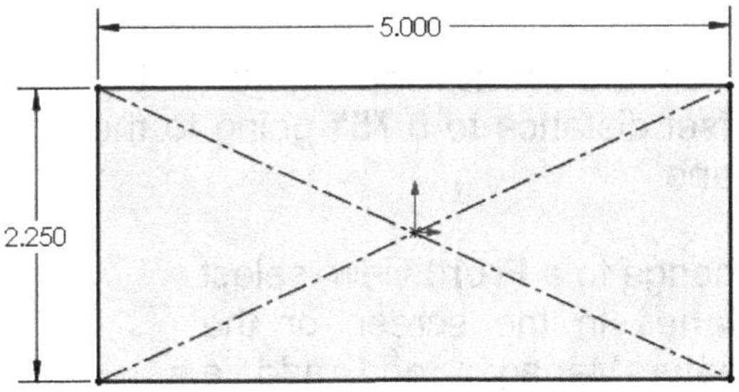

Start a new part using the **ANSI-INCHES** template, add a new sketch in the **Top Plane** and draw the profile shown. Extrude it upward **0.25"** and rename the feature "Base Plate."

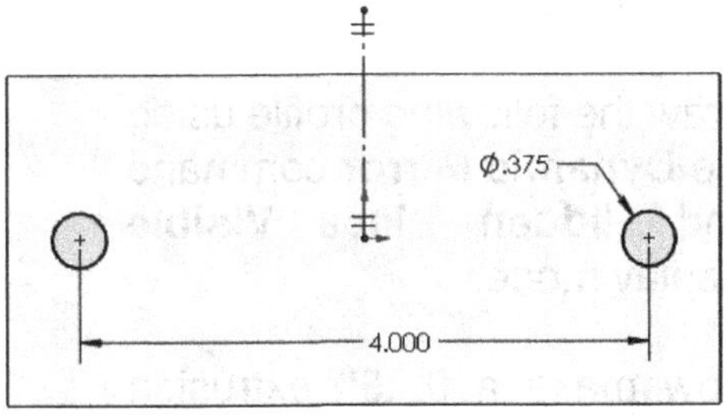

Now add a sketch in the top face of the Base Plate and draw the next sketch profile. Add a **Horizontal** relation between one hole and the origin to fully define the sketch. Use the **Dynamic Mirror** command to simplify your work.

Make an **Extruded Cut** using the Through All end condition, activate the **Draft** option and set the angle to **15°** to make two tapered holes.

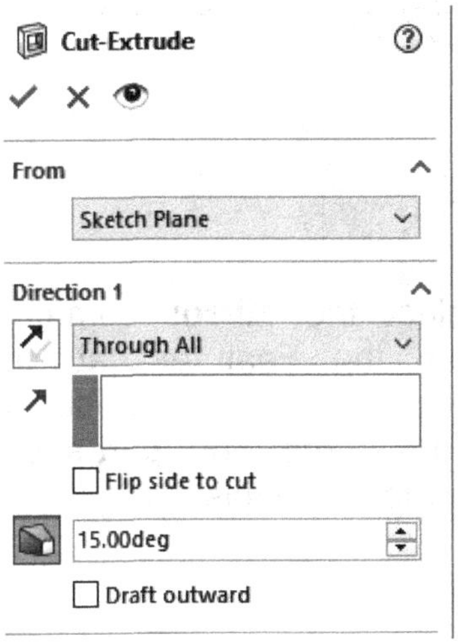

Now add a new sketch on the Top face. Draw a circle centered at the origin with a diameter of **1.75"**. Make an **Extruded Boss 0.875"** in Direction 1, and **0.875"** in Direction 2.

In the next step change to a Top view, add a sketch on the topmost face, draw a circle at the origin and dimension it **1.00"** in diameter.

Make an **Extruded Cut** using the Through All end condition, turn on the **Draft** option with **7.5°** and click **OK** to finish.

Change to a **Hidden Lines Visible** mode to see the tapered hole.

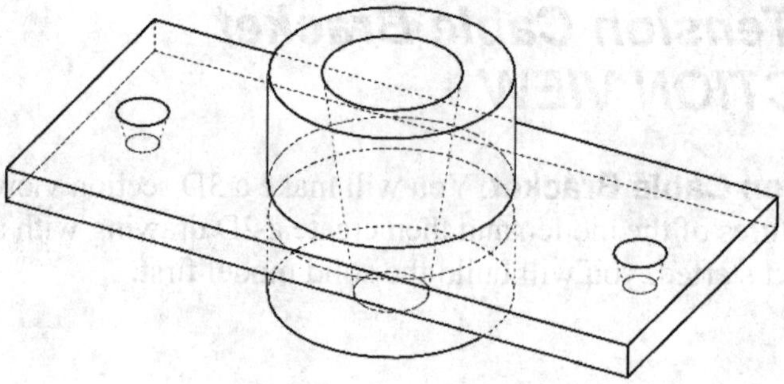

In the next step you will add two triangular supports. These supports will be offset from the center, and you will need to add a reference plane.

Select the **Reference Geometry** command and click on **Plane** from the drop-down list.

Select the **Front Plane** in the first reference from the fly-out FeatureManager, set the offset distance to **0.75"** going to the front of the part and click **OK** to create the new plane.

Change to a **Front** view, select **Plane1** in the screen or the FeatureManager and add a new sketch on it.

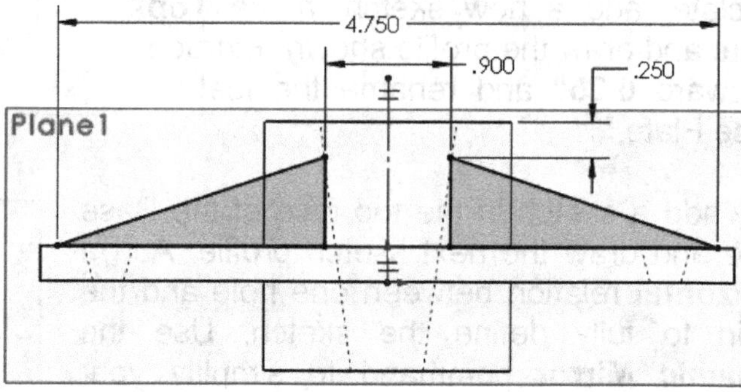

Draw the following profile using the **Dynamic Mirror** command and **Hidden Lines Visible** display mode.

Now make a **0.25"** extrusion using the **Blind** end condition going towards the center of the part.

Select the **Mirror** command from the Features tab and mirror the triangles about the **Front Plane**. Your part should now look like this.

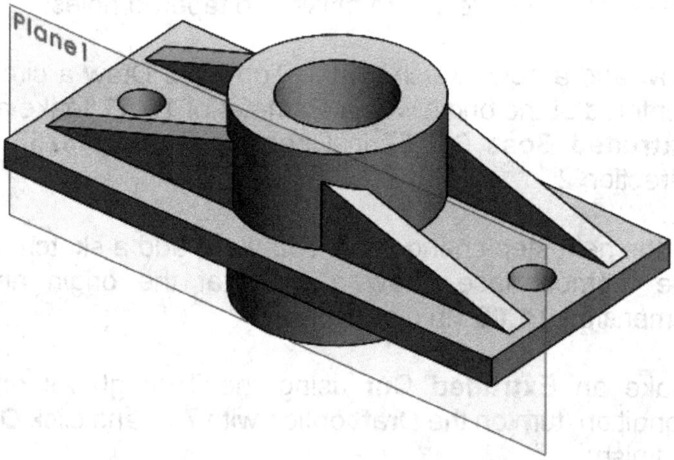

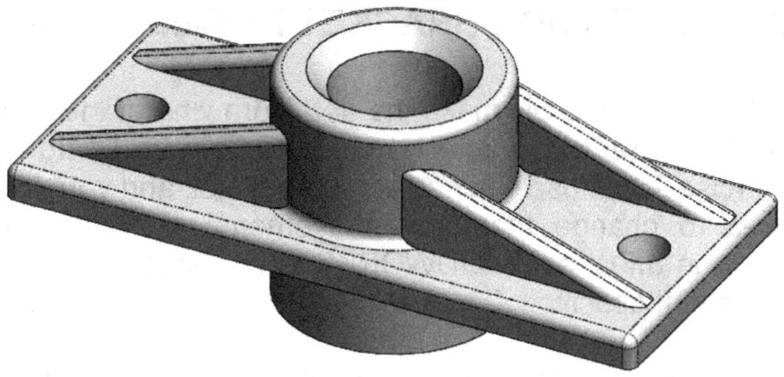

To finish the Tension Cable Bracket, add a **Chamfer** to the top inside edge of the cylindrical boss. Make the chamfer using the **Distance-Angle** option and size it **0.125" x 45°**.

Next, add a **0.0825" Fillet** to the following edges:

- Outer top edge of upright boss.
- All intersections of the upright boss with the base plate, above and below the base plate.
- The top outer edges of the base plate.
- The top edges of the four triangular supports.
- The vertical edges of the rectangular base.

When finished adding the fillets, set the material to **Cast Carbon Steel** and save your part as **Tension Cable Bracket.sldprt**.

You will now make a 3D section view of the **Tension Cable Bracket**. Select the menu **View, Display, Section View**, or the **Section View** command from the **View** toolbar.

The Front Plane is the default option for Section 1. Leave the Offset Distance at **0.00"** and click OK to continue. If needed, click on the **Reverse Direction** button to change the section view's orientation. Your sectioned model will look as follows.

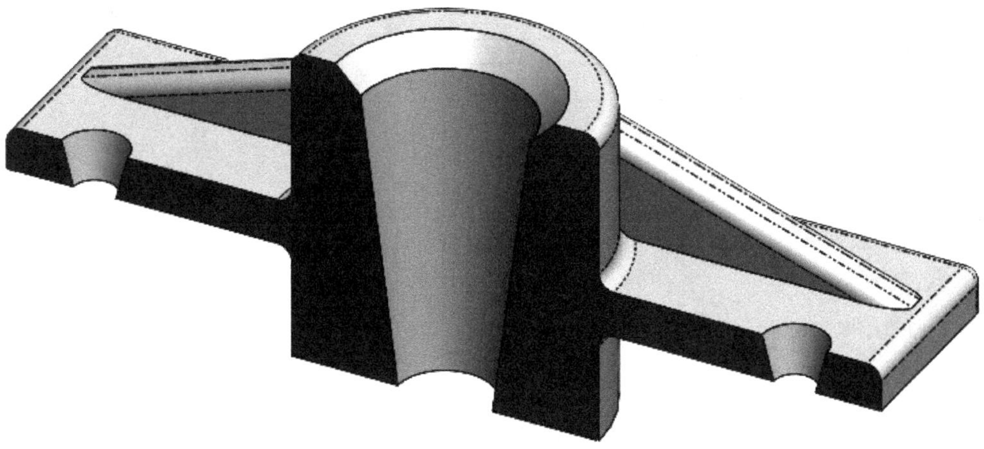

When ready, click on the **Section View** command again to turn it off and save your model before making the detail drawing.

CABLE TENSION BRACKET DETAIL DRAWING

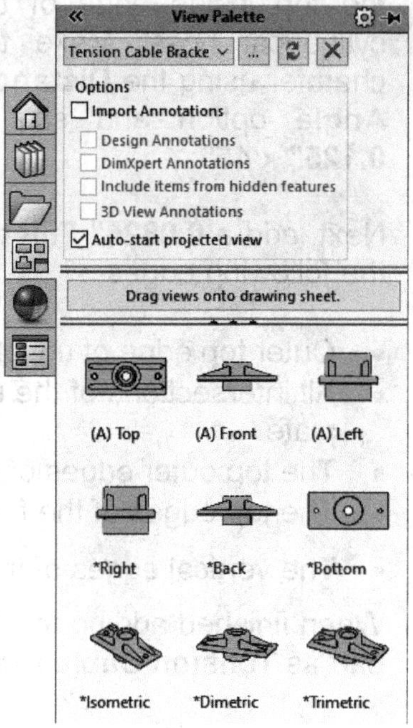

At this time, you will make the detail drawing of the **Cable Tension Bracket**. If you changed the **System Options** in exercise 9-1, those settings are already set, and you do not need to change them again. Remember that **System Options** affect all SOLIDWORKS documents and are not document specific.

Start a new drawing using the **Inches** template. Select the **View Palette** tab from the **Task Pane** on the right side of the screen and drag the **Top** and **Isometric** views separately onto the drawing.

Select the **Isometric** View and from the view's properties make sure the **Scale** is set to **1:2** and the Display Style to **Shaded with Edges** mode. Right click on the view and set the **Tangent Edges** to **Tangent Edges with Font**.

Now select the **Top view,** change its **Scale to 1:1** and change the Display Style to **Hidden Lines Visible**. Arrange the views to match the following image.

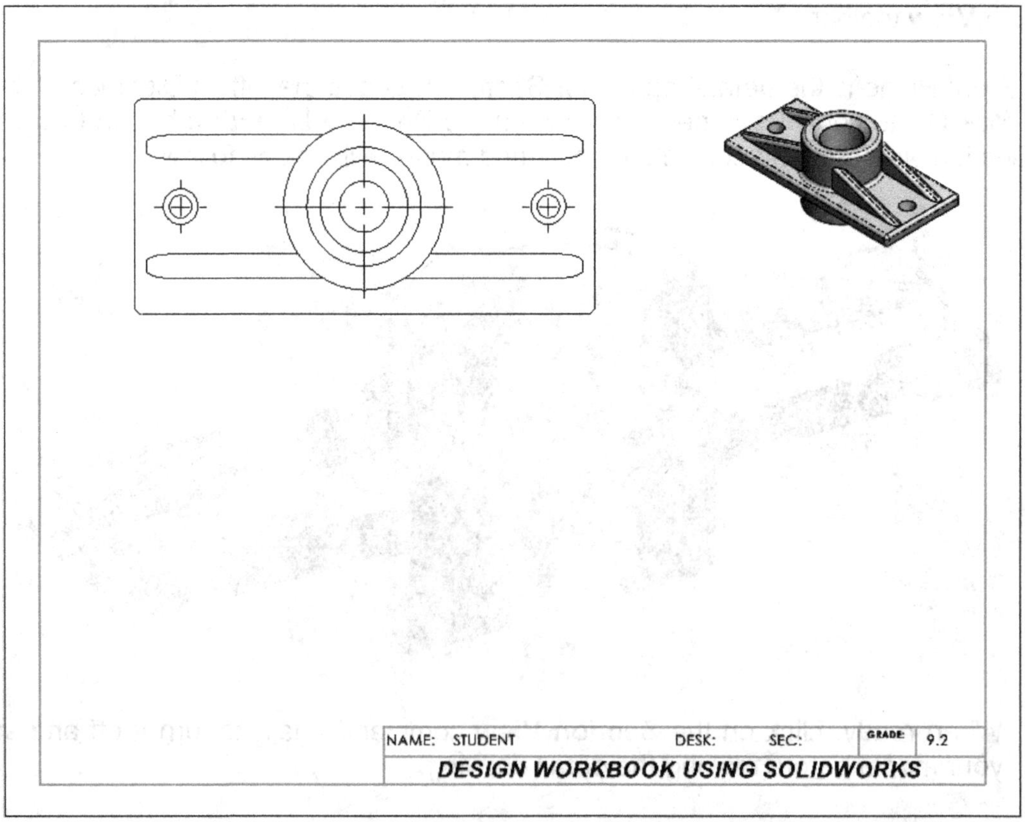

ADD A SECTION VIEW OF THE CABLE BRACKET

To add a section view, select the **Section View** command from the Drawing tab. In the Section View properties select the **Horizontal** section line option. Move your cursor to the center of the **Top View**; when the **Section Line** snaps to the center, click to locate the section line, move your cursor down to place the section view in place of the **Front** view, and click to locate it.

Remember the section view is aligned vertically with the top view, and parallel to the section line. The section line is automatically labeled **A-A,** and the section view is labeled **Section A-A**. If the section line arrows are pointing down, click in the **Flip Direction** button. Finally, use the **Centerline** command to add the missing centerlines to the Section View if needed.

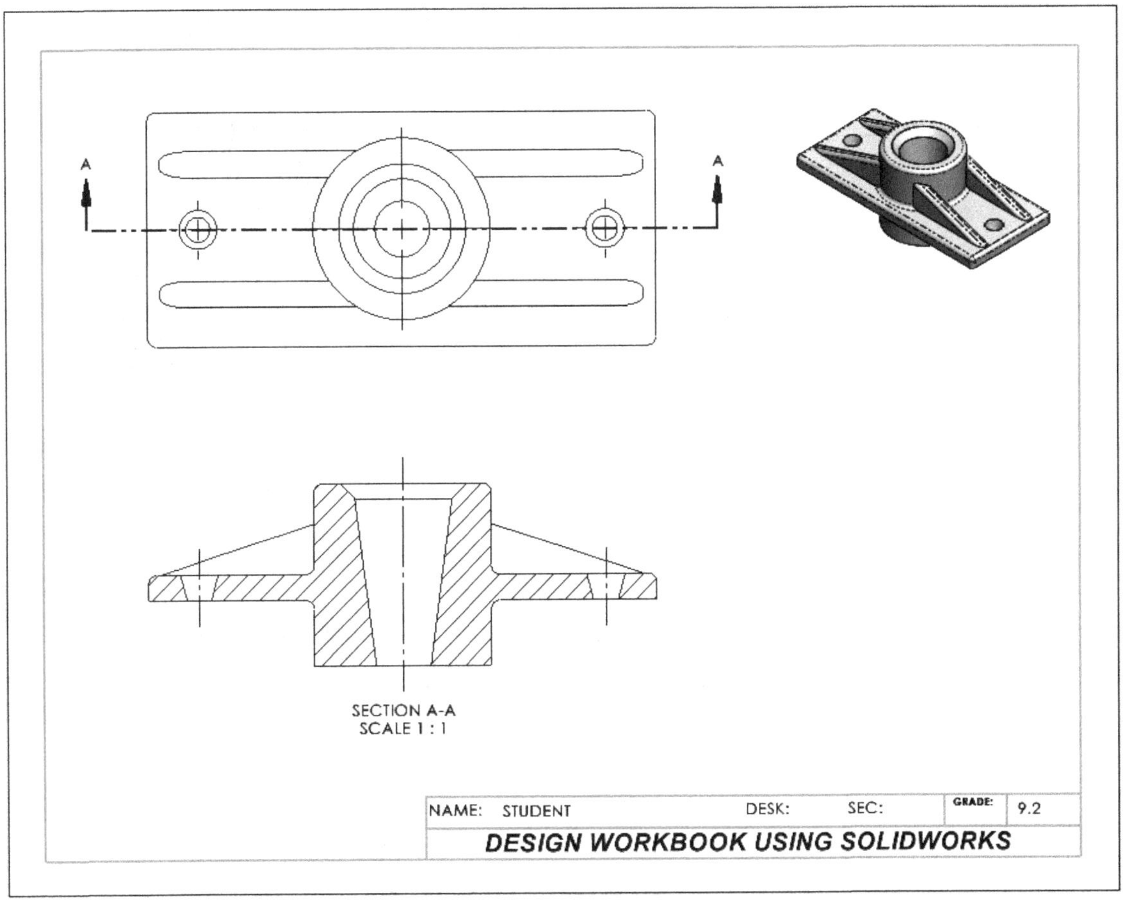

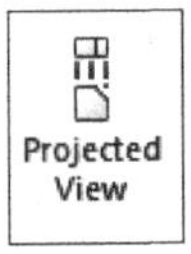

Now you need to add a **Right** view. From the Drawing tab, select the **Projected View** command. Select the section view and move to the right to project the **Right** view. When you reach the desired location, click to position the view. By projecting the view, the **Right** view will be aligned horizontally with the section view.

Change the **Right** view's properties to **Hidden Lines Visible**. From the Annotation tab select the **Note** command to add the drawing's title **TENSION CABLE BRACKET** and **SCALE 1:1**.

When finished save your drawing as **TENSION CABLE BRACKET.slddrw** and print a copy for your lab instructor.

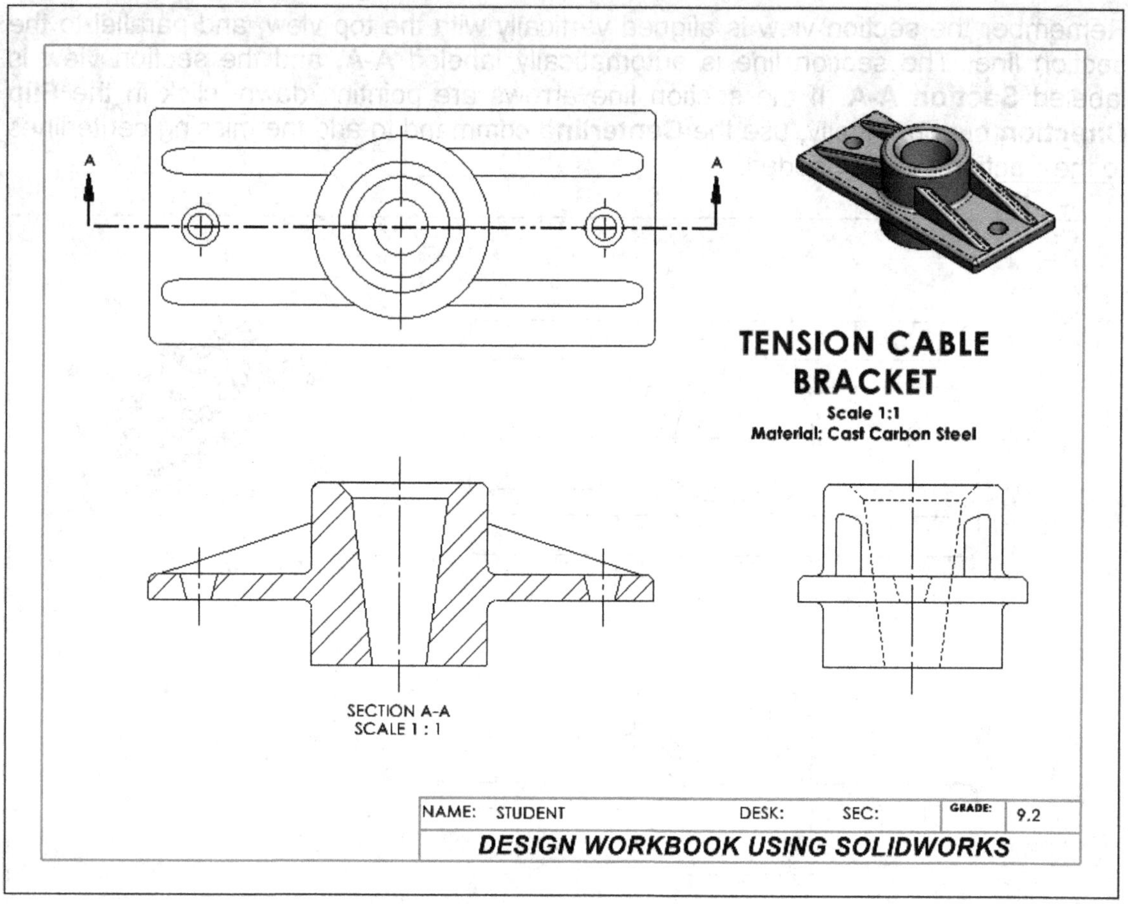

Exercise 9.3: Milling End Adapter Section Views

In Exercise 9.3, you will model a Milling End Adapter using the commands learned so far. You will make a section view of the 3D model to visualize the internal features of the model, and then you will create a detail drawing with a section view. To get started, you will first build the solid model part.

BUILDING THE MILLING END ADAPTER

Start a new part using the **ANSI-INCHES** template, add a new sketch in the **Front Plane**, and draw and dimension the following sketch. Remember to add a horizontal **Centerline** starting at the origin. You will need to add a horizontal relation between the top endpoints of the **0.125"** wide slots. Change the units of measure to four decimal places.

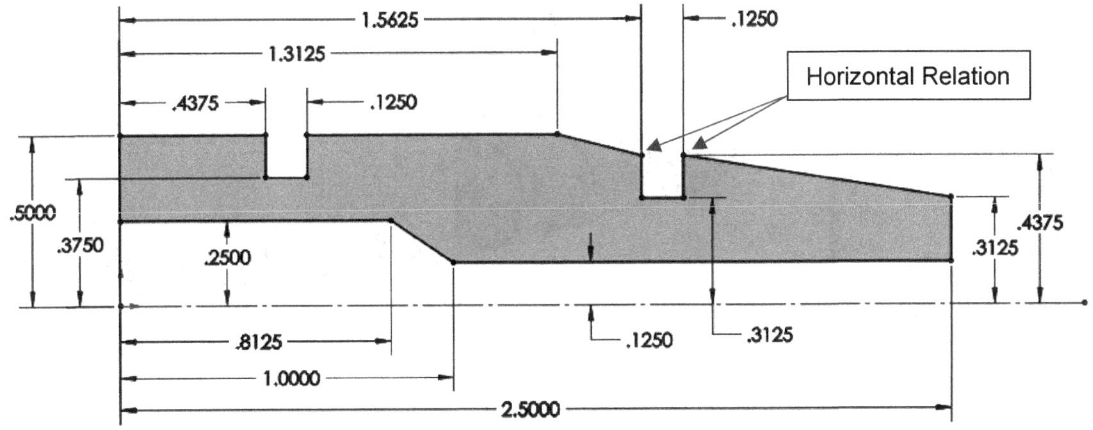

Select the **Revolved Base** command and revolve the profile **360°**. Now add a **Chamfer** to both outside edges of the part. Use the **Distance-Angle** option and make the chamfer **0.0625" x 45°**. Your part will look like the following.

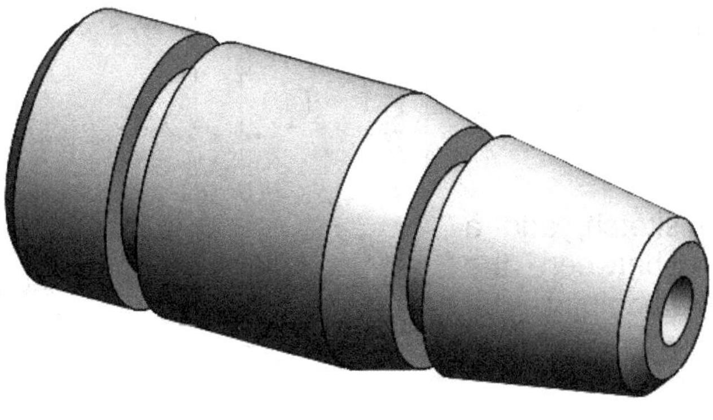

The adapter needs to have two screw holes on the outer surface. In order to locate the holes, you need to add a new plane first. Select the **Reference Geometry** command and click on **Plane**.

In the **First Reference** select the **Top Plane** and select the **Parallel** option. For the **Second Reference** select the outside face of the Milling Adapter. Using these options will create a new plane parallel to the **Top Plane** and **Tangent** to the selected cylindrical face.

If needed, use the **Flip offset** option to locate the new plane above the part, and not below. Click **OK** to finish.

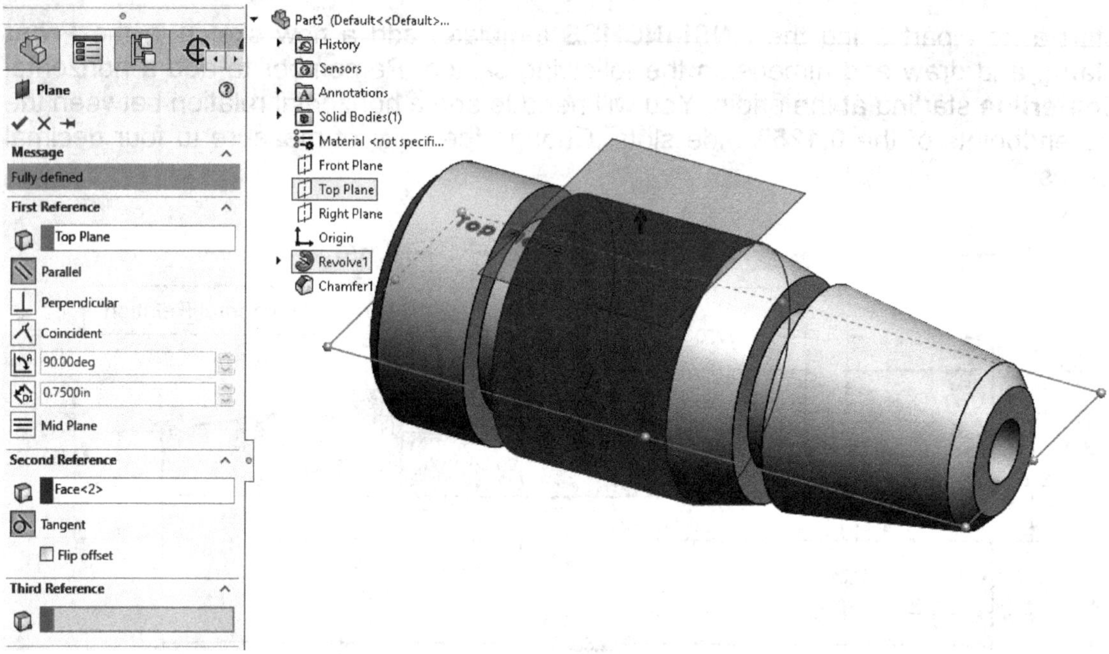

Change to a **Top** view, add a new sketch in the new plane (**Plane1**) and draw the profile indicated.

Add an **Equal** geometric relation to make both circles the same and dimension the diameter of either one **0.125"**.

To fully define the sketch, add a **Horizontal** relation between the circles' center and the origin.

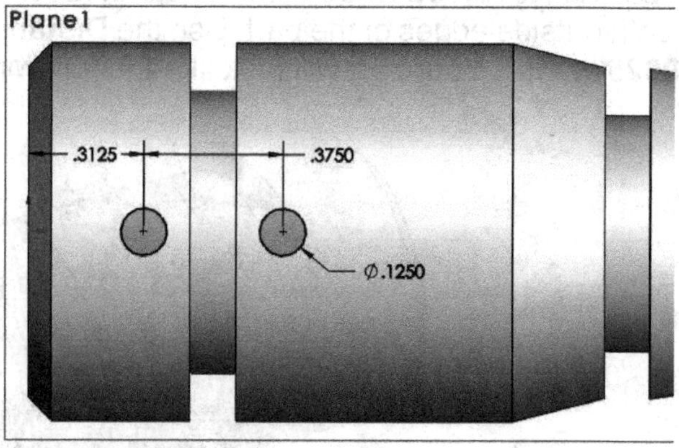

Now make an **Extruded Cut** down through the cylinder wall using an **Up to Next** end condition. The **Up to Next** end condition will make the cut until it finds the next model face, in this case the *inner* face of the Milling Adapter. This results in the holes only going through the top half of the part.

Now your model of the Milling End Adapter is complete.

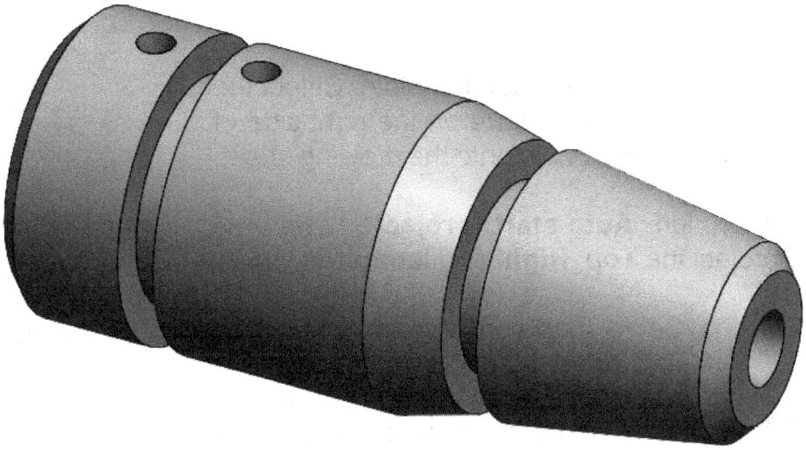

Optionally, you can add the sketch in the **Top Plane** and make the **Extruded Cut** going up, but we chose this option to learn a different approach and additional command options.

To complete the model, assign the **Manganese Bronze** material from the **Copper Alloys** library, and save your part as **Milling End Adapter.sldprt**.

Using the same process as the previous two exercises, select the **Section View** command using the **Front Plane**.

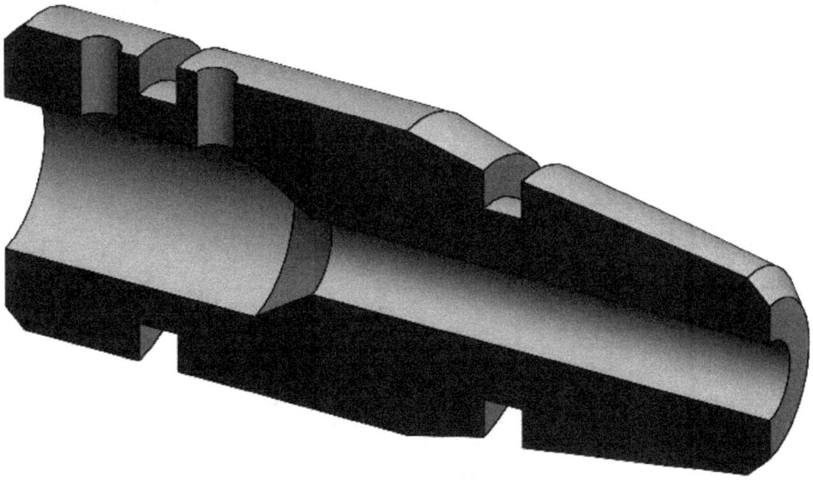

MILLING END ADAPTER DETAIL DRAWING

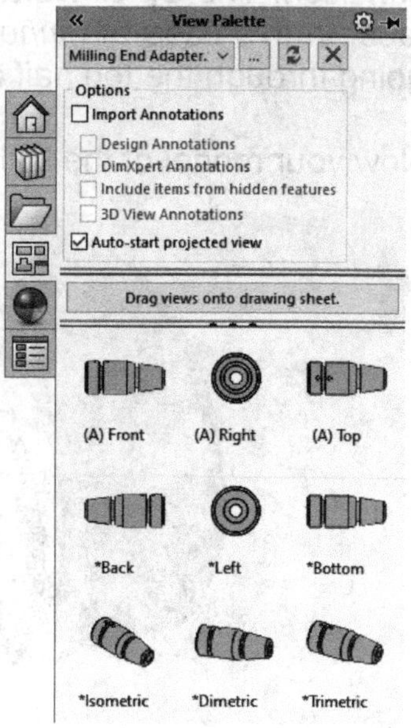

Now you will make the detail drawing of the **Milling End Adapter**. If you changed the **System Options** in exercise 9-1, those settings are already set, and you do not need to change them again. Remember that **System Options** affect all SOLIDWORKS documents and are not document specific.

Start a new drawing using the **Inches** template. Select the **View Palette** tab from the **Task Pane** on the right side of the screen and drag the **Front** view onto the drawing sheet.

Make sure the option **Auto-start projected view** is activated, and project the **Top**, **Right** and **Isometric** views.

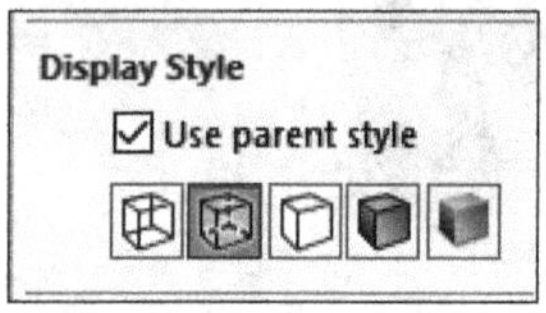

Select the **Front** view and change it to **Hidden Lines Visible** mode. The rest of the views will also change because they are set to use the parent view's style.

In the FeatureManager right-click on **Sheet1**, select **Properties** and change the sheet's scale to **2:1**. All four views will change scale. Now select the **Isometric** view, change its scale to **1:1** and display mode to **Shaded with Edges**. Finally, add the missing **Centerlines** to all views as shown.

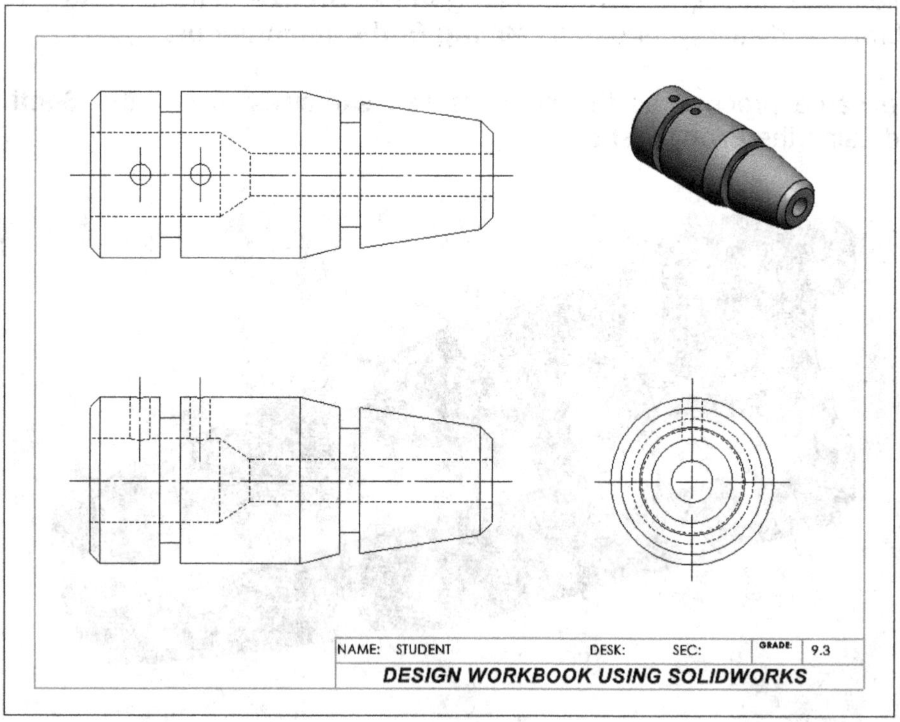

A **Broken-Out section** shows a partial section to a defined depth in a view. To create it, select the **Broken-Out Section** command from the Drawing tab. The **Spline** sketch tool will be automatically activated.

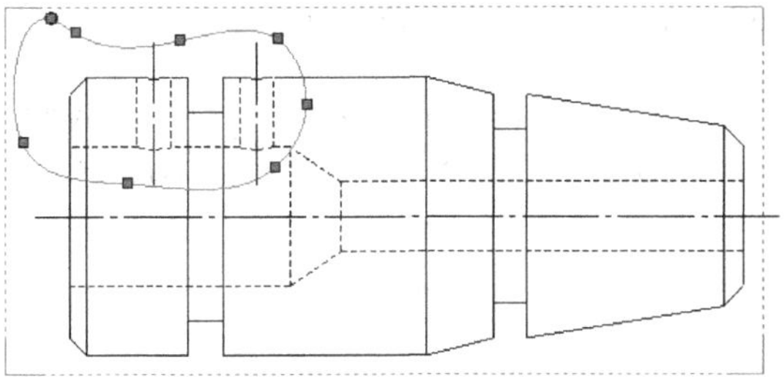

Go to the **Front** view and draw a closed profile to define the area for the section as suggested. When the spline (or any other closed profile) is completed, the **Broken-Out Section** properties are displayed.

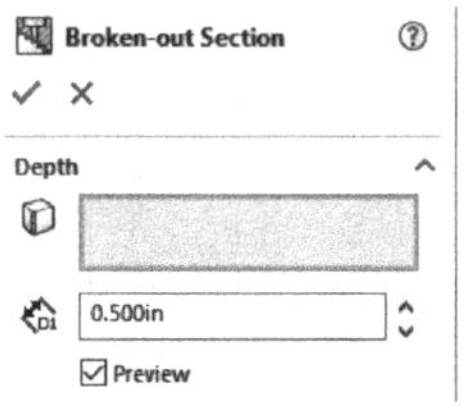

Turn on the **Preview** checkbox to preview the results. In the **Depth** distance box, enter **0.500** inches. This places the cut depth plane through the middle of the part. You will see the depth planes in the **Top** and **Right** views, including the arrows indicating the direction of cut, as shown. Click **OK** to finish.

Note: You can draw the closed spline before and drag the nodes to adjust the section area. When the closed profile covers the area of interest, pre-select the closed profile, and then select the **Broken-Out Section** view command.

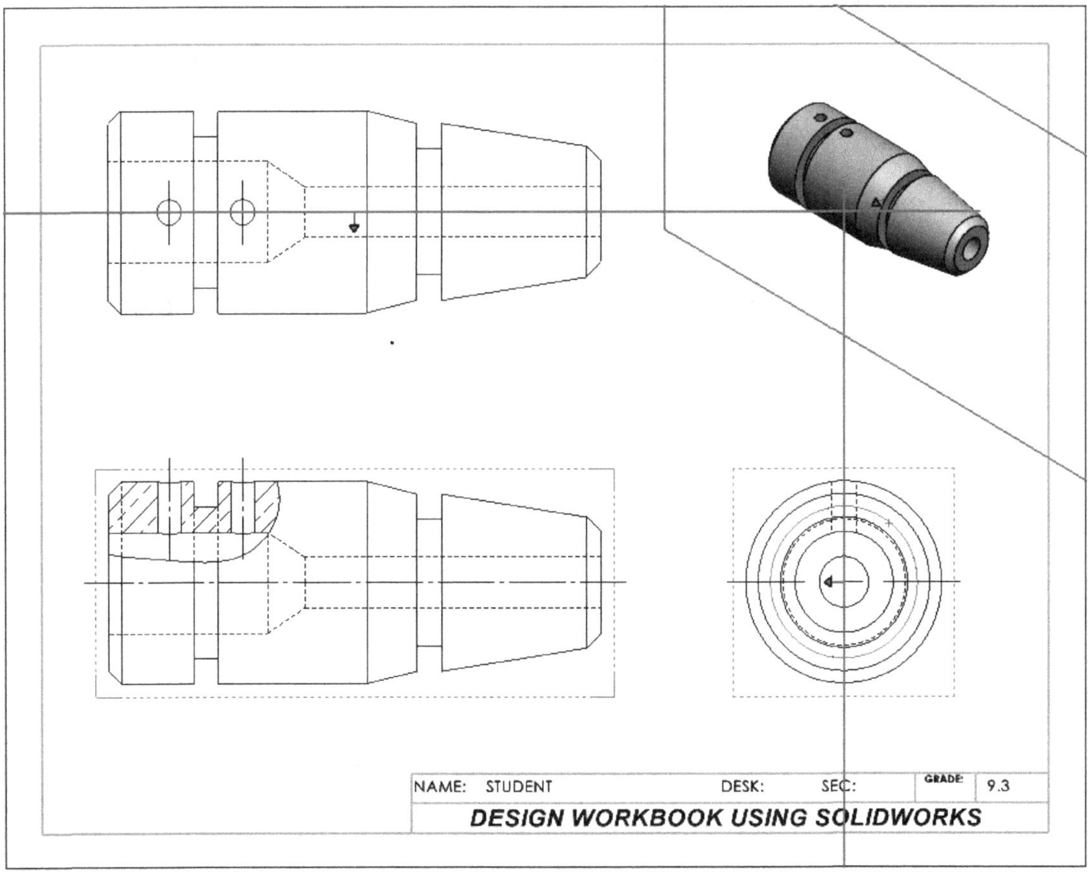

With the drawing views complete, use the **Note** command to add the part's name, drawing scale and material.

When you are finished, save your drawing as **MILLING END ADAPTER.slddrw** and print a copy to submit to your instructor.

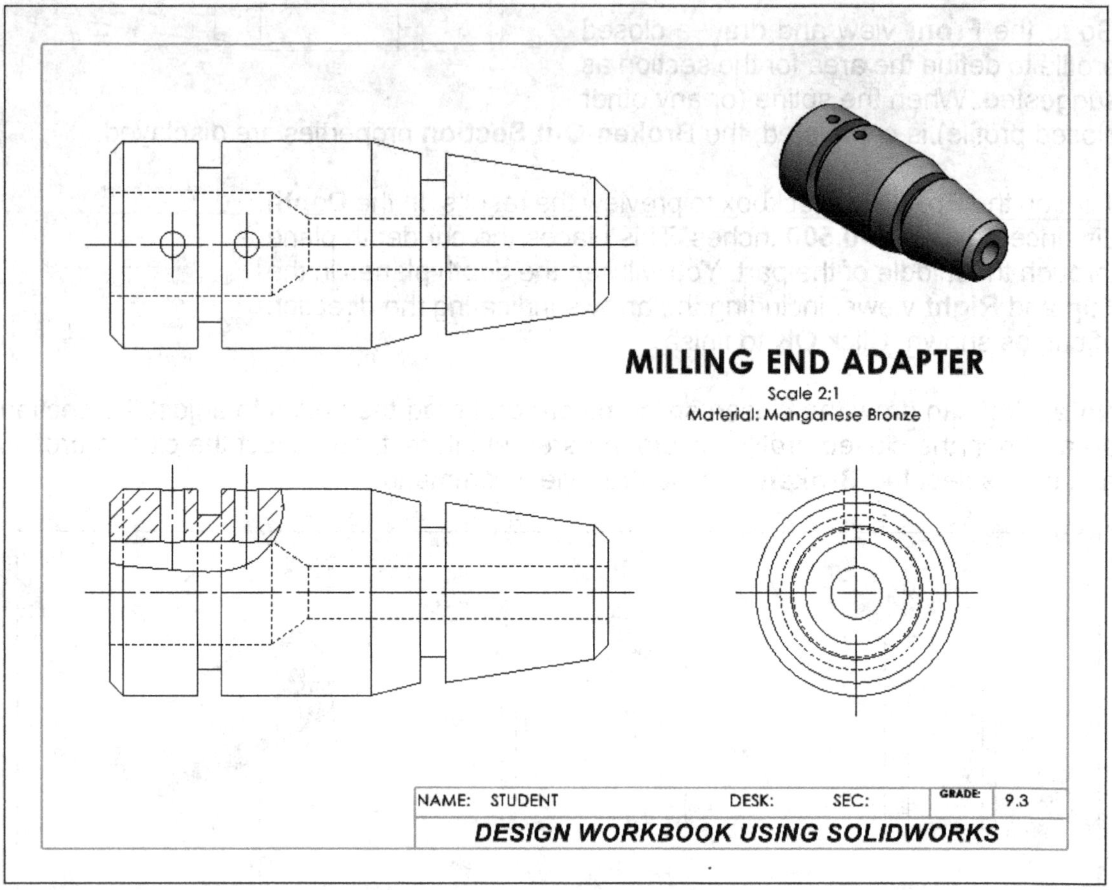

Exercise 9.4: Plastic Revolving Ball Assembly Section View

In Exercise 9.4, you will build the three parts of the Plastic Revolving Ball assembly. You will make a 3D section view of the model to visualize and inspect internal features of the assembly.

Next you will make an assembly drawing and create a 2D section view of the assembly. To get started, build the solid model of each part, and then create an assembly of the three parts.

Building the Ball

Use the **ANSI-INCHES** template and change the units of measure to four decimal places. Add a new sketch in the **Front Plane** and draw the sketch indicated. The arc center is in the **Centerline**.

Add a **Centerline** starting at the origin and dimension as indicated. Use the **Revolved Boss** command, make a full revolution and click **OK** to finish.

Set the material to **PF** from the **Plastics** library and save your part as **Plastic Ball.sldprt**.

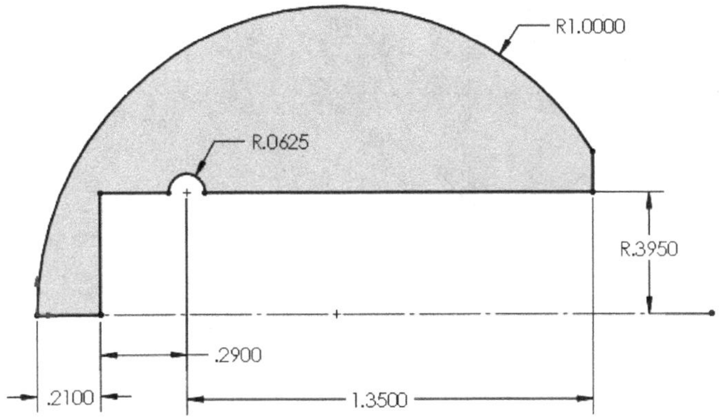

Building the Steel Shaft

Start a new part using the **ANSI-INCHES** template, add a new sketch in the **Front Plane** and draw the following sketch. Select the two horizontal lines at the top and add a **Collinear** relation between them to fully define the sketch.

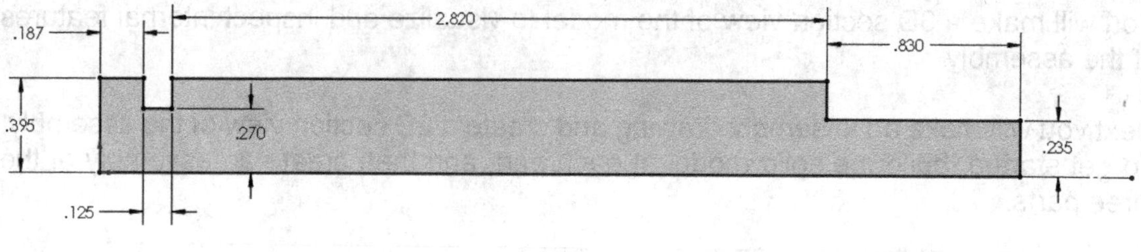

Select the **Revolved Boss** command and revolve the sketch **360°** and set the material to **AISI 316 Annealed Stainless Steel (SS)**, from the **Steel** library.

Change to a Right view, select the small end of the shaft, and add a new sketch. Draw a **Hexagon** and make a **Cut Extrude** using the **Blind** end condition **0.375"** deep. Finally add a **0.0375" x 45° Chamfer** on both ends of the shaft to finish. Save your part as **Steel Shaft.sldprt**.

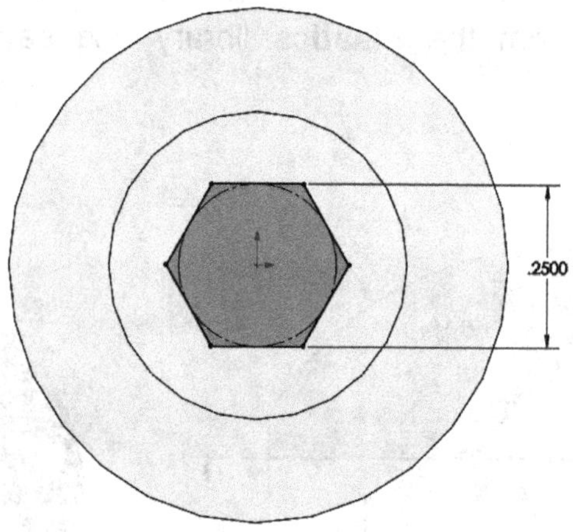

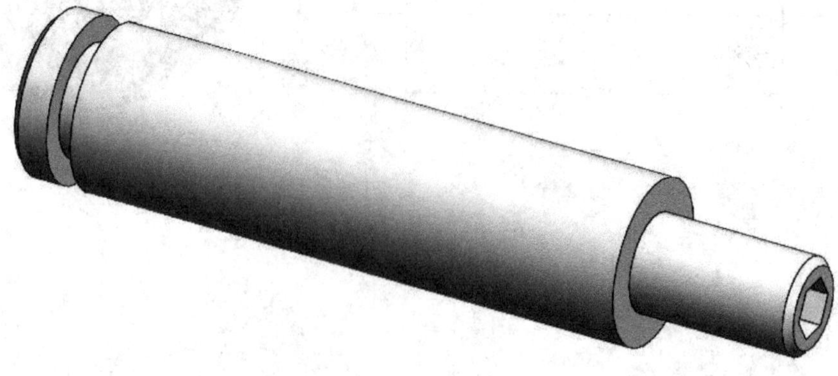

Building the Snap Ring

Start a new part using the **ANSI-INCHES** template, add a sketch in the **Front Plane**, draw the profile indicated, and change the units of measure to **four** decimal places. Remember to add a **Vertical** relation between the arc's center and the origin to fully define the sketch.

Select the **Revolved Boss** command and revolve the sketch **360°** to create the Snap Ring. Set the material to **Alloy Steel** from the **Steel** library.

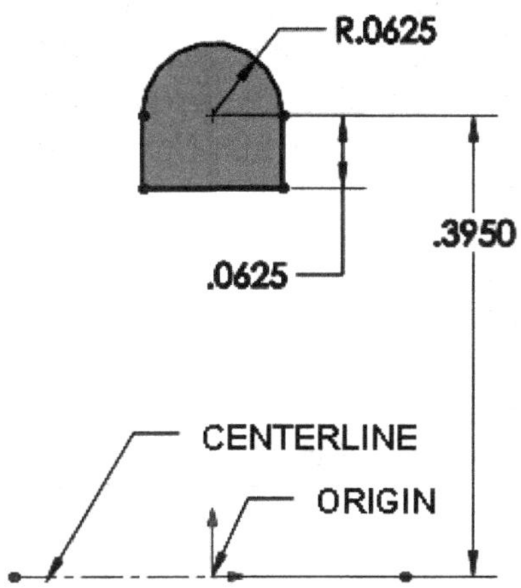

Next, change to a **Right** view, select the **Right Plane,** and start a new sketch. Draw the sketch profile shown using the Dynamic Mirror command and add a **Cut Extrude** using the End Condition **Through all - Both.**

Your part is complete and will look as indicated next.

Save your part as **Snap Ring.sldprt**.

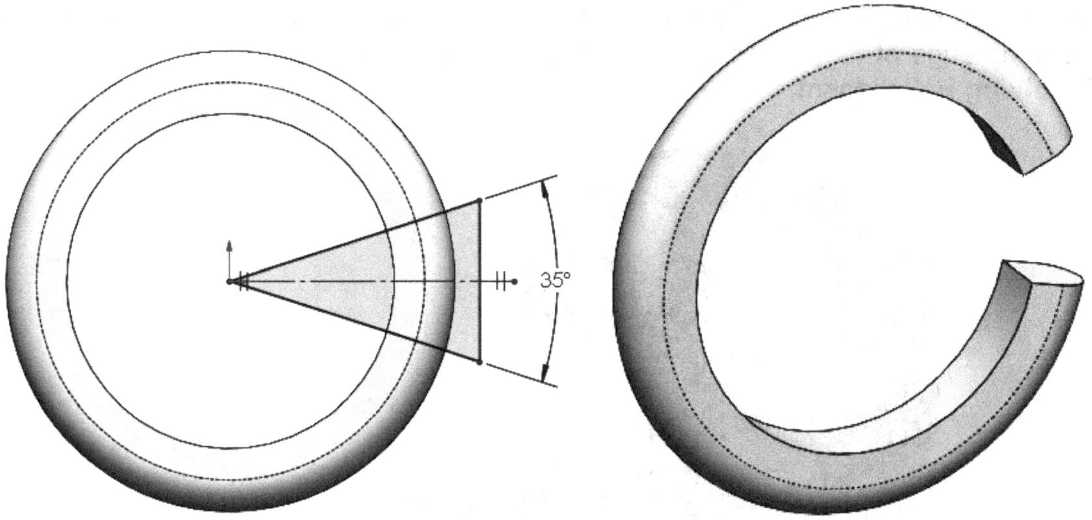

Building the Assembly

After the three parts are completed, make a new assembly and from the **Begin Assembly** command, select the **Steel Shaft** and click **OK** to add it at the assembly's origin. Next, use the **Insert Components** command, add the **Snap Ring** part and locate it near the shaft's end with the groove.

Use the **Mate** command and make the inside cylindrical face **Concentric** with the groove's cylindrical face, and finally add a **Coincident** mate between a flat face of the **Snap Ring** and the corresponding flat face of the shaft's groove, as shown below. At this point you can drag the **Snap Ring** and it will rotate about the shaft's axis.

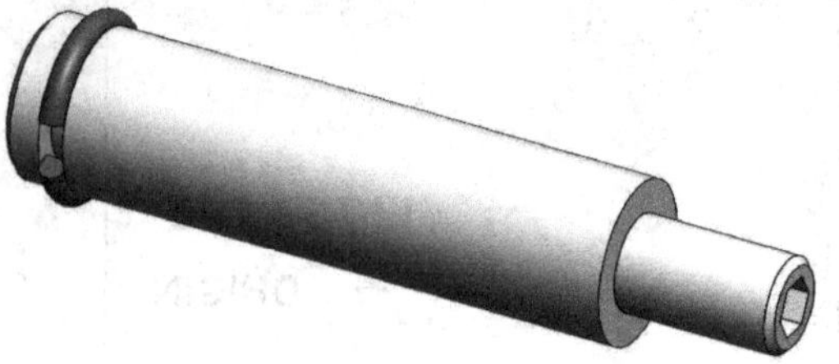

Now use the **Insert Components** command to add the **Plastic Ball** and locate it at the end of the shaft with the snap ring. Make the inside face of the **Plastic Ball Concentric** with the cylindrical face of the **Steel Shaft**.

If needed, move the ball away from the shaft, and rotate as shown next. Now add a new **Coincident** mate between the inside edge of the groove in the **Plastic Ball**, and the matching edge of the shaft's groove. When finished, save your assembly as **Plastic Revolving Ball.sldasm**.

After the assembly is complete, click on the **Section View** command and use the **Front Plane** as the section plane. Before you click **OK** to finish the section view, feel free to click and drag the distance arrow to see the effect in the model.

In the **Section View** options turn off the **Keep cap color** option to use the component's color in the section plane. Click OK to evaluate the assembly, and finally turn off the **Section View** to continue.

CREATING A DRAWING OF THE REVOLVING BALL ASSEMBLY

Make a new drawing using the **Inches** based template and select the **View Palette** tab. If needed, select the **Plastic Revolving Ball** assembly from the drop-down list, turn off the **Auto-Start projected view** option, and add the **Top** and **Isometric** views to your drawing. Select the **Isometric View**, set the view's scale to **1:2** and change the display mode to **Shaded with Edges**. Now select the **Top view** and set its scale **to 1:1** with **Hidden Lines Visible** mode.

To add the section view, select the **Section View** command, select the **Horizontal Section** option, and locate your cursor at the axis of the shaft in the **Top** view. When the section line snaps to the axis click to locate it. The **Section Scope** dialog is displayed, where you can select the assembly components that will *not* be cut by the section line. In this case you will not select any component and click **OK** to continue. Move the **Section** view below the **Top** view and click to locate it.

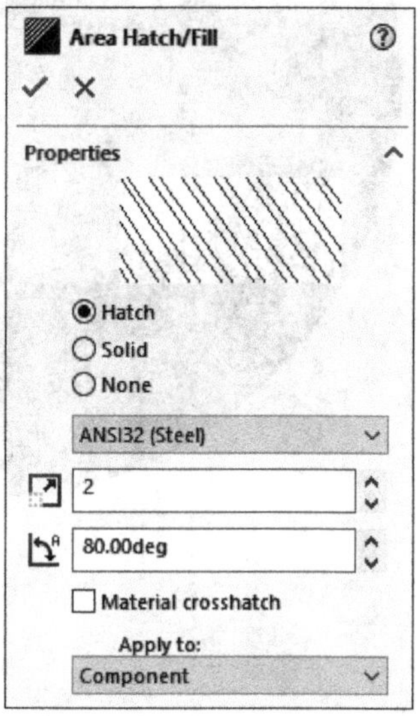

The section line will be automatically labeled **A-A** in the **Top** view and should be pointing up. If the arrows are pointed downward, select the flip direction box. The section view is automatically labeled **Section A-A**.

Once the section view is placed, you will notice that the **Area Hatch** lines may not be at different angles and the scale of the section lines may not be appropriate.

Select the section area of a component to view its properties, select the option **Apply to: Component**, turn off the **Material crosshatch** option, and change the **Hatch Pattern**, **Scale** and **Angle** to modify the density and direction of the crosshatch for each component.

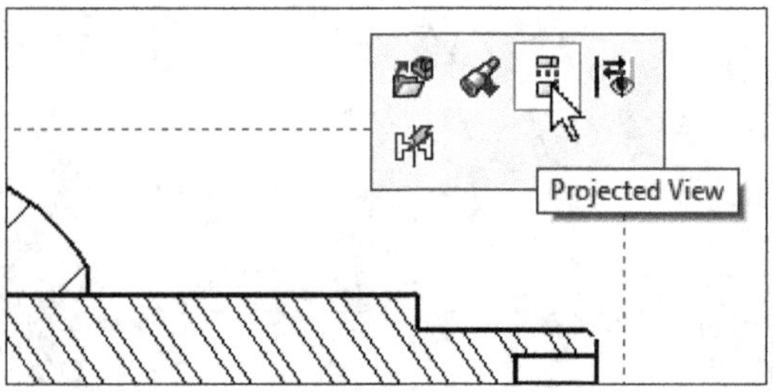

Now you need to add a **Right** view to your drawing. Select the **Projected View** command from the Drawing tab, select the Section view, and locate the **Right** view to the right. Optionally, select the Section view, and from the pop-up toolbar select the **Projected View** command.

When the **Right** view is projected, it will inherit the **Display Style** of the **Section** view. Select the **Right** view and change it to **Hidden Lines Visible**. Use the **Center Mark** command from the Annotations tab and add the missing center marks. Select the outer edge of the plastic ball to locate the center mark. If needed, use the **Centerline** command to add a centerline to the **Section** view as before.

Add a note to your drawing with the name and scale of the assembly with the same font settings used in previous drawings.

When finished, save your drawing as **Plastic Revolving Ball.slddrw** and print a copy for your instructor.

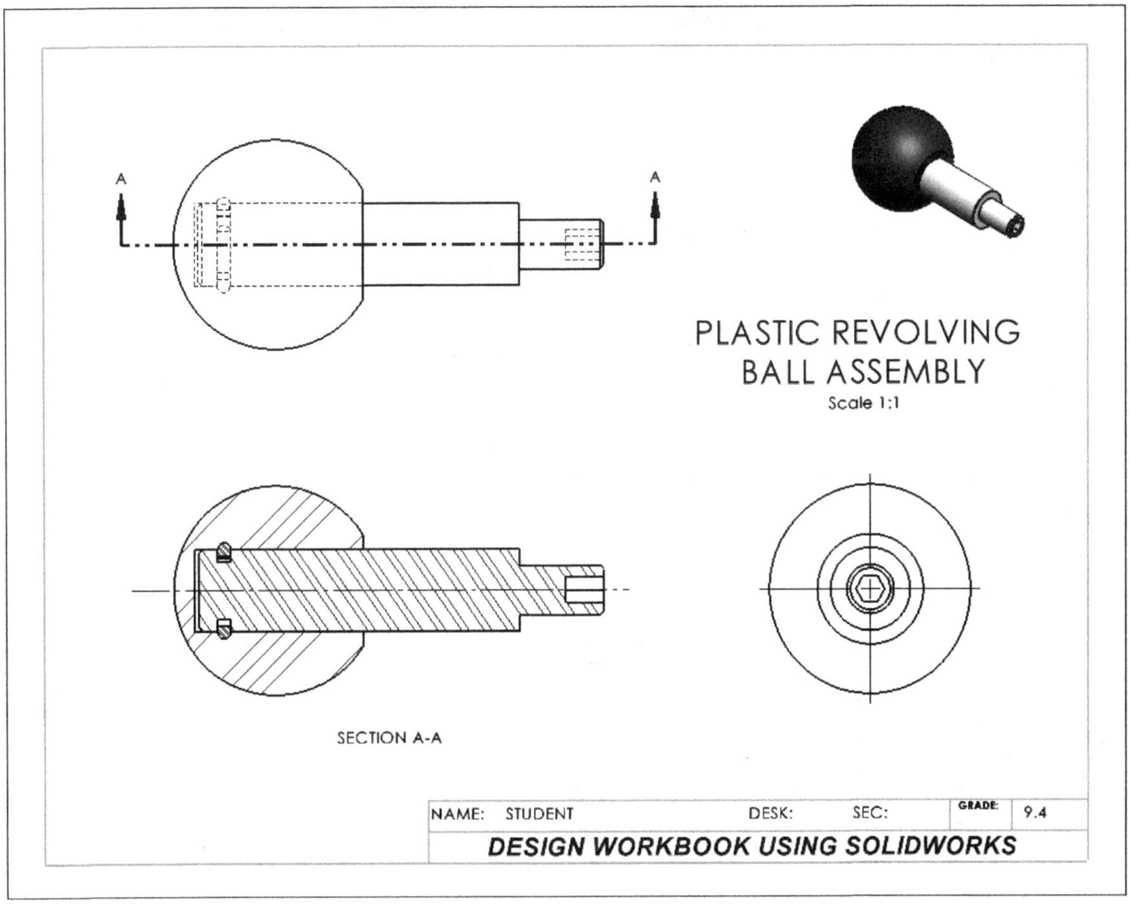

Supplementary Exercise 9-5: CLAMPING BLOCK

Build a solid model of the next drawing. Make a drawing and provide a **Section** in the place of the front view. Add a small isometric view in the upper right-hand corner of the sheet. Add the Title, Scale and Material annotations. The part's dimensions are given in millimeters.

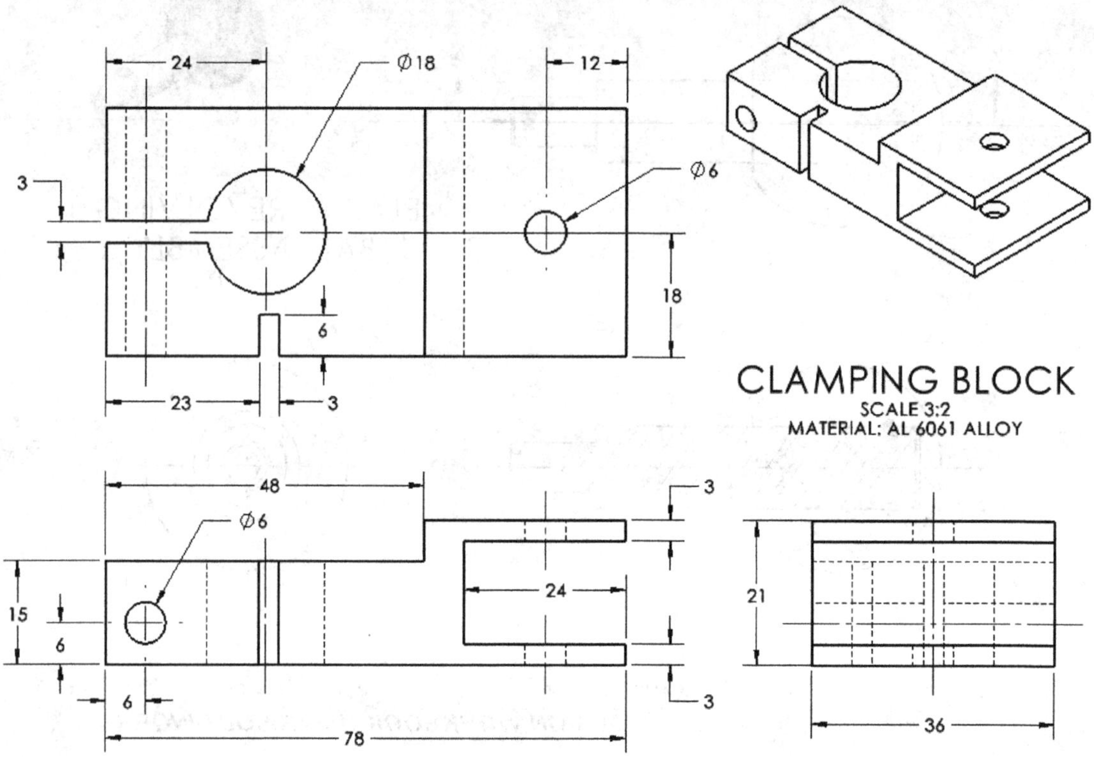

Supplementary Exercise 9-6: TWO WAY BENCH BLOCK

Build a solid model using the following image and make a drawing. In place of the **Front** view add a section view of the **Top** view. Draw an offset line through the two counterbored holes and the center hole. Next select the menu **Insert, Drawing View, Section View** and locate it in place of the **Front** view. From the **Section** view, project the **Right** view. Finally add an **Isometric** View in the upper right corner. Add notes with the Title and Scale as in previous exercises. Dimensions are given in Inches.

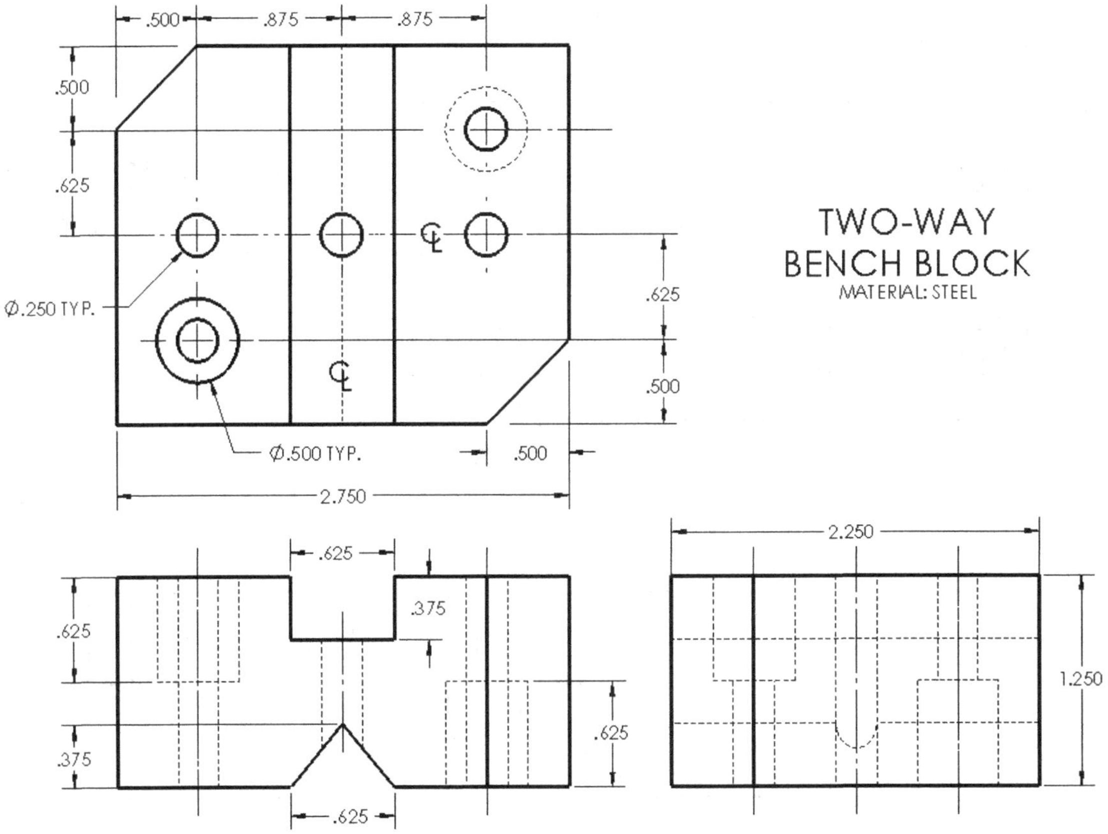

Supplementary Exercise 9-7: ELECTRICAL CONTACT PLATE

Build a solid model using the following image and make a drawing. In place of the Front view add a section view of the Top view. Draw an offset line through one of the small holes on the left, then through the slot with the two holes in the center and through one of the slots on the right. Next select the menu **Insert, Drawing View, Section View** and place it in place of the Front view. From the Section view, project the Right view. Finally add an Isometric View in the upper right corner. Add notes with the Title and Scale as in previous exercises. Dimensions are given in Inches.

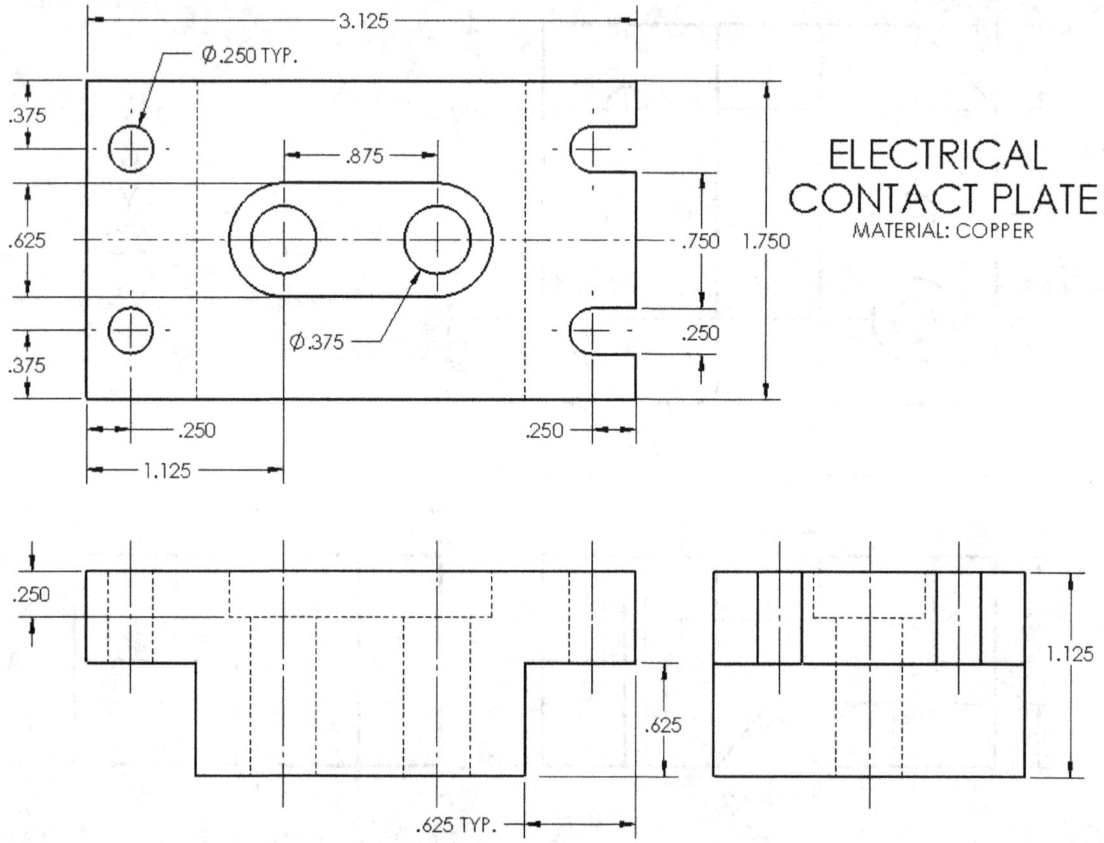

Supplementary Exercise 9-8:
DISC ASSEMBLY

Build solid models of the **Disk Assembly Pieces** below. Build a fully constrained Assembly of the parts and make a drawing of it.

Add a **Section View** in the place of the **Front** view. Finally, add an **Isometric** View in the upper right corner of the unexploded assembly in the upper right corner of your drawing. Add notes with the Title and Scale as in previous exercises. Dimensions are given in Inches.

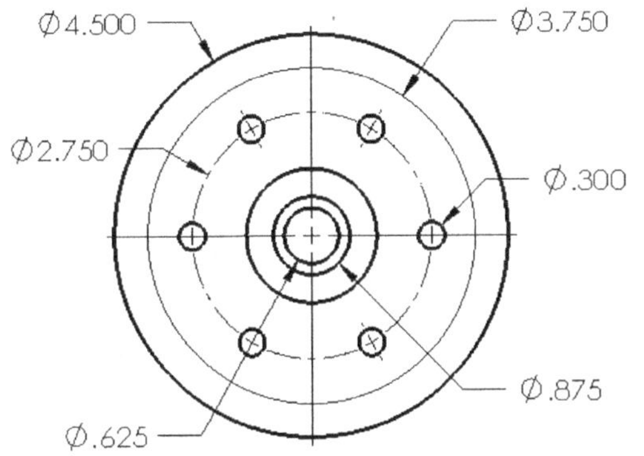

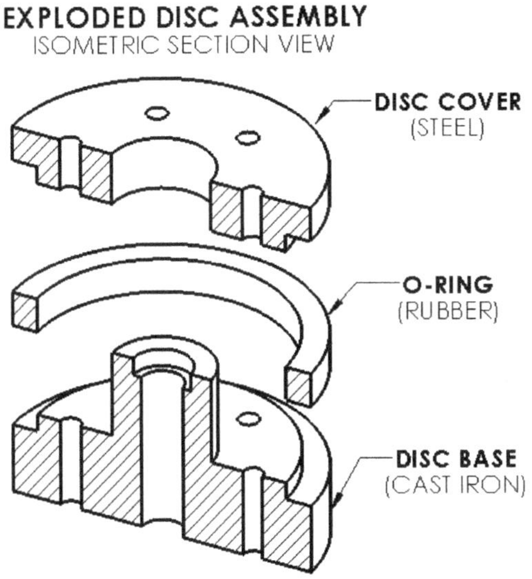

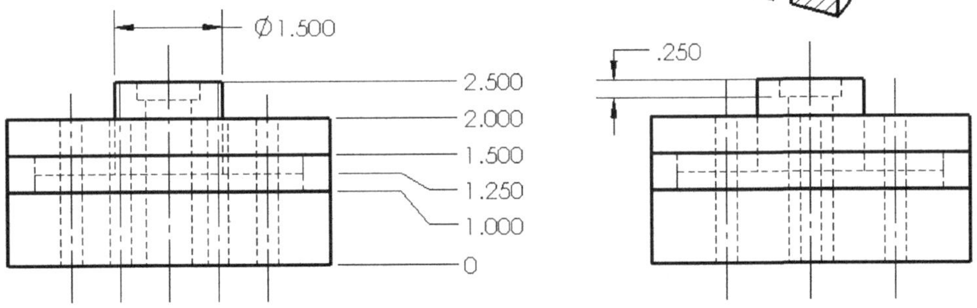

NOTES:

Design Workbook Lab 10
Manufacturing Detail Drawings

In this lab you will learn how to create a fully annotated 2D detail drawing of a part or assembly, including all the necessary views required in a manufacturing environment. In each exercise you will build the part(s), make the detail drawing, import the solid model dimensions into the drawing, and manually add any missing dimensions and annotations.

The following introduction will teach you how to use the necessary tools to make a complete part or assembly drawing using SOLIDWORKS.

THE DRAWING SHEET AND SHEET FORMAT

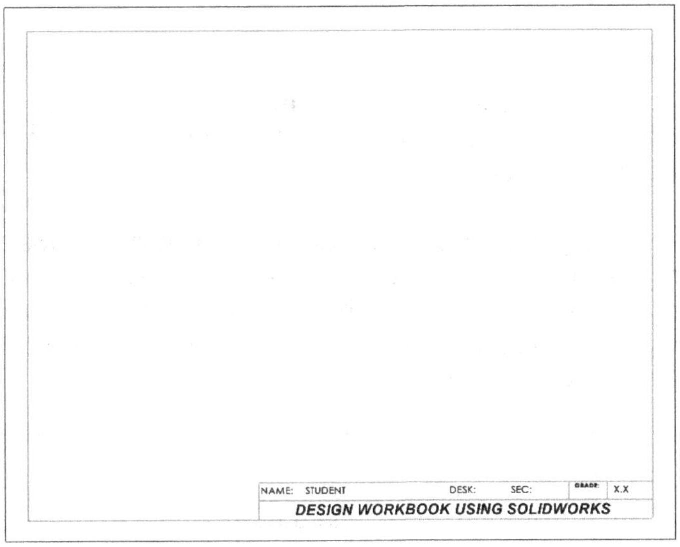

In chapter 1 you created a 2D drawing template called **TITLEBLOCK-INCHES.drwdot**.

In this drawing template you added a Title Block with information about the drawing such as student name, desk, section, exercise number, etc.

In industrial and manufacturing drawings, a title block usually has more information including the drawing's scale, material, weight, vendor, part number, notes, revision level, etc. to correctly fabricate the part or assembly, and it is common practice to add multiple drawing Sheets as needed.

The **Sheet Format** is in a "locked" layer in the drawing. The purpose of this is to prevent you from accidentally modifying it while making a 2D drawing. The **Sheet Format** can be easily accessed to make changes if needed.

Note: When you modify the **Sheet Format**, the drawing views are automatically hidden.

DETAIL DRAWING OPTIONS

As previously covered, the options set in the **System Options** tab control the behavior and functionality of all the SOLIDWORKS documents regardless of the model open, including Sketch options, display style, file locations, performance settings, etc.

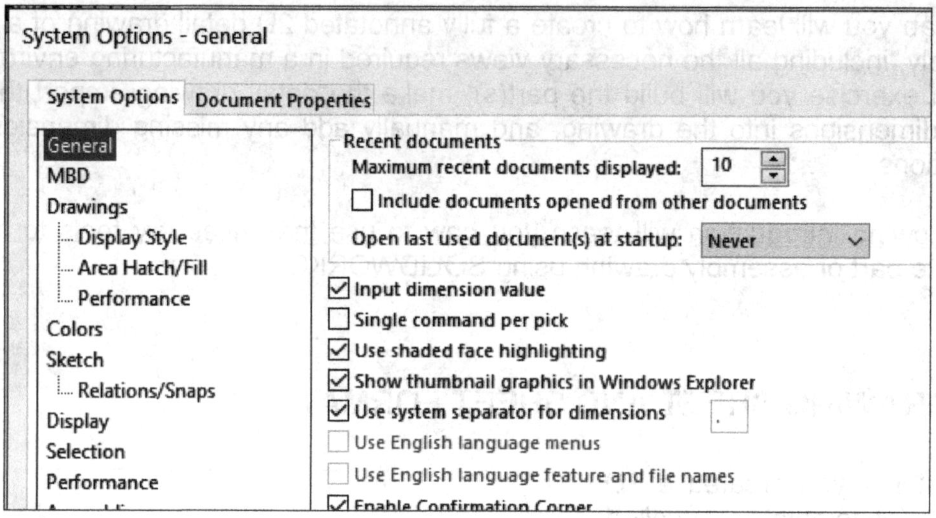

The options set in the **Document Properties** tab are specific to the model currently open in SOLIDWORKS, including the units of measure, material, and the **Drafting Standard** used.

Selecting a **Drafting Standard** in a document will automatically adjust all annotation styles, line types and weights, arrow styles and size, symbols, etc. Since most companies use one of the multiple drafting standards available in SOLIDWORKS, making changes to these options is not a common practice in order to prevent deviations from industry standards.

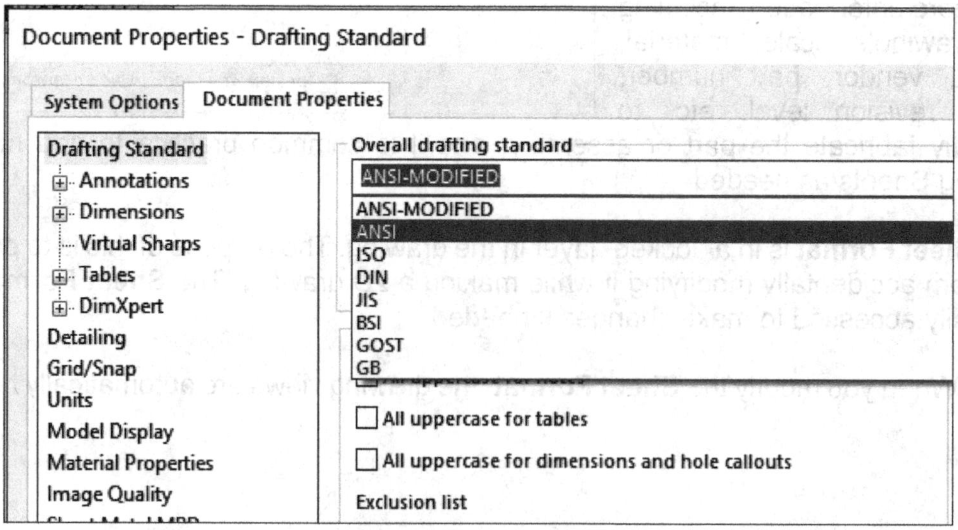

ADDING MODEL VIEWS TO A DRAWING

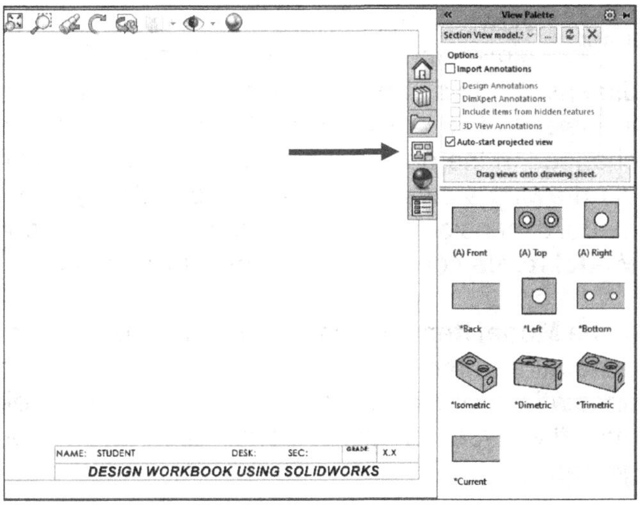

When you have a part or assembly open, and start a new drawing, select the **View Palette** tab where you can see a preview of each orientation of the model currently open in **SOLIDWORKS**, or you can select a different model from the drop-down list at the top of the **View Palette**.

It is advisable to turn on the **Auto-Start Projected View** option at the top of the **View Palette**. From here you can drag any view orientation needed, and automatically project the other views from it. If this option is not activated, you will need to add the different views individually, or use the **Projected View** command as before.

From the **View Palette** drag the **Front** view onto the drawing and move the cursor around to see the other projected views. Click above to add a **Top** view, on the upper right to add an **Isometric** view, and to the right side to add a **Right** view.

The projected views will inherit the Display Style of the original view. Select the Front view and change it to Hidden Lines Visible or Hidden Lines Removed to see the effect.

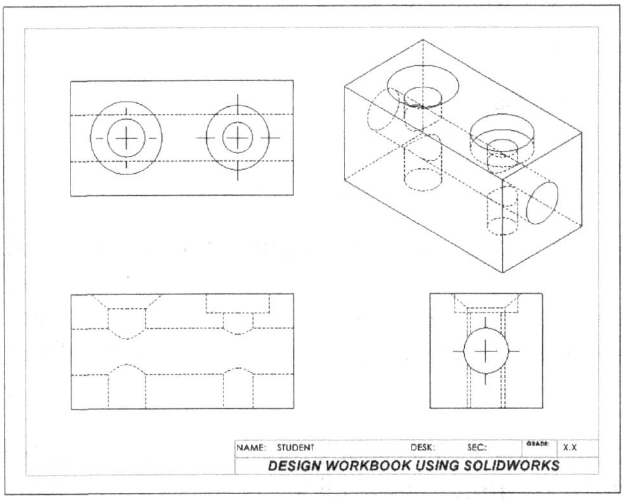

If the drawing views are too big for the sheet, right click on *Sheet1* in the FeatureManager, or in the drawing (*Not on a model view*) and select **Properties** to change the scale of the entire sheet.

Select the **Isometric** view. In the view properties change the view's scale to make it smaller and change the Display Style to **Shaded with Edges** mode.

Remember that you can click-and-drag a view to move it in the drawing sheet, in order to have enough space for dimensions and annotations.

If the orthographic views are projected from one view, they will remain aligned vertically or horizontally.

After adding the necessary views to your drawing, you need to add dimensions and annotations to the different views. When you make a 3D part, sketch and feature dimensions were added. These are called **Parametric** dimensions and can be imported into a detail drawing, making it easier to fully annotate it. When parametric dimensions are changed in the part *or* the drawing, both the 3D model and the 2D drawing views are updated accordingly. To import the 3D model dimensions and annotations into a 2D drawing, select the **Model Items** command from the **Annotation** tab.

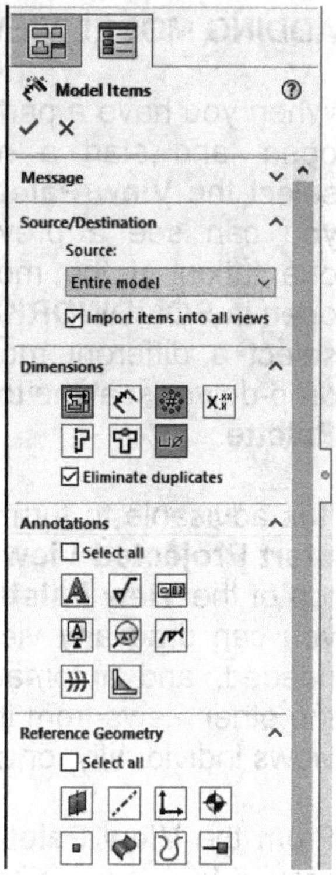

In the **Model Items** command options, you can choose to import annotations into a selected view or all drawing views, the annotations of the entire 3D model or only a selected feature, and the type of dimensions, annotations, and reference geometry.

Some of the annotations available to import into a drawing include:

	Add all dimensions added in a sketch and those created by 3D features.
	Dimensions that had been manually excluded to be imported into a drawing.
	The number of instance copies in patterns.
	Dimensions with tolerance values.
	Dimensions of Hole Wizard profiles.
	Dimensions for Hole Wizard locations.
	Hole specifications and count.

If the imported 3D model dimensions do not fully describe the detail drawing, you can manually add the missing necessary dimensions. By default, manually added dimensions have a parenthesis, which is interpreted as a reference dimension, meaning its value is not exact. A dimension's parentheses can be turned on or off individually in the dimension's properties.

To turn off the parentheses for all manually added dimensions, go to the menu **Tools**, **Options**, **Document Properties**, select the **Dimensions** section and turn off the "**Add Parentheses by default**" option.

Remember this is a document option, not a system setting. You must set this option for each drawing individually, or if you want it in all drawings, it has to be changed in the drawing's template.

After the model dimensions and hole callout annotations are imported, the dimensions can be arranged in the different views for visibility.

To **Move** a dimension to a different model view that would help us to better describe the model, you must drag the dimension to the other view while holding the **Shift** key.

After you add the **Center Mark** and **Centerlines** to the drawing, and arrange the dimensions, your finished drawing will look as follows.

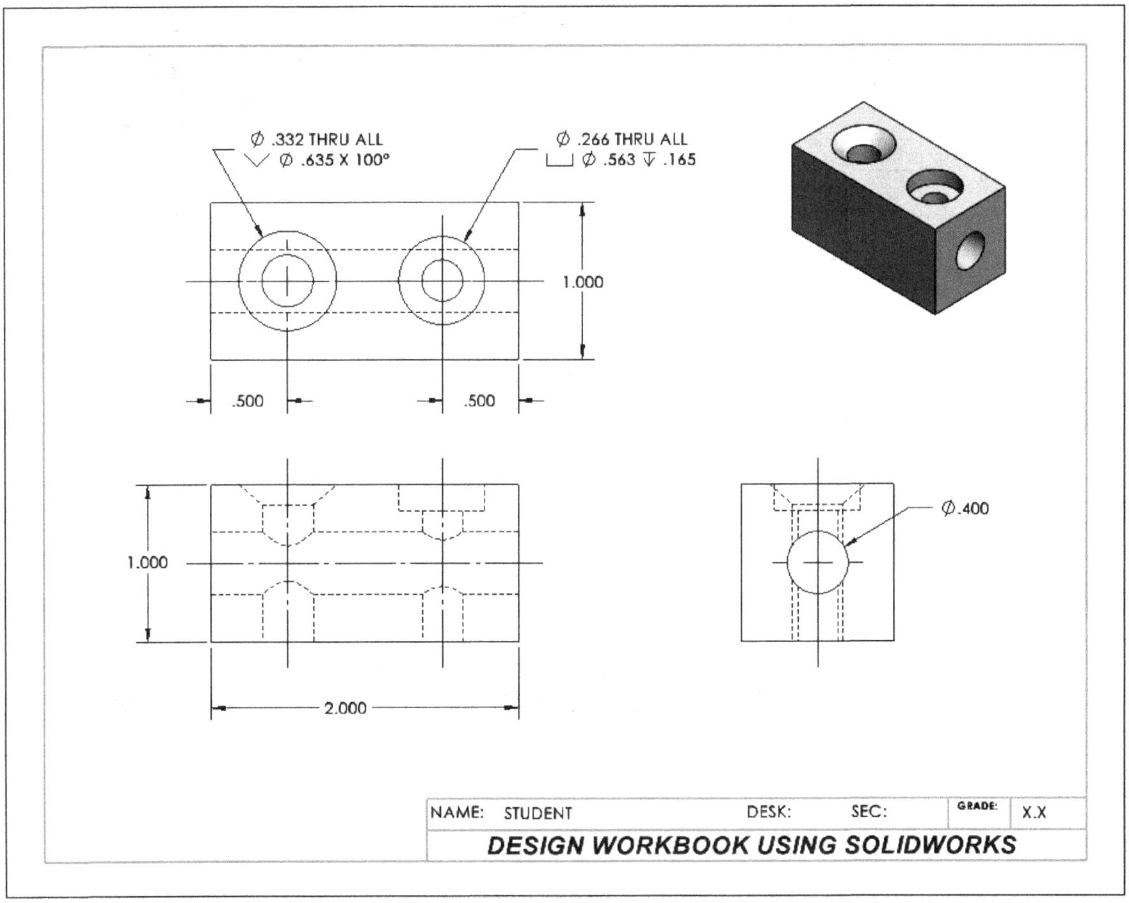

Now you can manually add the vertical location dimensions of the holes using the **Smart Dimension** tool and turn off the parentheses. Using the **Note** command add the **Title**, **Scale** and **Material** for the part.

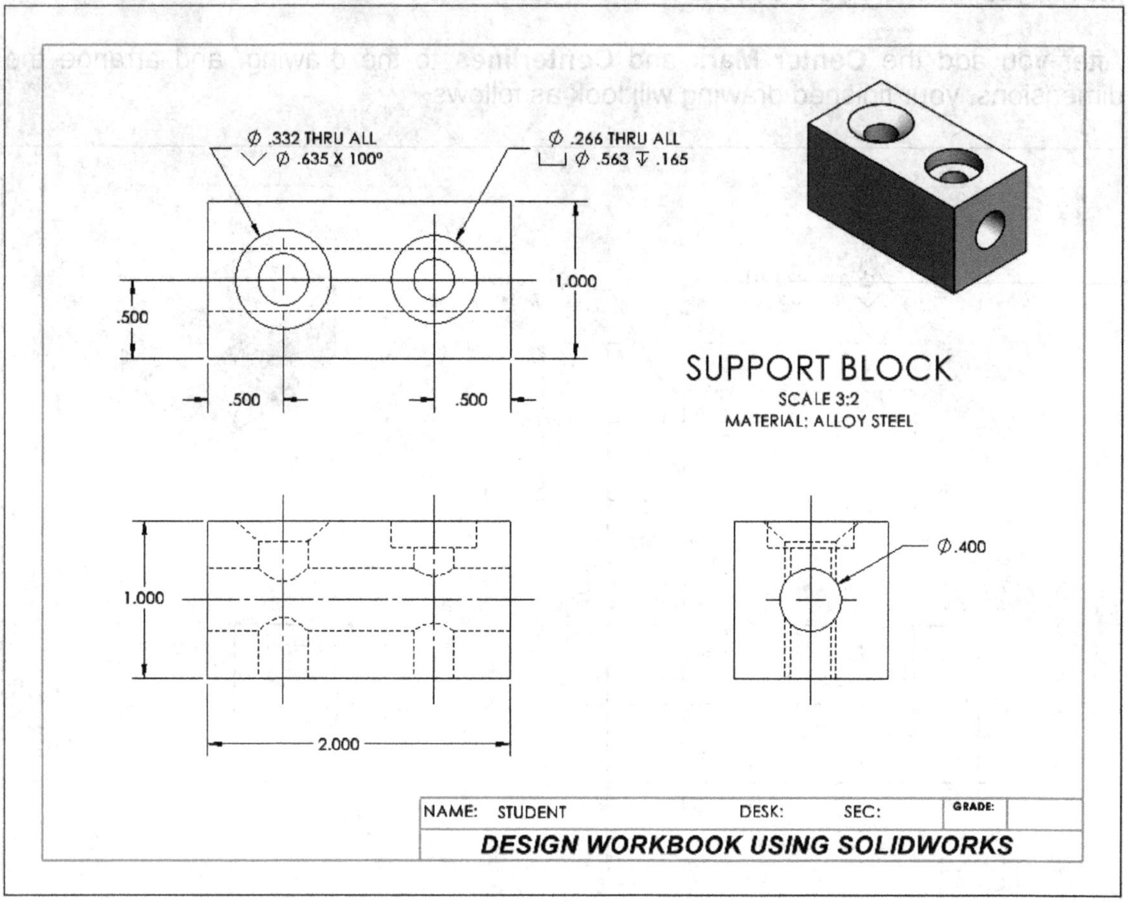

Exercise 10.1: GUIDE BLOCK Drawing

In Exercise 10.1, you will make a solid model of the Guide Block using previously learned commands. Next, you will make a drawing with three orthographic views and an Isometric of the part. You will import the 3D model dimensions and complete the drawing with the title annotations.

BUILDING THE GUIDE BLOCK PART

Start a new part using the **ANSI-INCHES** template, add a new sketch in the **Top Plane** and draw the profile shown here. Use the **Dynamic Mirror** command to maintain symmetry.

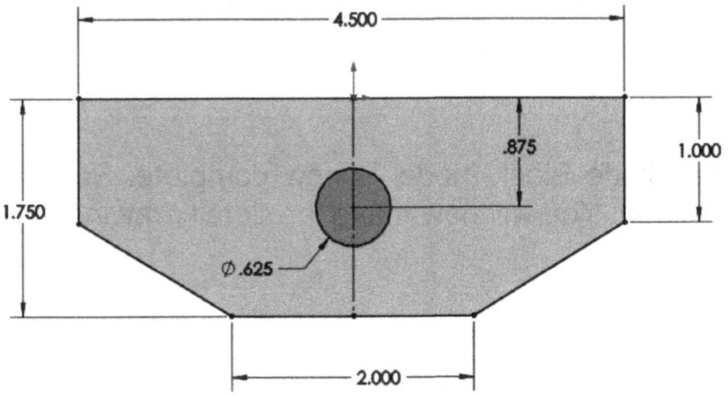

When finished, make a **0.75" Extruded Boss** in the upward direction.

Now change to a **Back**-view orientation, select the back face, and add a new sketch in it. Draw a rectangle and dimension it as indicated.

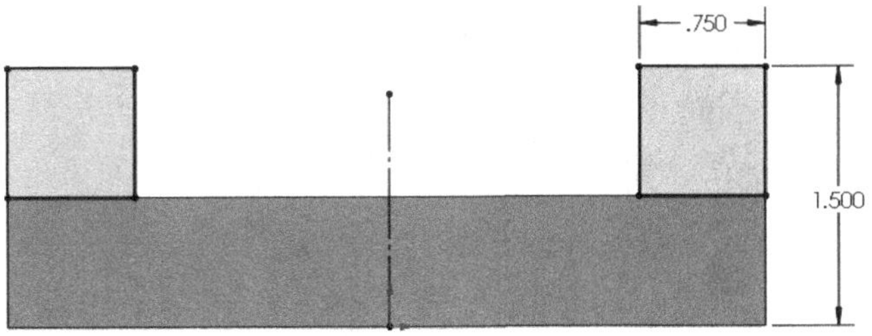

Add a vertical **Centerline** through the origin and **Mirror** the rectangle to the other side. Change to an **Isometric** view orientation and **Extrude** the sketch **1.00"** towards the front of the part.

The last feature is a slot that goes all the way through from front to back. Change to a **Front** view orientation, select the front face of the part, and add a new sketch.

Add a rectangle and dimension it as shown in the next page. To maintain symmetry, add a vertical centerline and add a **Midpoint** relation between the vertical centerline's top endpoint and the lower horizontal line of the rectangle.

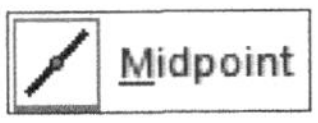

Now use the **Extruded Cut** command to cut the sketch using a **Through All** condition. When finished, change the part's material to **AISI 1020** from the **Steel** library.

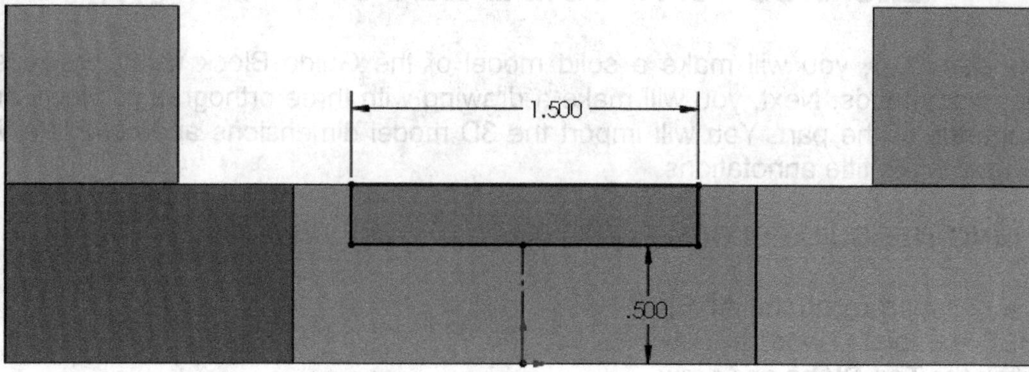

The Guide Block model is now complete. Save your part as **Guide Block.sldprt** to continue. You will now make the detail drawing.

MAKE A MULTI-VIEW DRAWING

With the Guide Block still open, start a new drawing using the Inches template. Open the **View Palette**, make sure the **Auto-Start Projected view** option is turned on and the **Guide Block** is selected in the drop-down list at the top.

Select the **Front** view and drag it into the drawing sheet, move up to add the **Top** view, to the top-right to add an **Isometric** view, and to the right to add a **Right** view. If needed, right click in **Sheet1** on the FeatureManager, select **Properties** and change the scale to 1:1. Select the Front view and change it to **Hidden Lines Visible** mode; the rest of the projected views will change at the same time.

Select the **Isometric** view and change its scale to 1:2 and **Shaded with Edges** display mode. Use the Centerline command to add the missing centerlines to the Front and Right views.

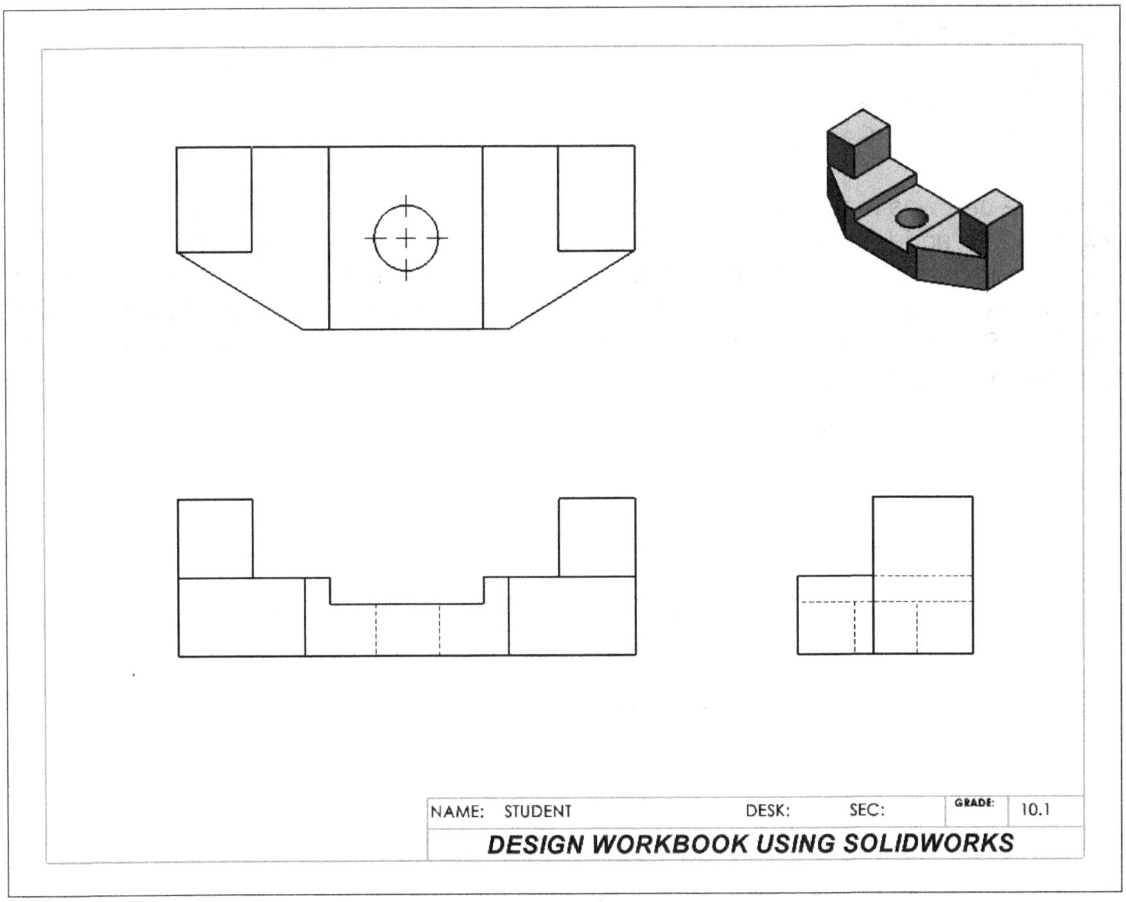

DIMENSIONING THE DRAWING

From the **Annotations** tab, select the **Model Items** command. In the Source section select **Entire Model,** turn on the option **Import items into all views** and click **OK**. Your 2D drawing with the imported 3D model dimensions is shown next.

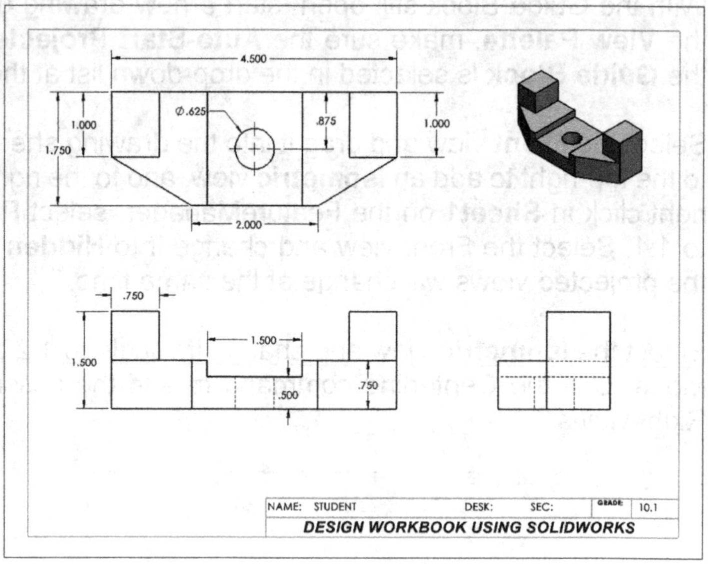

To distribute the dimensions in the drawing, select the **0.750"** vertical dimension in the **Front** view, hold down the **Shift** key and drag it to the **Right** view. In the **Top** view, there are two **1.00"** dimensions, and one of them is redundant. Select one and delete it.

Since the part was made symmetric about the origin, some dimensions are not present, but can be added manually using the **Smart Dimension** tool. Add the missing dimensions to match the completed drawing. Remember to turn off the parentheses to the manually added dimensions in the dimension's properties. Finally add a **Note** with the Title, Scale and Material. Save your drawing as **GUIDE BLOCK** and **Print** a copy to submit to your instructor.

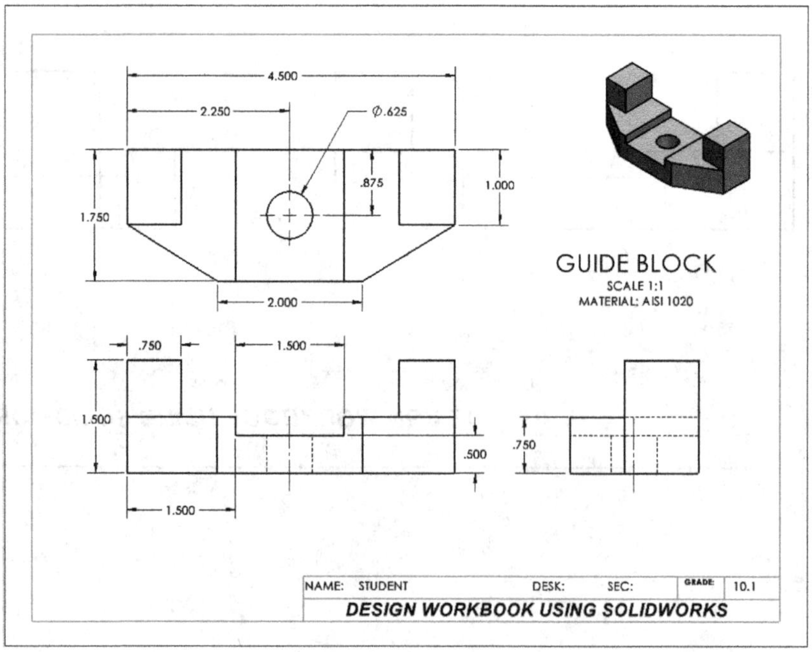

Exercise 10.2: PIPE JOINT Drawing

In this Exercise 10.2, you will build a solid model of a Pipe Joint. This exercise is dimensioned using Millimeters.

Next, you will make a drawing with three orthographic views and an Isometric of the part. You will import the 3D model dimensions and complete the drawing with annotations.

BUILDING THE PIPE JOINT MODEL

Start a new part using the **ANSI-METRIC** template, add a new sketch in the **Top Plane** and draw the next sketch profile.

Draw the two large circles, add a relation to make their centers **Horizontal**, and then draw the two lines. Add a **Tangent** relation between both circles and each line.

Add a vertical centerline that extends beyond the large diameter circle and dimension the sketch.

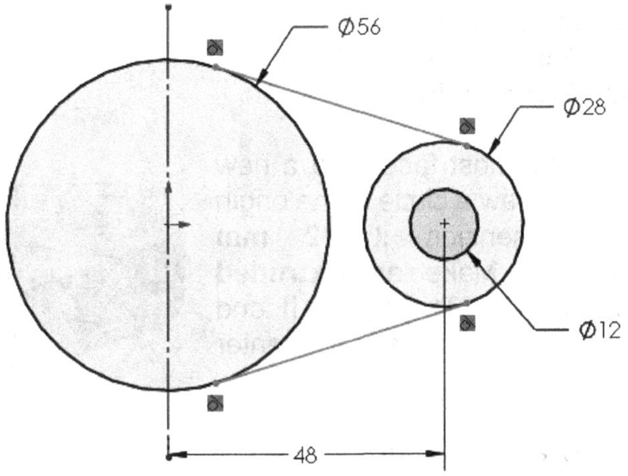

Next use the **Trim Entities** tool and remove the left and right side of the **56 mm** diameter circle, and the left side of the **28 mm** diameter circle.

Use the **Mirror Entities** command to mirror the entire sketch about the vertical centerline.

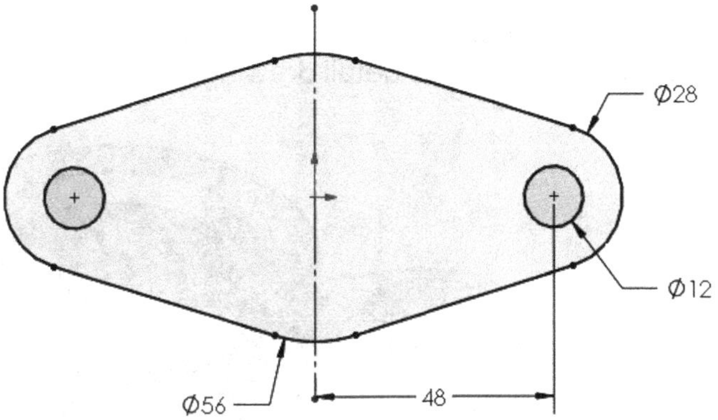

Finally **Extrude** the Sketch **16** mm up to complete the base feature for the Pipe Joint.

Next, add a new sketch in the top face of the part, draw two circles **concentric** to the holes in the first feature, add an **Equal** sketch relation and dimension one of them **18 mm** in diameter. Make an **Extruded Cut 6 mm** deep into the base to make the counterbore holes.

Add a new sketch in the top face and draw a circle at the origin. Add an **Equal** relation to the **56 mm** diameter edge and add a **24 mm Extruded Boss**.

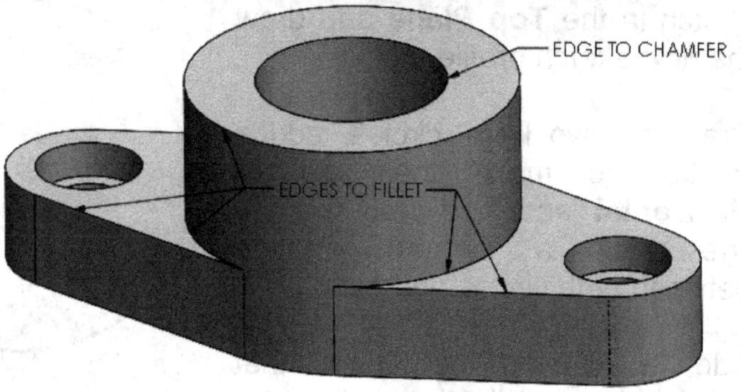

In the topmost face, add a new sketch, draw a circle at the origin and dimension it **32 mm** diameter. Make an **Extruded Cut** using a **Through All** end condition to make the center hole.

Finally add a **3 mm** Fillet to the edges indicated, and a **3 mm x 45° Chamfer** to the top inside edge. Set the part's material to **Wrought Stainless Steel**.

The Pipe Joint part is now finished. Save your part as **PIPE JOINT.sldprt** to continue. Now you will make the detail drawing.

MAKE A MULTI-VIEW DRAWING

With the Pipe Joint open, start a new drawing using the **Metric** template. Open the **View Palette**, and for this exercise make sure the **Auto-Start Projected view** option is turned off and the **Pipe Joint** is selected in the drop-down list.

From the **View Palette** drag the **Top** and **Isometric** views. Select the **Isometric** view and change it to **Shaded with Edges** mode.

If needed, right click on the **Top** view, select **Tangent Edge**, **Tangent Edges Removed**, and change the **Top** view's scale to **1:1.**

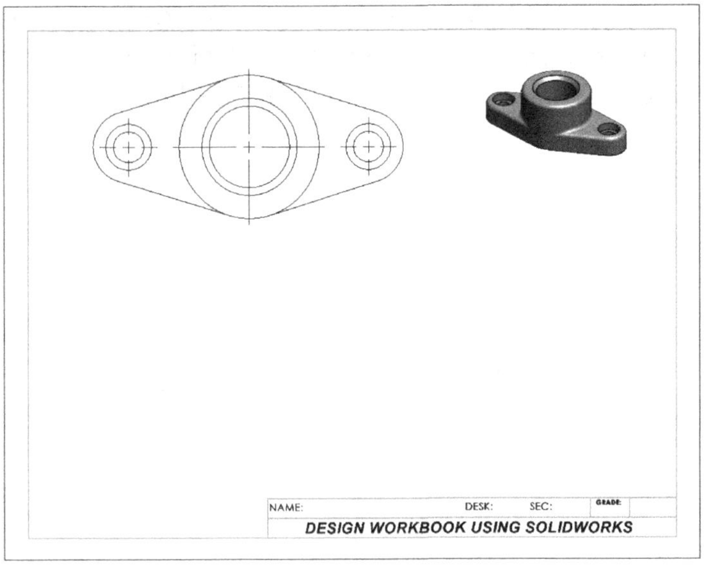

From the Drawing tab select the **Section View** command.

Activate the **Horizontal** section view option and click in the **Top** view when the section line snaps in the middle of the part.

Immediately locate the section view below in place of the **Front** view.

Now select the **Section** view, from the Drawing tab select the **Projected View** command and locate the projected **Right** view on the right side of the sheet.

Select the **Right** view and set the Display Style to **Hidden Lines Visible**. Add the **Centerlines** to all views to continue. The final drawing views layout is shown here.

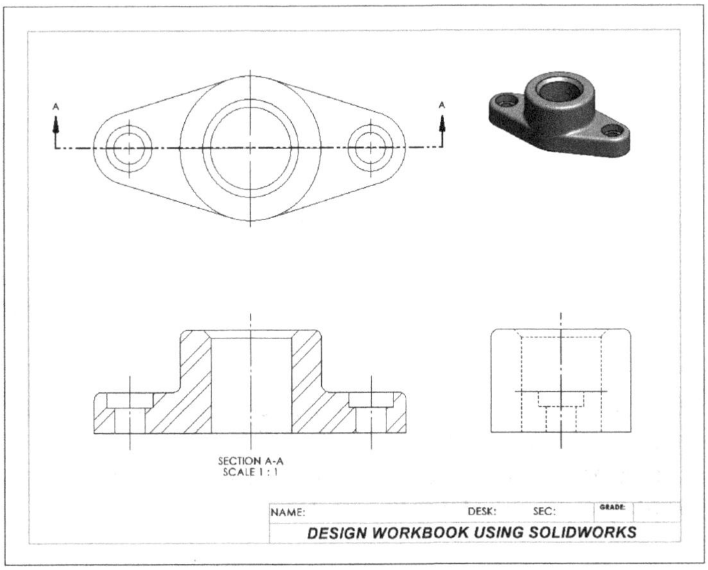

DIMENSIONING THE DRAWING

From the Annotations tab, select the **Model Items** command. In the Source section select **Entire Model,** turn on the option **Import items into all views** and click **OK**.

When you import dimensions from a 3D model, dimensions are first added to the **Detail** views, then **Section** views, and then to main views. Move the **48 mm** dimension to the **Top** view. Remember to hold down the **Shift key** when you move a dimension to a different view. Delete the dimensions of the counterbore hole (both diameters and the depth) and the chamfer. Your drawing should look like this.

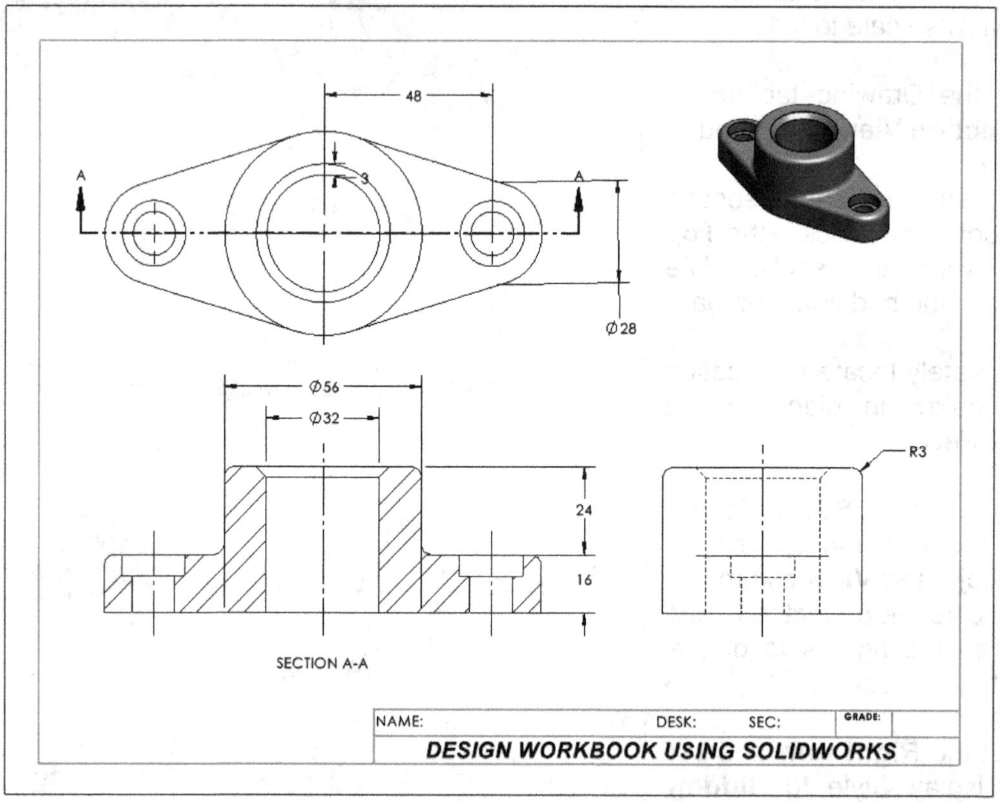

After moving the **28mm** Diameter dimension it will be shown as a linear dimension. To change it to a radial dimension right click on it, and select **Display Options**, **Display as Diameter**. To show it as a **Radius** dimension, right click on it again and select **Display Options**, **Display as Radius**.

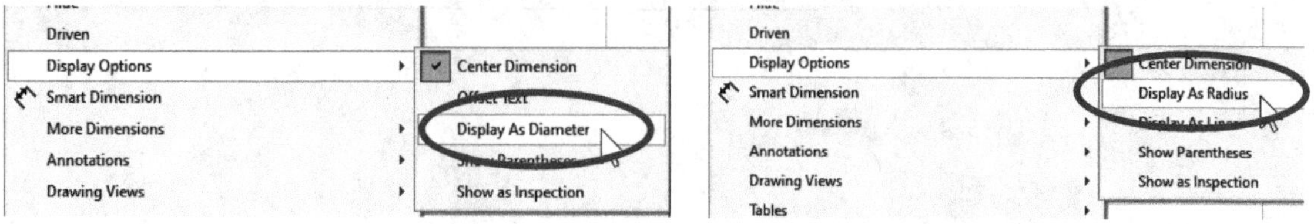

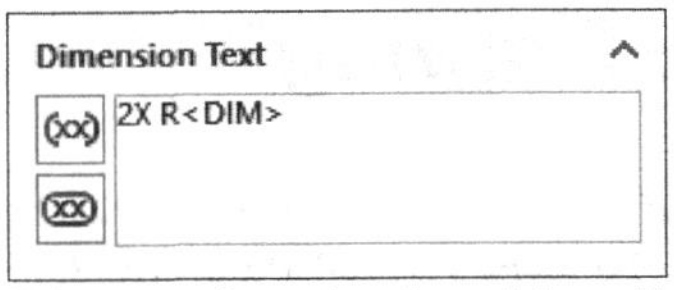

Since this dimension references both sides of the part, select the **R14.00**, and in the dimension's properties add a "**2X**" before "**R<DIM>**" in the **Dimension Text box**. The final dimension should read "**2X R14**". With the **Smart Dimension** tool, add a dimension between the two counterbore holes. Use the **Smart Dimension** tool to add a dimension between the two counterbore holes.

From the **Smart Dimension** command Drop-down menu, select **Chamfer Dimension**. In the **Section** view select the chamfer edge, then the vertical edge of the hole and locate the Chamfer dimension.

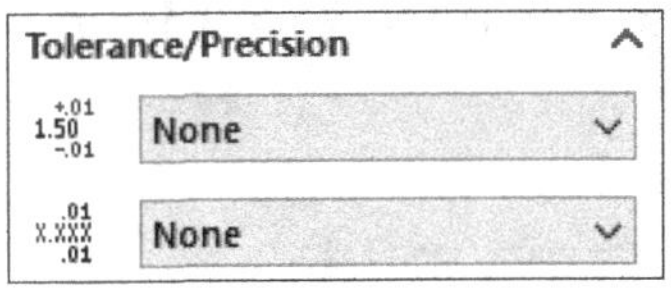

To change the tolerance and/or precision of a dimension, in the properties, under Tolerance/Precision change the required Tolerance and Precision as needed.

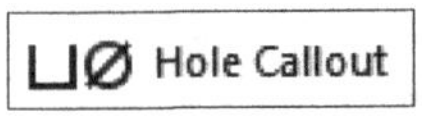

Finally, from the Annotation tab select the **Hole Callout** command and click on the outer edge of the counterbore hole in the **Top** view. Locate the annotation to finish. As with the radial dimension, add a "**2X**" in the **Dimension Text box**.

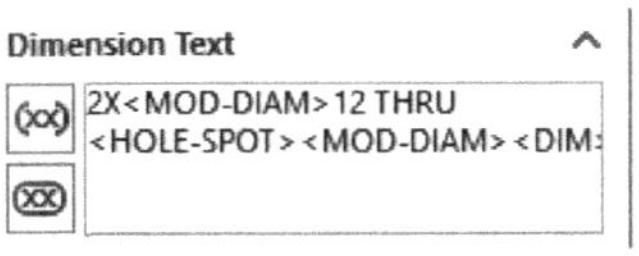

Next use the **Note** command to add a note with title, scale, material, "**ALL FILLETS & ROUNDS -R3**", and "**UNITS MM**". Change the font and size as needed.

Save your drawing as **PIPE JOINT.slddrw** and print a copy to submit to your lab instructor.

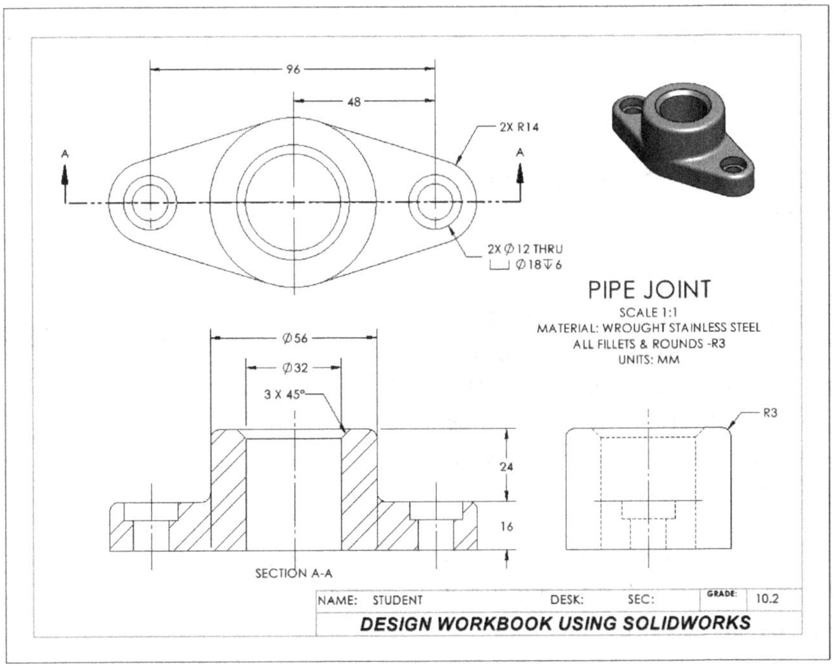

Exercise 10.3: PEDESTAL BASE Drawing

Quite often in mechanical design, features such as holes are repeated in a regular circular pattern. Examples include flanged pipe joints, inspection cover plates, motor housings, and many others. Circular and rectangular patterns have been introduced in earlier exercises; here, you will create a part with multiple bolt circle patterns and discuss dimensioning practices for the part in a multi-view drawing.

BUILDING THE PEDESTAL BASE

Start a new part using the **METRIC** template and add a new sketch in the **Front Plane** and draw the sketch profile shown here.

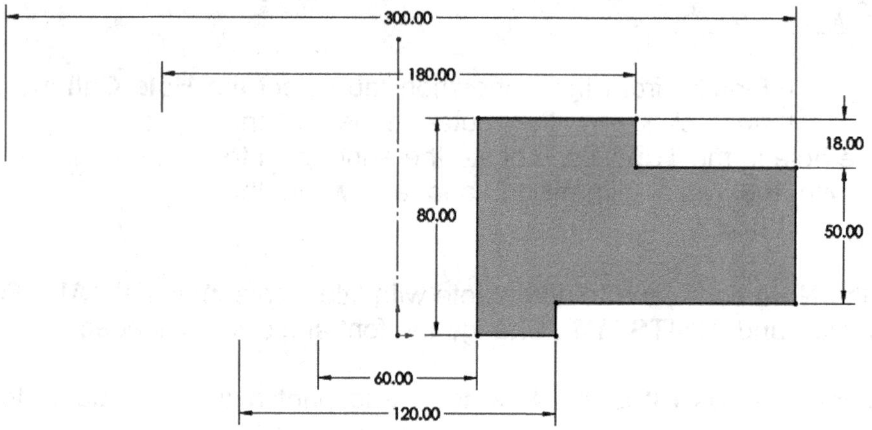

Since this sketch will be used to make a **Revolved Boss**, it is helpful to add diameter dimensions to the sketch. To add a diameter dimension, select the centerline and the entity to dimension, and then cross the centerline ***BEFORE*** locating the dimension; it will be automatically doubled.

While you add the **doubled dimensions**, the mouse cursor will look as: 

When you finish adding the doubled dimensions, press the **ESC** key to return to the **Smart Dimension** tool and finish adding the height dimensions. When the sketch is complete, add a **360° Revolved Base** and click **OK**.

Next, create the circular hole patterns in the base. The holes needed can be made using the **Hole Wizard**, which will be covered in the next lab. For this exercise we will use a regular **Cut Extrude** feature.

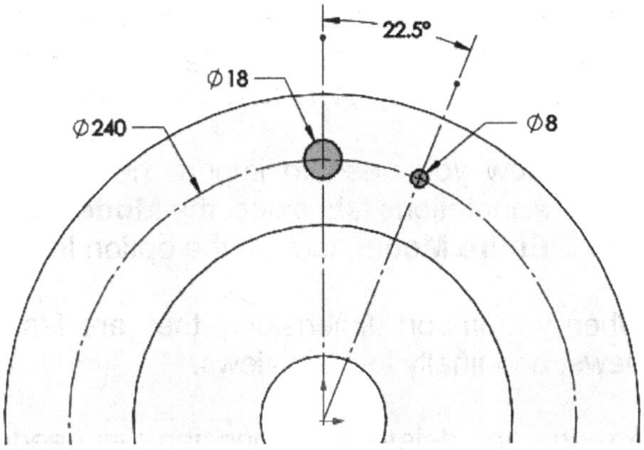

Change to a **Top** view and add a new **Sketch** in the outer top face of the base feature. Draw two construction lines and a construction circle.

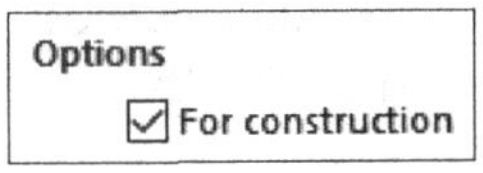

After adding the **240 mm** circle select it and activate the checkbox **For Construction**. Finally dimension it as indicated.

When the sketch is complete select the **Cut Extrude** command and use the **Through All** end condition.

From the Linear Pattern command's drop-down list, select the **Circular Pattern** command, select the circular outer edge for the direction, add the **Cut Extrude** feature in the **Features to pattern** selection box, set the number of copies to **8** and use the equal spacing option in **360°**.

To finish the part, add a **5mm Fillet** to the outer edges, except for the holes.

Set the part's material to **Copper** and save your part as **PEDESTAL BASE.sldprt**.

MAKE A MULTI-VIEW DRAWING

Using the **METRIC** drawing template start a new drawing. Open the **View Palette**, make sure the **Auto-Start Projected view** option is turned **off** and the **Pedestal Base** is selected in the drop-down list.

From the **View Palette** drag a **Top** view and then an **Isometric** view. Select the **Isometric** view and change it to **Shaded with Edges** mode. If needed, right click on the **Top** view, and select **Tangent Edge**, **Tangent Edges Removed**, and change the Top view's scale to 1:4.

From the **Drawing** tab select the **Section View** command, select the **Horizontal** section view option, click in the **Top** view when the section line snaps in the middle of the part, and locate the section view in place of the **Front** view.

Now select the **Section** view, click on the **Projected View** command from the Drawing tab, and locate the projected **Right** on the right side of the sheet. Select the **Right** view and set the Display Style to **Hidden Lines Visible**. Add the **Centerlines** to continue.

Now you need to import the part's dimensions into the drawing. From the Annotations tab, select the **Model Items** command. In the Source section select **Entire Model,** turn on the option **Import items into all views** and click **OK**.

When you import dimensions, they are first added to the **Detail** views, then **Section** views, and finally to main views.

Arrange and delete the dimensions as needed to match the next drawing. Remember to hold down the **Shift key** to move dimensions to a different view. Delete the diameter dimensions of the two holes and change the **60 mm** dimension to display as diameter.

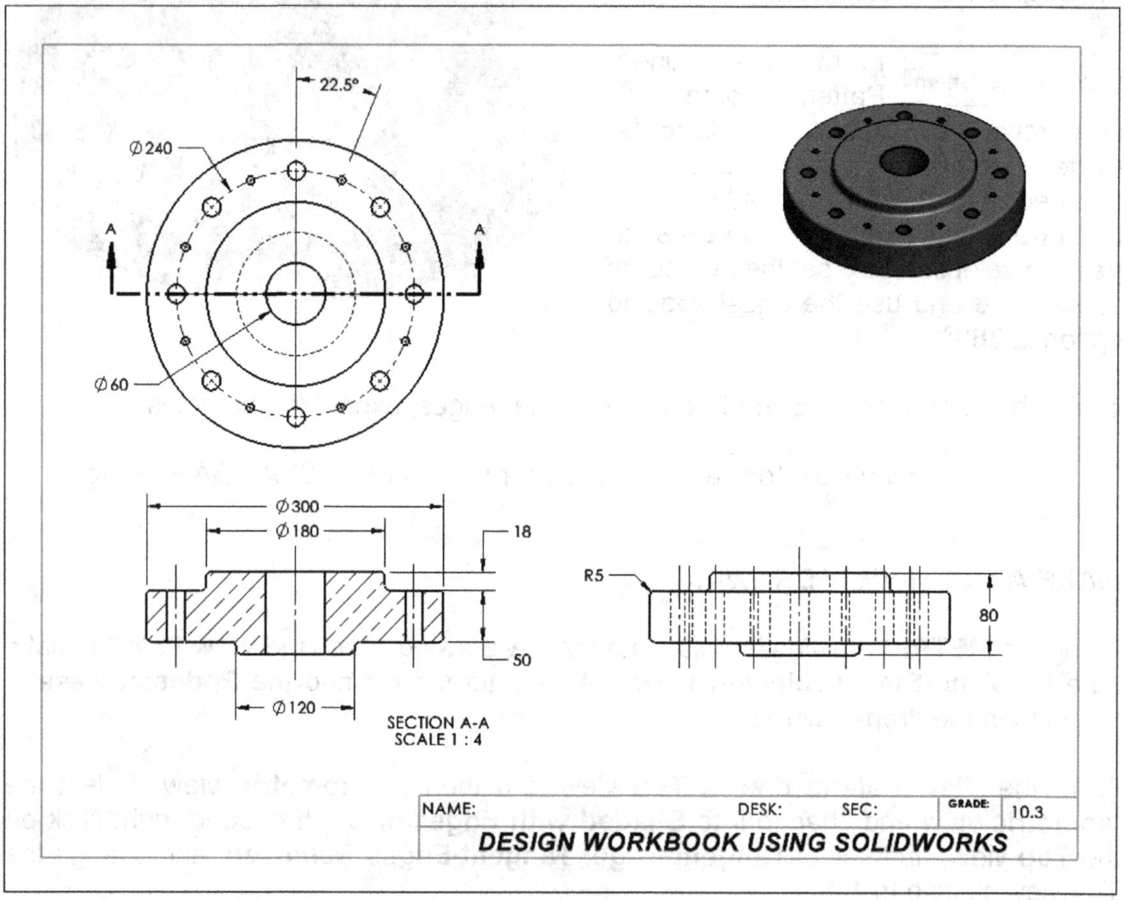

Now, add a **centerline** in the **Top** view from the origin to the first small hole to reference the imported **22.5°** dimension.

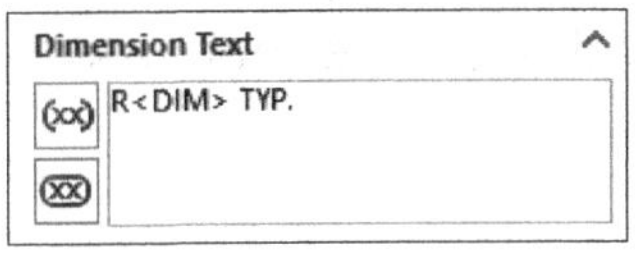

The **Fillet** dimension is added to a single fillet. To communicate the fillet's dimension is the same for all rounded edges—select the dimension, and in the **Dimension Text** box type "**TYP.**" at the end of the dimension text.

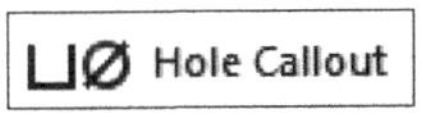

To complete the drawing, select the **Hole Callout** tool from the Annotations tab, and click in both patterned holes; the correct annotation will be added. In the **Dimension Text** of both hole annotations type "**8X**" before the dimension text code.

Next use the **Note** command to add a note with title, scale, material, and "**UNITS MM**". Change the font and size as needed.

Save your drawing as **PEDESTAL BASE.slddrw** and print a copy to submit to your lab instructor.

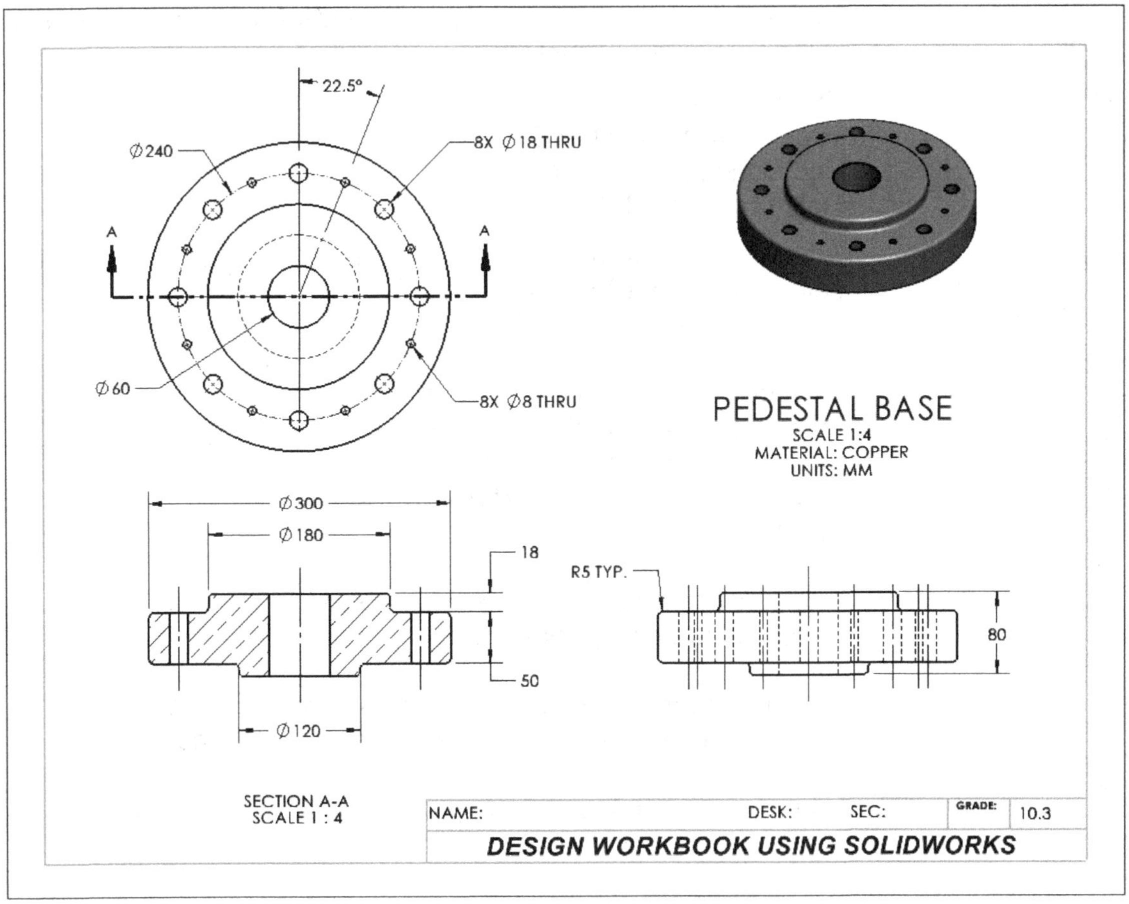

Exercise 10.4: TOOLING PAD Drawing

In Exercise 10.4, you will build a solid model of the Tooling Pad, make a drawing with three orthographic views, import the model dimensions, and add the necessary annotations to complete the drawing.

BUILDING THE TOOLING PAD MODEL

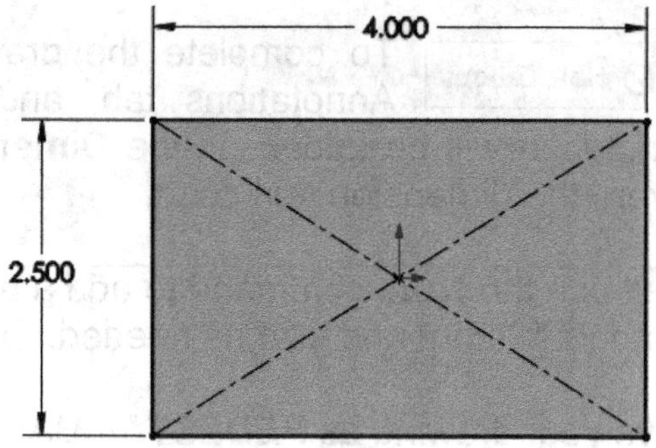

Start a new part using the INCH template, add a new sketch in the **Top** plane, draw the sketch indicated using the **Center Rectangle** tool starting at the origin, and dimension as shown. Finally **Extrude** the profile **0.725"** up.

Using the Hole Wizard

Now you will learn how to use the **Hole Wizard** to add holes on the Tooling Pad. Select the **Hole Wizard** command from the Features tab. In the **Hole Specification** select the Hole option, and set the following parameters as indicated:

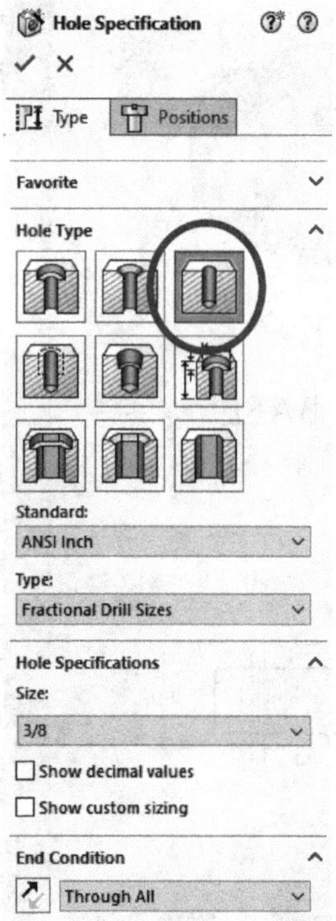

Standard: **ANSI Inch**
Type: **Fractional Drill Sizes**
End Condition: **Through All**
Hole Diameter: **3/8 in**

After the hole parameters have been set, select the **Positions** tab; this is the second part of the Hole Wizard.

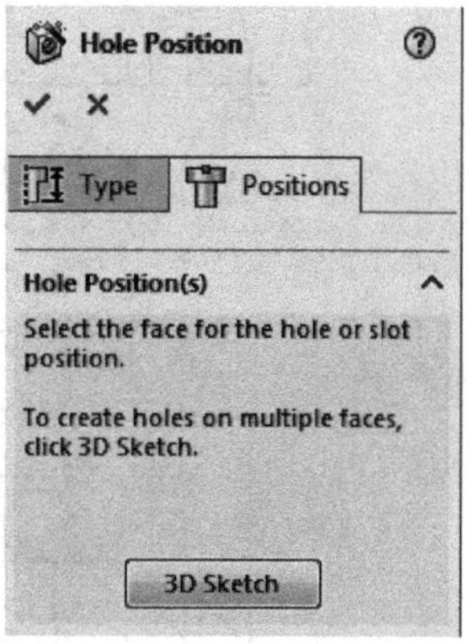

Change to a **Top** view and select the top face of the base feature. In this step of the **Hole Wizard** we are editing a sketch, and you will use a sketch **Point** to define the location of each hole.

After selecting the face, the sketch **Point** tool is automatically selected.

Click to add three points in the top face and dimension them. At each **Point** location you will see a preview of the hole.

When you finish dimensioning the locations click **OK** to finish.

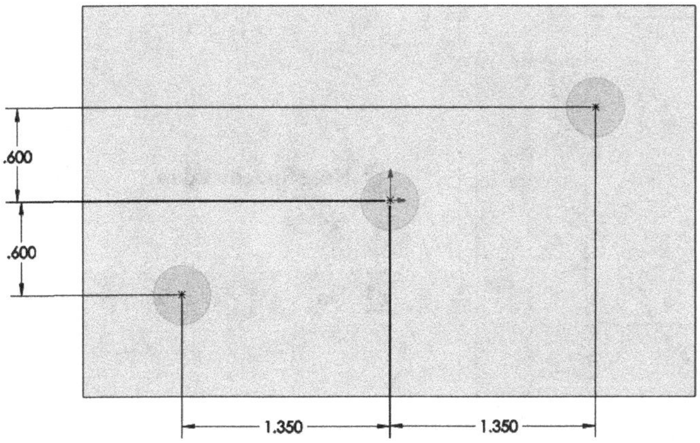

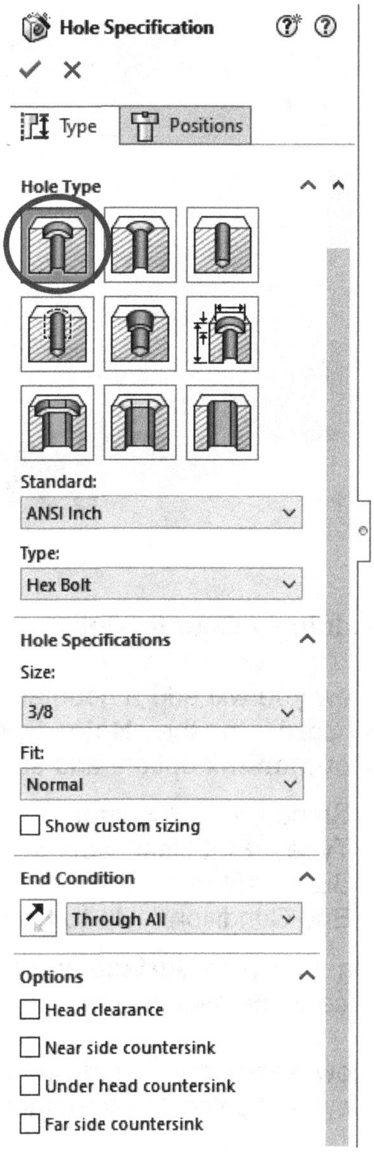

Add the Counterbore Holes

Now you will add counterbore holes using the **Hole Wizard**. In the **Hole Specification** select the **Counterbore** option, and set the parameters indicated.

Standard = **ANSI Inch**
Type = **Hex Bolt**
Size = **3/8 in**
End Condition = **Through All**

Note: The option **Show Custom Sizing** allows you to customize the counterbore hole dimensions if needed.

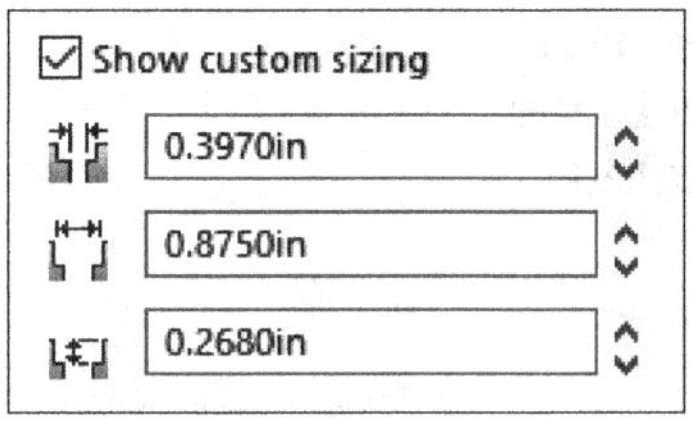

Now select the **Positions** tab to locate the counterbore holes. Change to a **Top** view and select the top face of the base feature as before.

After the top face is selected, the sketch **Point** tool is activated. Add two points in the top face, dimension them as indicated above and click **OK** to finish.

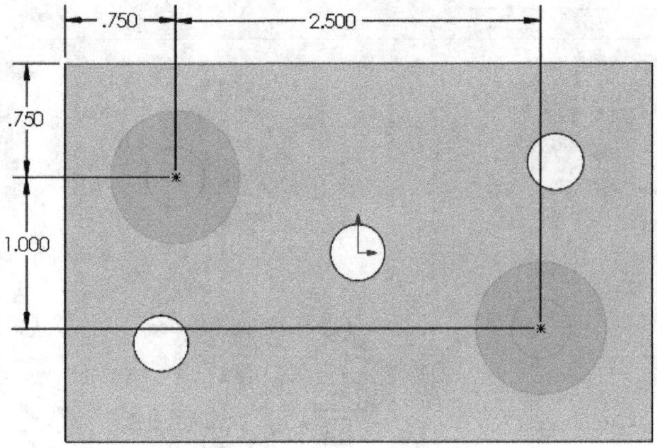

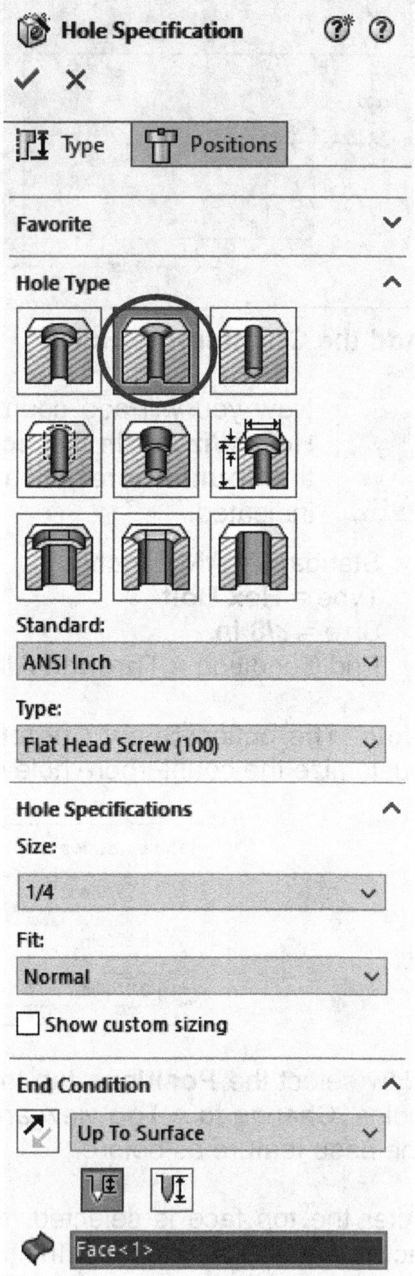

Adding a Countersink

Now you will add a countersink hole using the **Hole Wizard.** In the **Hole Specification** select the **Countersink** option, and set the options indicated.

Standard = **ANSI Inch**
Type = **Flat Head Screw**
Size = **1/4 in**
End Condition = **Up to Surface**

*In the **Up to Surface** end condition select the inner face of the middle hole.

Now select the **Positions** tab. Change to a **Front** view and select the front face of the base feature.

Click to add a **Point** in the **Front** face, dimension it and add a **Vertical** relation to the origin to fully define the sketch. Click **OK** to complete the **Hole Wizard**.

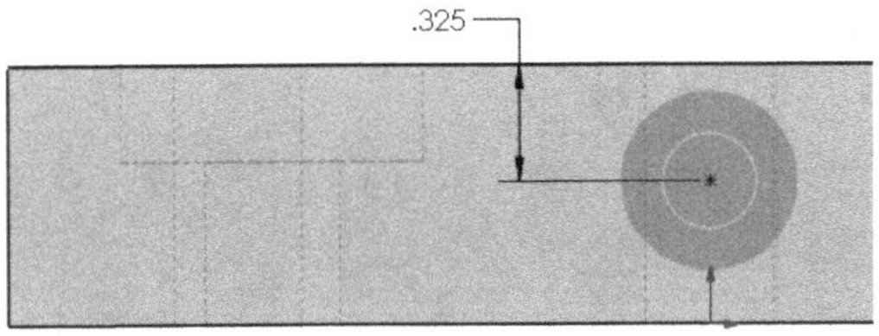

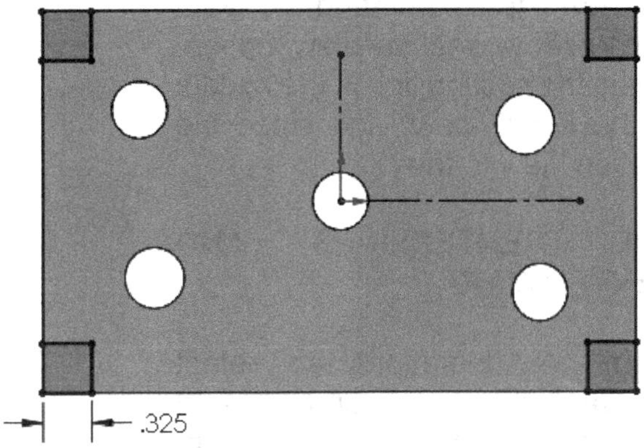

Adding the Bottom Base Feet

Change to a **Bottom** view to add the four corner feet to the Tooling Pad. Add a sketch in the bottom face and draw the profile shown.

Draw a rectangle in a corner and make two perpendicular lines **Equal** to make a square.

Use the **Mirror Entities** command about a Vertical and Horizontal centerline to add the other three corners.

Extrude the sketch **0.125"**, add a **0.0625" Fillets** to the outer top edges and the outer vertical edges.

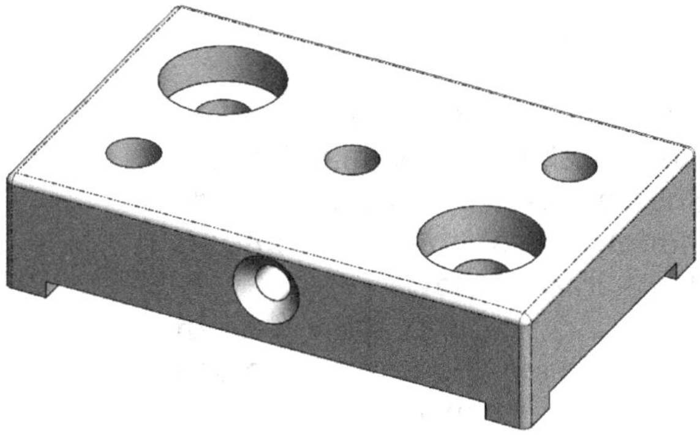

Set the material to **Titanium Ti-8Mn, Annealed** from the **Titanium Alloys** library, and save your part as **Tooling Pad.sldprt**.

MAKE A MULTI-VIEW DRAWING

Start a new drawing using the **Inch** template and change the sheet's scale to **1:1**. Open the **View Palette**, make sure the **Auto-Start projected view** option is turned **on** and the **Tooling Pad** is selected in the drop-down list.

From the **View Palette** drag the **Front** view, and project a **Top** and an **Isometric** view. Select the **Isometric** view, change it to **Shaded with Edges** mode and change the scale to **1:2**.

If needed, right click on the **Front** and **Top** view and select **Tangent Edge, Tangent Edges Removed.**

ADD A SECTION VIEW

From the **Drawing** tab select the **Section View** command. Activate the **Vertical** section view option, locate the section line in the middle of the **Front** view, and place the section view to the right.

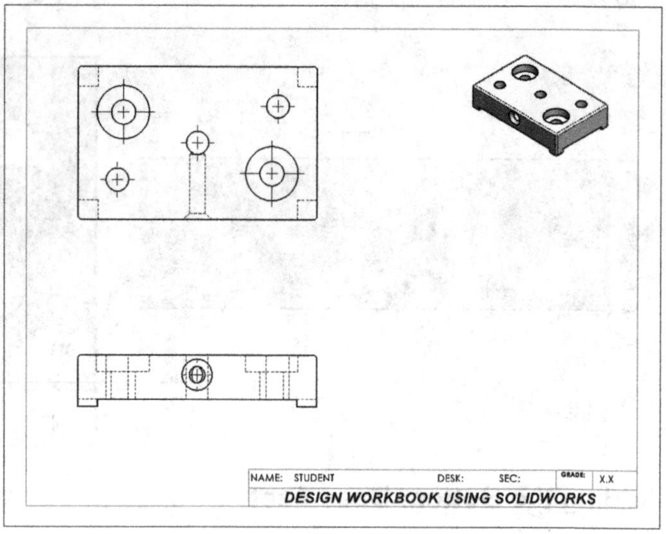

ADD CENTERLINES AND CENTER MARKS

From the Annotations tab, select the **Centerline** command and add the missing centerlines to the Top, Front, and Section views. If missing, also add the missing **Center Marks**.

DIMENSIONING THE DRAWING

Now you need to import the part's dimensions into the drawing.

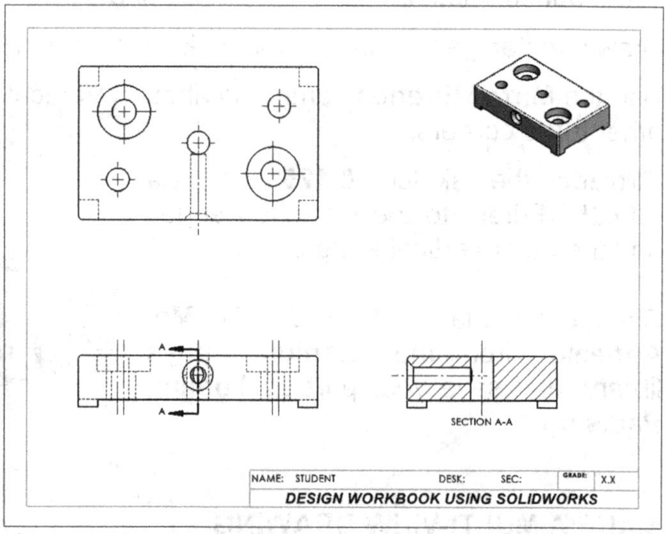

From the **Annotations** tab, select the **Model Items** command. In the Source section select **Entire Model,** turn on the option **Import items into all views.**

Activate the **Hole Callout** option and click **OK** to add the dimensions.

The **Hole Callout** option adds the correct annotations to **Hole Wizard** features, including the number of instances.

In this case, the dimensions added fully annotate the drawing. There are only a few changes to make, like moving dimensions to a different view (hold **Shift** to move dimensions). In the bottom feet, since all sides are equal, only one dimension is needed; in this case add a **TYP** text after the **0.325"** dimension. Also, a dimension's arrows can be **reversed.** Select the dimension and click in the arrow to reverse it, as shown. Add any missing dimensions as needed to finish.

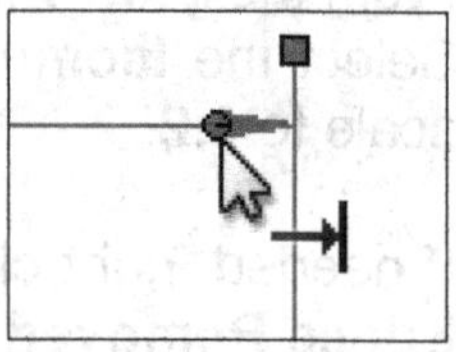

Finally add a note to the drawing including the title, scale, material, and "**ALL FILLETS AND ROUNDS R-0.0625"**". Change the font and size as needed.

Save your drawing as **TOOLING PAD.slddrw** and print a copy to submit to your lab instructor.

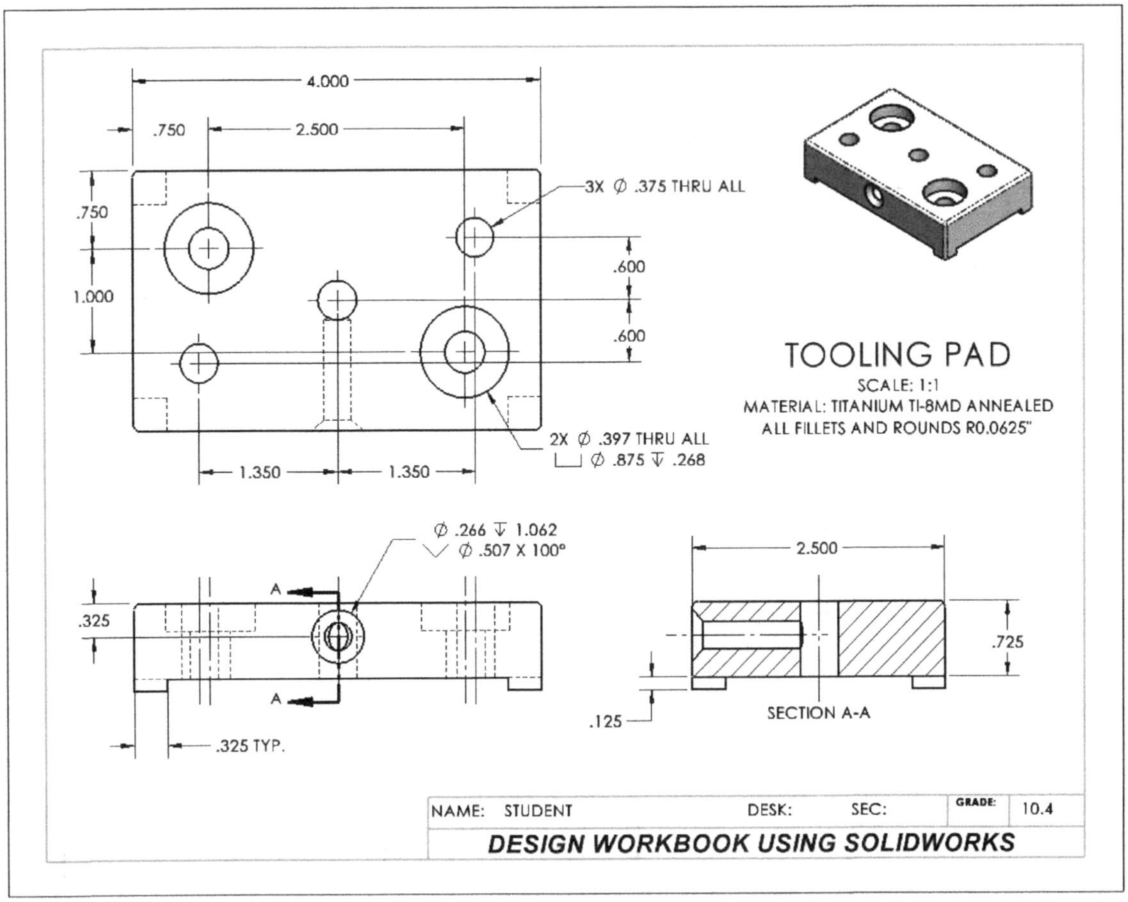

Supplementary Exercise 10-5: PILLOW BLOCK

Build a solid model of the image below. Make a detail drawing and Dimension it. Insert a small Isometric of the part in the upper right-hand corner of the sheet. Provide the proper Titles, Scale and other pertinent notes. Drawing dimensions are in Millimeters.

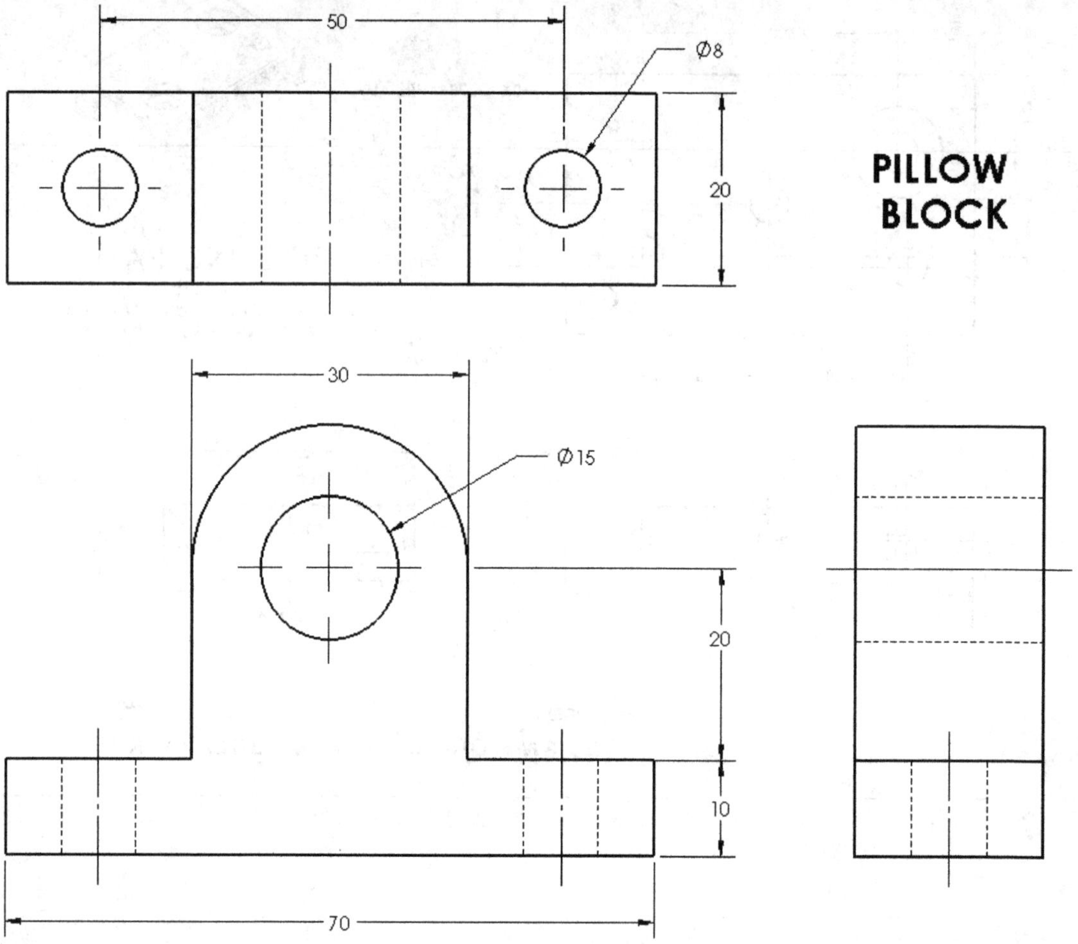

Supplementary Exercise 10-6: CHASSIS BOX

Build a solid model of the image below. Make a detail drawing and Dimension it. Insert a small Isometric of the part in the upper right-hand corner of the sheet. Provide the proper Titles, Scale and other pertinent notes. Drawing dimensions are in Inches.

CHASSIS BOX

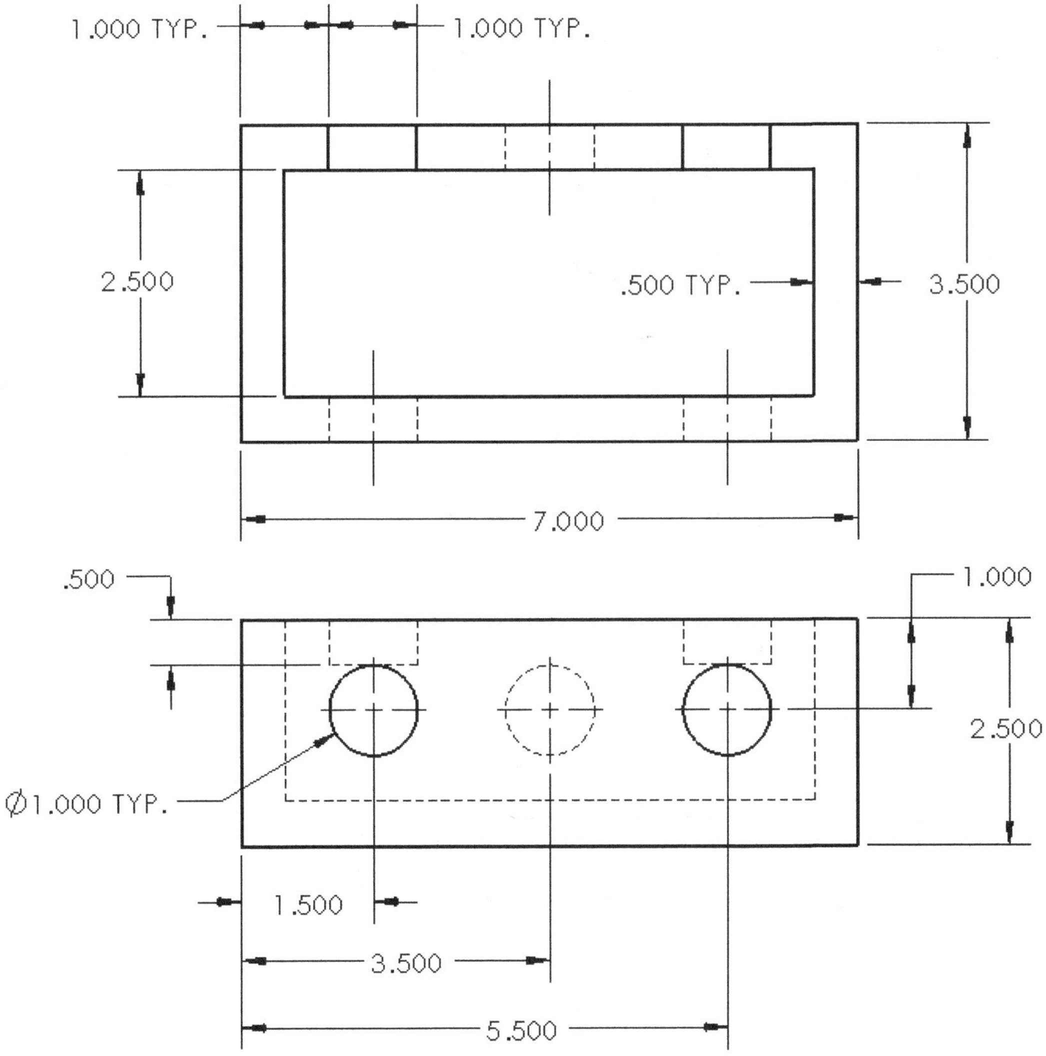

Supplementary Exercise 10-7: COLUMN BASE

Build a solid model of the image below. Make a detail drawing and Dimension it. Instead of a Front view, add a Horizontal section view of the Top view. Add an Isometric of the part in the upper right-hand corner of the sheet. Provide the proper Titles, Scale, and other pertinent notes. Drawing dimensions are in Inches.

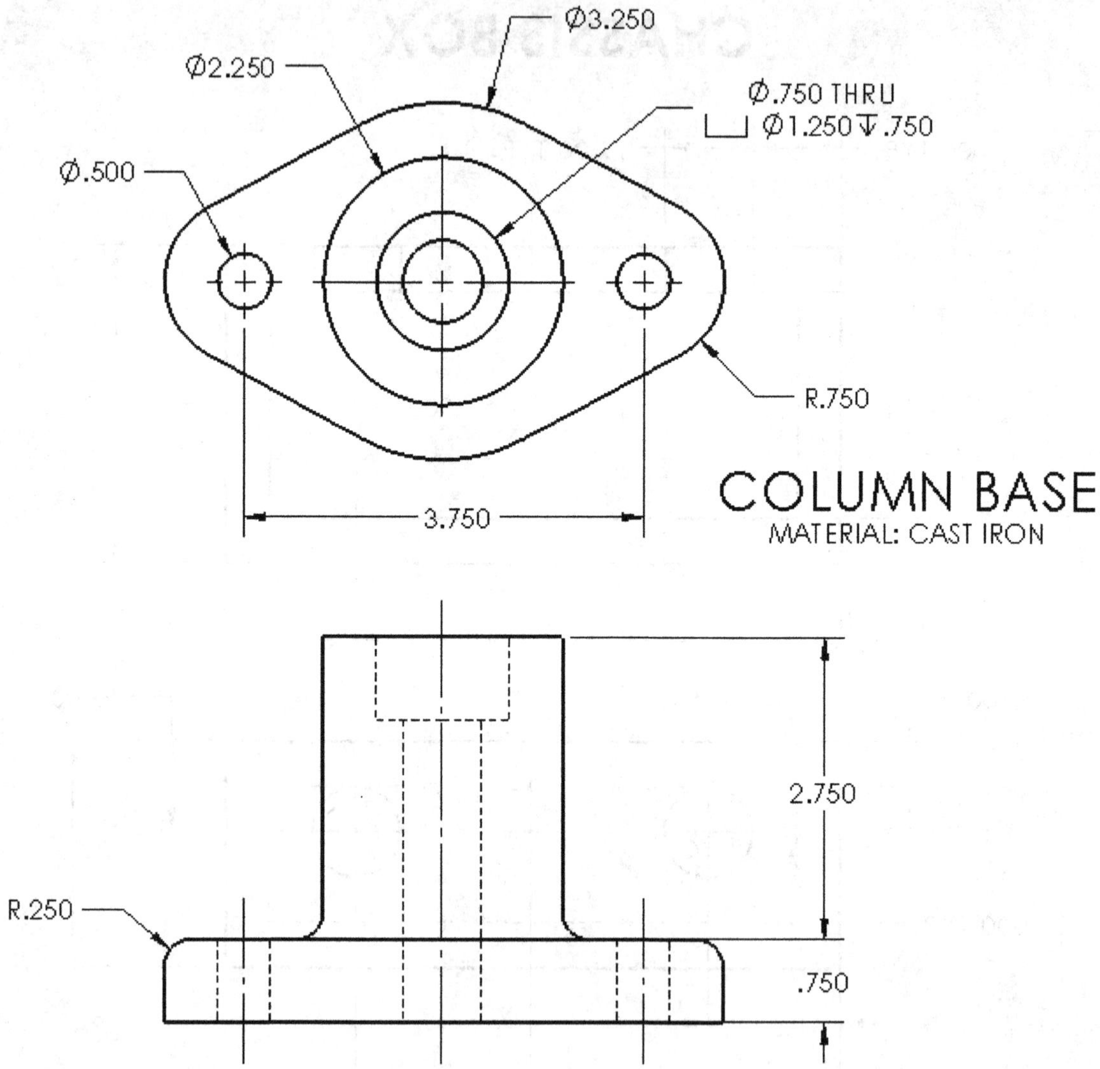

Supplementary Exercise 10-8: VALVE HOUSING

Build a solid model of the image below. Make a detail drawing and Dimension it. Add a Section view in place of the Front view, and project the Right view from the Section view. Add an Isometric view in the upper right-hand corner of the sheet. Provide the proper Titles, Scale, and other pertinent notes. Drawing dimensions are in Inches.

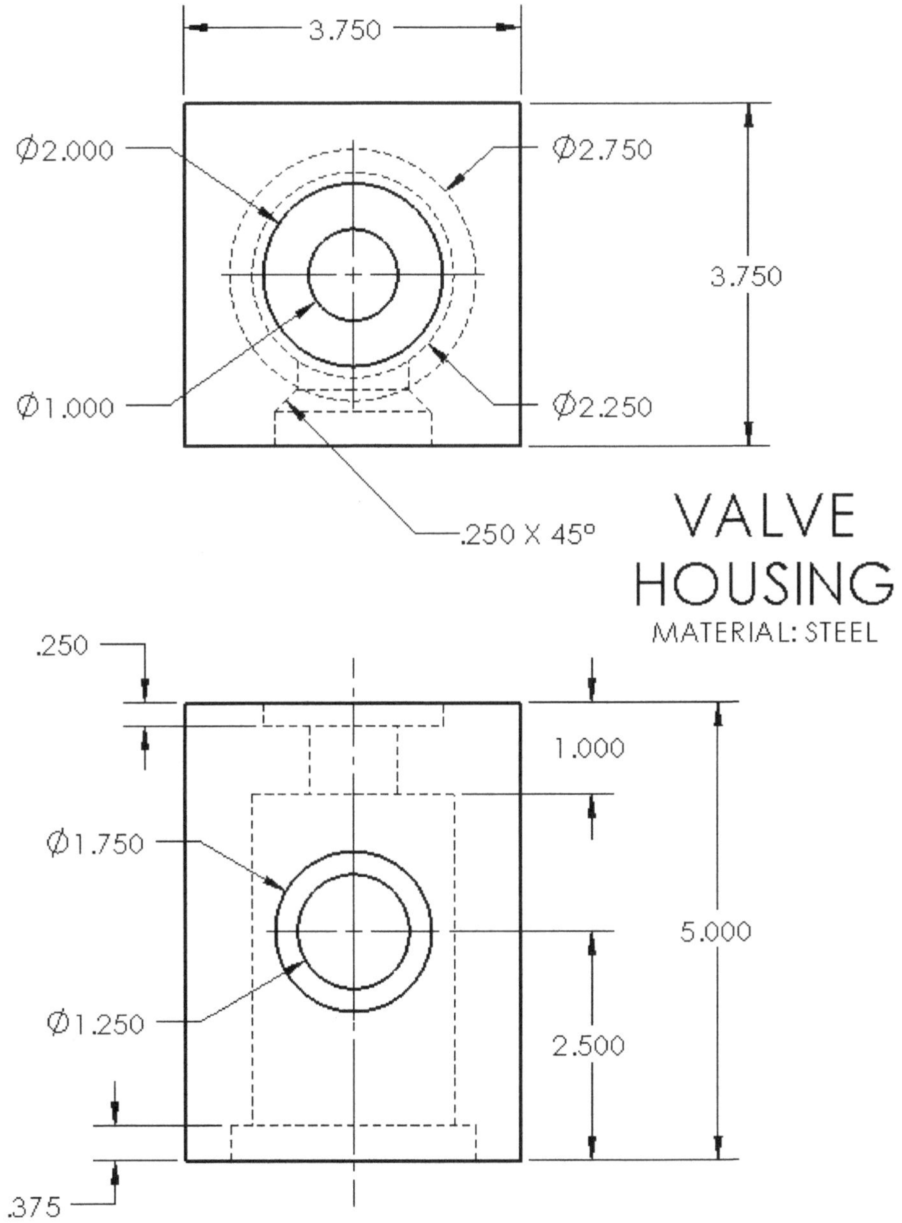

NOTES:

APPENDIX A
DRAWING SHEET TEMPLATE

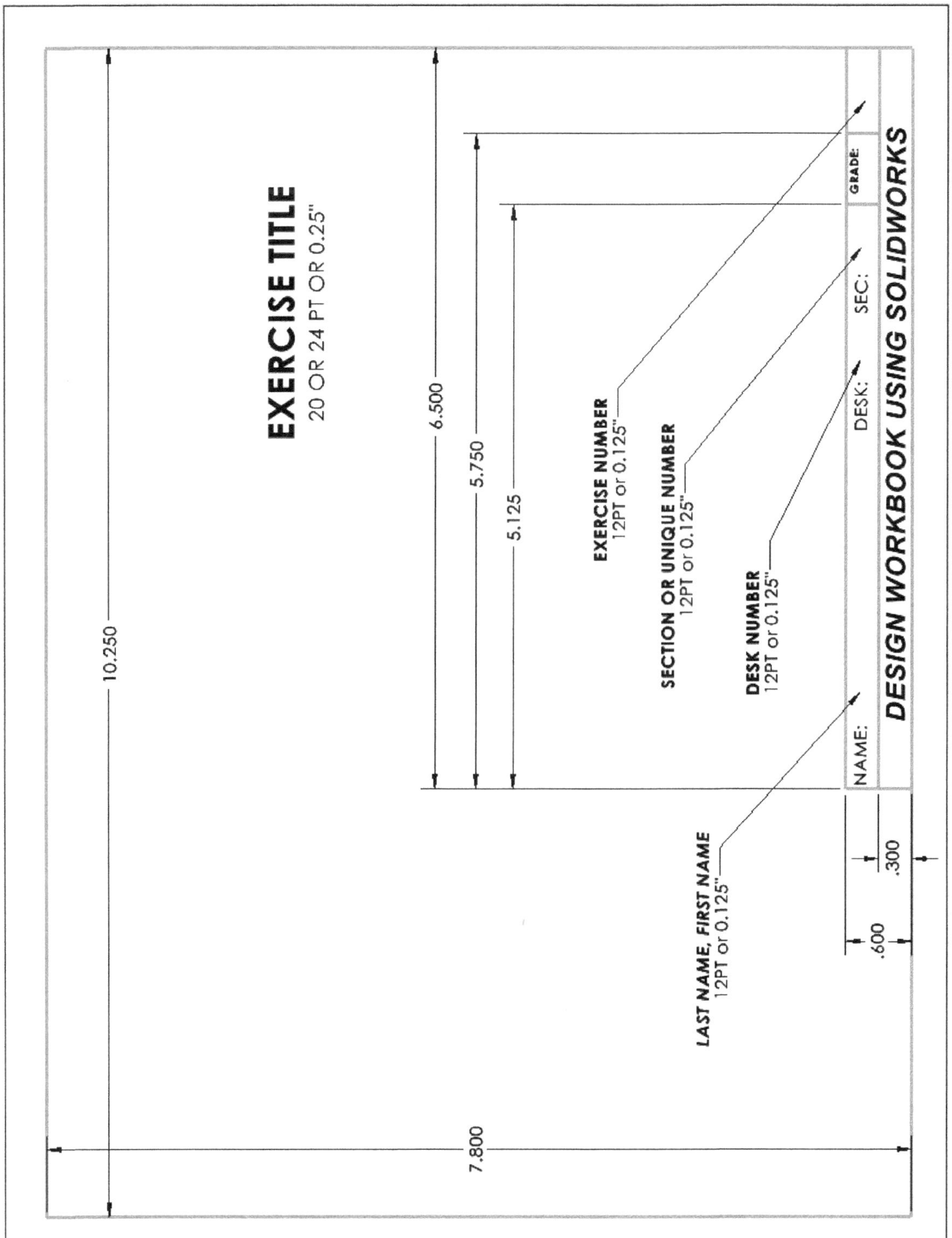

NOTES:

9781630577810